Linear Algebra

Linear Algebra

A geometric approach

E. Sernesi
Professor of Geometry
La Sapienza
Rome
Italy

Translated by
J. Montaldi
Mathematics Institute
Warwick University
UK

CHAPMAN & HALL
London · Glasgow · New York · Tokyo · Melbourne · Madras

Published by Chapman & Hall, 2–6 Boundary Row, London SE1 8HN

Chapman & Hall, 2–6 Boundary Row, London SE1 8HN, UK

Blackie Academic & Professional, Wester Cleddens Road, Bishopbriggs, Glasgow G64 2NZ, UK

Chapman & Hall, 29 West 35th Street, New York NY10001, USA

Chapman & Hall Japan, Thomson Publishing Japan, Hirakawacho Nemoto Building, 6F, 1–7–11 Hirakawa-cho, Chiyoda-ku, Tokyo 102, Japan

Chapman & Hall Australia, Thomas Nelson Australia, 102 Dodds Street, South Melbourne, Victoria 3205, Australia

Chapman & Hall India, R. Seshadri, 32 Second Main Road, CIT East, Madras 600 035, India

English language edition 1993

Printed in Great Britain by T.J. Press, Padstow

ISBN 0 412 40670 5 (HB) 0 412 40680 2 (PB)

A catalogue record for this book is available from the British Library

Library of Congress Cataloging-in-Publication data available

Contents

Appendices

Preface

This book is written primarily for Mathematics students, and it covers the topics usually contained in a first course on linear algebra and geometry. Adopting a geometric approach, the text develops linear algebra alongside affine and Euclidean geometry, in such a way as to emphasize their close relationships and the geometric motivation.

In order to reconcile as much as possible the need for full rigour with the equally important need not to weigh the treatment down with too abstract and formal a theory, the linear algebra is developed gradually and in alternation with the geometry. This is done also to give due prominence to the geometric aspects of vector spaces.

The exposition makes use of numerous examples to help the reader acquire the more delicate concepts. Some of the chapters contain 'Complements' which develop further more specialized topics. Many of the exercises appearing at the end of each chapter have solutions at the end of the book. The structure of the book is designed to allow as much flexibility as possible in designing a course, both by reordering or omitting chapters, and, within individual chapters, by omitting examples or complements.

E.S.

Notes for the reader

This text book assumes a knowledge of the basics of set theory and the principal properties of the fundamental sets of numbers, for which we use the following notation.

$\mathbf{N}$: the set of natural numbers (including 0);
$\mathbf{Z}$: the set of integers;
$\mathbf{Q}$: the set of rational numbers;
$\mathbf{R}$: the set of real numbers;
$\mathbf{C}$: the set of complex numbers.

On a first reading of this text, it is not strictly necessary to be familiar with complex numbers.

The notation and symbols used are those most commonly used in the mathematical literature. For the reader's benefit, they are listed here.

The empty set is denoted $\emptyset$.

$A \subset B$ and $B \supset A$ mean that the set A is a subset of the set B.

$a \in A$ means that a is an element of the set A.

If $A \subset B$, then $B \setminus A$ denotes the difference B minus A, consisting of the elements of B which do not belong to A.

If $n \geq 1$ is an integer and A is a set, then A^n denotes the Cartesian product of A with itself n times.

The expression

$$\begin{aligned} f : A &\rightarrow B \\ a &\mapsto b \end{aligned}$$

means that the map f from the set A to the set B sends the element $a \in A$ to the element $b \in B$.

If $f : A \to B$ and $g : B \to C$ are two maps, then their composite is denoted $g \circ f$.

For every positive integer k the symbol $k!$ means the product $1.2.3 \ldots k$ and is called k factorial. By definition, one takes $0! = 1$.

Given $a, b \in \mathbf{R}$, with $a < b$, the symbols (a, b), $[a, b]$, $(a, b]$ and $[a, b)$ mean the intervals with end points a and b which are, respectively, open, closed, open at the left and open at the right.

The conjugate $a - ib$ of the complex number $z = a + ib$ is denoted $\bar{z}$. The modulus of z is $|z| = \sqrt{a^2 + b^2}$.

For other symbols, we refer the reader to the list at the end of the book.

The notions introduced in the appendices are used liberally throughout the text.

Part I
Affine Geometry

1

Vectors and vector spaces

The study of geometry in secondary schools is based upon Euclid's axiomatic system in the modern formulation given at the end of the nineteenth century by David Hilbert. For plane geometry, this system considers points and lines as the primitive objects. It also takes as primitive the notions of a point belonging to a line, a point lying between two points, equality of segments and equality of angles (the notions of segment and angle are defined in terms of the axioms).

There is an analogous system of axioms for space geometry.

We will adopt a different point of view, founding geometry on the concept of 'vector'. The axiomatics based on this concept are not only very simple but are also of great importance throughout Mathematics.

To motivate the definitions needed, we begin by introducing the concept of vectors in the Euclidean plane and in Euclidean space (which henceforth we refer to as the *ordinary plane* and *space*), and then highlight those properties which will subsequently be used to formulate the axioms. For now, we limit ourselves to an intuitive approach without being too concerned about giving complete proofs.

A *based vector* (or *oriented segment*) in ordinary space is specified by a *base point* A and an *end point* B and is denoted by the symbol (A, B). The point A is also called the *initial point* of the vector. A based vector is represented by an arrow which joins the points A and B as in Fig. 1.1.

Two based vectors (A, B) and (C, D) are said to be *equivalent* if they have the same *direction* and the same *length*. That is, they are equivalent if they lie on two parallel (possibly coincident) lines

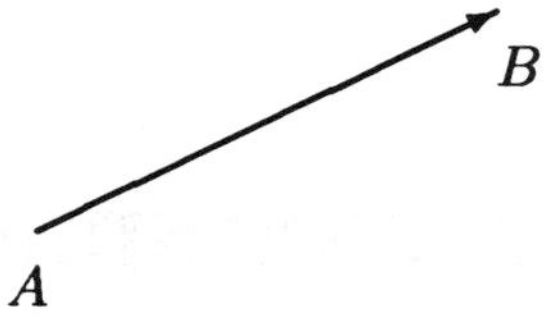

Fig. 1.1

and if moving one of the lines in such a way that it always remains parallel to the other it is possible to move one segment so that both the initial and the end points coincide. In the set of all based vectors this equivalence is indeed an equivalence relation because it satisfies for obvious reasons the three properties of reflexivity, symmetry and transitivity. A *geometric vector* (or simply *vector*) is by definition an equivalence class of based vectors, that is, it is the set of all oriented segments equivalent to a given oriented segment (Fig 1.2). Vectors will usually be denoted by letters in bold face $\mathbf{a}, \mathbf{b}, \mathbf{v}, \mathbf{w}$, etc.

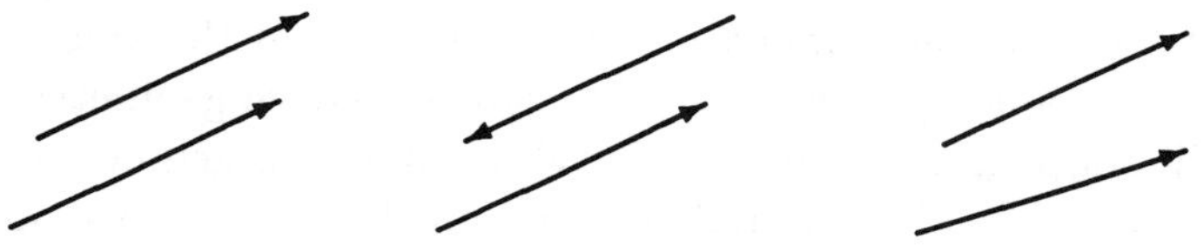

Fig. 1.2

Every based vector that determines the geometric vector $\mathbf{a}$ is said to be a *representative* of $\mathbf{a}$. The vector having representative (A, B) will also be denoted $\overrightarrow{AB}$. Given a point A each vector $\mathbf{a}$ has one and only one representative whose base point is A.

In the definition, we did not exclude the possibility that $A = B$. The vector determined by (A, A) is called the *zero vector*: it has zero length and undefined direction and is denoted by $\mathbf{0}$.

The *sum of two vectors* can be defined using their representatives in the following way (Fig. 1.3).

Let $\mathbf{a} = \overrightarrow{AB}$ and $\mathbf{b} = \overrightarrow{BC}$, then $\mathbf{a} + \mathbf{b} = \overrightarrow{AC}$.

If instead the vectors $\mathbf{a}$ and $\mathbf{b}$ are given by representatives based

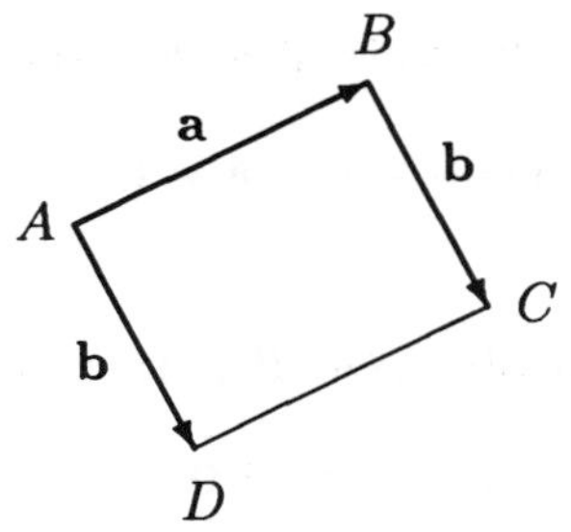

Fig. 1.3

at the same point, that is $\mathbf{a} = \overrightarrow{AB}$ and $\mathbf{b} = \overrightarrow{AD}$ then $\mathbf{a} + \mathbf{b} = \overrightarrow{AC}$, where C is the fourth vertex of the parallelogram whose other three vertices are A, B, and D with C lying opposite the common base point A. This method of constructing $\mathbf{a} + \mathbf{b}$ is called the *parallelogram rule.*

The operation of adding two vectors is associative, that is,

$$\mathbf{a} + (\mathbf{b} + \mathbf{c}) = (\mathbf{a} + \mathbf{b}) + \mathbf{c} \tag{1.1}$$

for every triple of vectors $\mathbf{a}, \mathbf{b}, \mathbf{c}$. This is verified immediately using Fig. 1.4.

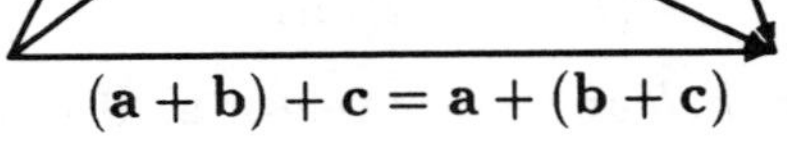

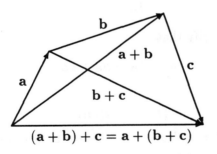

Fig. 1.4

From (1.1) it follows that when adding three vectors it is possible to omit the brackets because the expression $\mathbf{a} + \mathbf{b} + \mathbf{c}$ has only one meaning. A similar property holds for the sum of any finite number of vectors — see Observation 1.3(2).

From the way addition of vectors is defined, it is obviously commutative, that is,

$$\mathbf{a} + \mathbf{b} = \mathbf{b} + \mathbf{a}$$

for every pair of vectors $\mathbf{a}, \mathbf{b}$. Note also that the vector $\mathbf{0}$ satisfies

$$\mathbf{a} + \mathbf{0} = \mathbf{0} + \mathbf{a} = \mathbf{a}$$

for every vector $\mathbf{a}$.

If $\mathbf{a} = \overrightarrow{AB}$ then $-\mathbf{a}$ denotes $\overrightarrow{BA}$. It is then obvious that

$$\mathbf{a} + (-\mathbf{a}) = \mathbf{0}.$$

We now define the product of a vector $\mathbf{a}$ by a scalar k (in the context of vectors, real numbers are often called scalars). This is, by definition, the vector $k\mathbf{a}$ which is parallel to $\mathbf{a}$, its length is that of $\mathbf{a}$ multiplied by $|k|$, and its direction is either the same as that of $\mathbf{a}$ if $k > 0$, or opposite to that of $\mathbf{a}$ if $k < 0$; if $k = 0$ or $\mathbf{a} = \mathbf{0}$ then $k\mathbf{a} = \mathbf{0}$.

The operation of multiplication of a vector by a scalar is compatible with the operation of summing two vectors, and with the operations of addition and multiplication of scalars. For example, the following identity is easy to establish:

$$n\mathbf{a} = \mathbf{a} + \mathbf{a} + \cdots + \mathbf{a} \quad (n \text{ times})$$

for every vector $\mathbf{a}$ and every positive integer n. In particular,

$$1\mathbf{a} = \mathbf{a}.$$

One can easily verify that

$$(h + k)\mathbf{a} = h\mathbf{a} + k\mathbf{a}$$

and that

$$(kh)\mathbf{a} = k(h\mathbf{a})$$

for every pair of scalars k, h and every vector $\mathbf{a}$. It is also easy to verify the following identity geometrically:

$$k(\mathbf{a} + \mathbf{b}) = k\mathbf{a} + k\mathbf{b}$$

for every scalar k and every pair of vectors $\mathbf{a}, \mathbf{b}$.

In a similar way, one can introduce vectors in a line, and vectors in the ordinary plane.

It is important to notice that in order to define vectors and the operations of addition and scalar multiplication, we have only used the notion of two lines being parallel and the possibility of comparing the lengths of two segments on parallel lines (i.e. that it is possible to find a real number k which represents the ratio of the lengths of two segments, and conversely to associate to any segment and any scalar k a second segment whose length is k times that of the first segment). The axioms of Euclidean geometry guarantee that these are indeed possible.

In our definitions we have not needed to compare two arbitrary segments or to measure angles between two lines. In particular, it is not necessary to have at our disposal an absolute unit for measuring distance, nor do we need the concept of perpendicularity.

Now that we have verified, in an intuitive geometric way, the properties of vectors, we will turn this point of view upside down and take these properties as the axioms for the definition of 'vector space'.

A vector space is always defined with respect to a *field* of scalars (cf. Appendix A), which in the preceeding examples was $\mathbf{R}$ but which can be chosen completely generally. We will not retain the maximum possible generality, but will take our field to be a subfield of $\mathbf{C}$.

Consequently, with the exception of Appendix A, we will from now on *denote by* $\mathbf{K}$ *a subfield of* $\mathbf{C}$, *which we suppose to be fixed once and for all.*

At first reading it is sufficient to limit oneself to the case $\mathbf{K} = \mathbf{R}$. Nevertheless, it is important to bear in mind that what follows is valid in far greater generality.

Definition 1.1

A vector space over $\mathbf{K}$, or a $\mathbf{K}$-vector space, is a non-empty set $\mathbf{V}$ such that,

1) for every pair of elements $\mathbf{v}, \mathbf{w} \in \mathbf{V}$ there is defined a third element of $\mathbf{V}$ called the *sum* of $\mathbf{v}$ and $\mathbf{w}$, which is denoted $\mathbf{v} + \mathbf{w}$,
2) for every $\mathbf{v} \in \mathbf{V}$ and every $k \in \mathbf{K}$, there is defined an element of $\mathbf{V}$ called the product of $\mathbf{v}$ and k, denoted $k\mathbf{v}$,

in such way that the following properties are satisfied:

[VS1] (Associativity of vector addition) For every $\mathbf{u}, \mathbf{v}, \mathbf{w} \in \mathbf{V}$,

$$(\mathbf{u} + \mathbf{v}) + \mathbf{w} = \mathbf{u} + (\mathbf{v} + \mathbf{w}).$$

[VS2] (Existence of zero) There exists an element $\mathbf{0} \in \mathbf{V}$, called the *zero vector*, such that,

$$\mathbf{0} + \mathbf{v} = \mathbf{v} + \mathbf{0} = \mathbf{v}$$

for every $\mathbf{v} \in \mathbf{V}$.

[VS3] (Existence of opposites) For every $\mathbf{v} \in \mathbf{V}$ the element $(-1)\mathbf{v}$ satisfies the identity,

$$\mathbf{v} + (-1)\mathbf{v} = \mathbf{0}.$$

[VS4] (Commutativity) For every $\mathbf{u}, \mathbf{v} \in \mathbf{V}$ one has,

$$\mathbf{u} + \mathbf{v} = \mathbf{v} + \mathbf{u}.$$

[VS5] (Distributive property over sum of vectors) For every $\mathbf{u}, \mathbf{v} \in \mathbf{V}$ and every $k \in \mathbf{K}$, one has,

$$k(\mathbf{u} + \mathbf{v}) = k\mathbf{u} + k\mathbf{v}.$$

[VS6] (Distributive property over sum of scalars) For every $\mathbf{v} \in \mathbf{V}$ and every $h, k \in \mathbf{K}$, one has,

$$(h + k)\mathbf{v} = h\mathbf{v} + k\mathbf{v}.$$

[VS7] (Associativity of scalar multiplication) For every $\mathbf{v} \in \mathbf{V}$ and every $h, k \in \mathbf{K}$, one has,

$$(hk)\mathbf{v} = h(k\mathbf{v}).$$

[VS8] For every $\mathbf{v} \in \mathbf{V}$ one has,

$$1\mathbf{v} = \mathbf{v}.$$

We will say that the two operations above define on $\mathbf{V}$ the *structure of a* $\mathbf{K}$*-vector space.*

It may be possible to define on a set the structure of a $\mathbf{K}$-vector space in more than one way. In other words there may be more than one way to define on $\mathbf{V}$ the operations that make it into a vector space, and possibly over different fields — see Example 1.2(4).

The elements of a vector space $\mathbf{V}$ are called *vectors*, and the elements of $\mathbf{K}$ are called *scalars*.

The vectors of the form $k\mathbf{v}$, $k \in \mathbf{K}$ are said to be *proportional to* or *multiples of* $\mathbf{v}$.

For every $\mathbf{v} \in \mathbf{V}$, the vector $(-1)\mathbf{v}$, called the opposite of $\mathbf{v}$, is written $-\mathbf{v}$, and one writes $\mathbf{u} - \mathbf{v}$ rather than $\mathbf{u} + (-\mathbf{v})$.

From the 8 axioms, there follow several elementary properties of vector spaces which will be discussed at the end of this chapter (in Observations 1.3), and which from then on we will assume the reader knows.

When $\mathbf{K} = \mathbf{R}$ or $\mathbf{K} = \mathbf{C}$, $\mathbf{V}$ is often called a *real vector space* or a *complex vector space*, respectively.

A set with just one element is in a trivial way a $\mathbf{K}$-vector space whose only element is the zero vector. We will now see a few non-trivial examples of vector spaces.

Examples 1.2

1. Let $\mathbf{V}$ be the set of geometric vectors in the plane. The properties that we derived at the beginning of this chapter show that the operations of addition and scalar multiplication define on $\mathbf{V}$ the structure of a vector space over $\mathbf{R}$. In a similar way, the sets of geometric vectors in space or in a line form vector spaces over $\mathbf{R}$.

2. Let $n \geq 1$ be an integer and $\mathbf{V} = \mathbf{K}^n$, the set of ordered n-tuples of elements in $\mathbf{K}$.

   Define the sum of two n-tuples $(x_1, \ldots, x_n)$ and $(y_1, \ldots, y_n)$ by

   $$(x_1, \ldots, x_n) + (y_1, \ldots, y_n) = (x_1 + y_1, \ldots, x_n + y_n),$$

   and for each $k \in \mathbf{K}$ and $(x_1, \ldots, x_n) \in \mathbf{K}^n$, define

   $$k(x_1, \ldots, x_n) = (kx_1, \ldots, kx_n);$$

   in particular $-(x_1, \ldots, x_n) = (-x_1, \ldots, -x_n)$.

   It is easy to check that with these operations $\mathbf{K}^n$ satisfies axioms VS1, ..., VS8, and so is a $\mathbf{K}$-vector space. It is usually called the *numerical n-space over* $\mathbf{K}$.

   The numerical 1-space over $\mathbf{K}$ is $\mathbf{K}$ itself, which with the usual definitions of sum and product, is a vector space over itself.

   If $(x_1, \ldots, x_n) \in \mathbf{K}^n$, the scalars $x_1, \ldots, x_n$ are called the *components*, and x_i the *i-th component*, of $(x_1, \ldots, x_n)$.

3. Let I be any non-empty set and let $\mathbf{V}$ be the set whose elements are the maps $f : I \to \mathbf{K}$. For every $f, g \in \mathbf{V}$ define $f + g : I \to \mathbf{K}$ by putting

   $$(f + g)(x) = f(x) + g(x)$$

for every $x \in I$. We thus obtain an element $f + g \in \mathbf{V}$. If $k \in \mathbf{K}$ and $f \in \mathbf{V}$, define $kf : I \to K$ by

$$(kf)(x) = kf(x)$$

for every $x \in I$. We thus obtain an element $kf \in \mathbf{V}$.

It is easy to verify that with these two operations $\mathbf{V}$ is a $\mathbf{K}$-vector space.

4. Let $\mathbf{F}$ be a subfield of $\mathbf{K}$ and let $\mathbf{V}$ be a vector space over $\mathbf{K}$. Then $\mathbf{V}$ has induced on it the structure of an $\mathbf{F}$-vector space by the same operations that define the $\mathbf{K}$-vector space structure. In other words, the sum of two vectors remains the same as before, and multiplication by a scalar $\alpha \in \mathbf{F}$ is defined by treating α as an element of $\mathbf{K}$, and so by using the definition of multiplication by scalars in $\mathbf{K}$.

   For example the field $\mathbf{C}$ of complex numbers can be considered not only as a complex vector space (the numerical 1-space over $\mathbf{C}$), but also as a real vector space because $\mathbf{R}$ is a subfield of $\mathbf{C}$.

Observations 1.3

1. Some of the properties that follow from the axioms VS1, ..., VS8 are obvious for geometric vectors, but require a proof for more general vector spaces.

   Let $\mathbf{V}$ be a $\mathbf{K}$-vector space.

   In $\mathbf{V}$ *there is a unique zero vector*. That is, if $\mathbf{0}_1$ and $\mathbf{0}_2$ are two vectors satisfying $\mathbf{0}_1 + \mathbf{v} = \mathbf{v}$ and $\mathbf{0}_2 + \mathbf{v} = \mathbf{v}$ for every $\mathbf{v} \in \mathbf{V}$, then $\mathbf{0}_1 = \mathbf{0}_2$.

   To see this, first put $\mathbf{v} = \mathbf{0}_1$ in $\mathbf{0}_2 + \mathbf{v} = \mathbf{v}$ to obtain $\mathbf{0}_2 + \mathbf{0}_1 = \mathbf{0}_1$. Next put $\mathbf{v} = \mathbf{0}_2$ in the other equation, to obtain $\mathbf{0}_2 + \mathbf{0}_1 = \mathbf{0}_2$. Thus

$$\mathbf{0}_1 = \mathbf{0}_2 + \mathbf{0}_1 = \mathbf{0}_1 + \mathbf{0}_2 = \mathbf{0}_2.$$

   For every $\mathbf{v} \in \mathbf{V}$ *there is only one opposite*, that is, if $\mathbf{v} + \mathbf{v}_1 = \mathbf{0} = \mathbf{v} + \mathbf{v}_2$ then $\mathbf{v}_1 = \mathbf{v}_2$.

   Indeed, one has,

$$\begin{aligned} \mathbf{v}_1 &= \mathbf{0} + \mathbf{v}_1 = (\mathbf{v} + \mathbf{v}_2) + \mathbf{v}_1 = \mathbf{v} + (\mathbf{v}_2 + \mathbf{v}_1) \\ &= \mathbf{v} + (\mathbf{v}_1 + \mathbf{v}_2) = (\mathbf{v} + \mathbf{v}_1) + \mathbf{v}_2 = \mathbf{0} + \mathbf{v}_2 = \mathbf{v}_2. \end{aligned}$$

For every $\mathbf{a}, \mathbf{b} \in \mathbf{V}$, *the equation* $\mathbf{x}+\mathbf{a} = \mathbf{b}$ *has a unique solution* $\mathbf{x} = \mathbf{b} - \mathbf{a}$. In fact, $(\mathbf{b} - \mathbf{a}) + \mathbf{a} = \mathbf{b}$, and, as $\mathbf{x} + \mathbf{a} = \mathbf{b}$, one has,

$$\mathbf{x} = (\mathbf{x} + \mathbf{a}) - \mathbf{a} = \mathbf{b} - \mathbf{a}.$$

For every $\mathbf{v} \in \mathbf{V}$, *one has* $0.\mathbf{v} = \mathbf{0}$, where $0 \in \mathbf{K}$ is the zero scalar and $\mathbf{0}$ is the zero vector. This is deduced as follows: $0\mathbf{v} = (0+0)\mathbf{v} = 0\mathbf{v} + 0\mathbf{v}$, thus $0\mathbf{v} = 0\mathbf{v} - 0\mathbf{v} = \mathbf{0}$.

Analagously, $k\mathbf{0} = \mathbf{0}$ for every $k \in \mathbf{K}$. This follows from a similar argument: $k\mathbf{0} = k(\mathbf{0} + \mathbf{0}) = k\mathbf{0} + k\mathbf{0}$, so $k\mathbf{0} = k\mathbf{0} - k\mathbf{0} = \mathbf{0}$.

2. Let $\mathbf{V}$ be a $\mathbf{K}$-vector space. Given three vectors $\mathbf{u}, \mathbf{v}, \mathbf{w} \in \mathbf{V}$ one writes $\mathbf{u} + \mathbf{v} + \mathbf{w}$ for the sum calculated in one of the two possible ways, which, by axiom VS1, give the same result.

   Suppose now we are given $n \geq 3$ vectors $\mathbf{v}_1, \ldots, \mathbf{v}_n \in \mathbf{V}$. We want to show that calculating their sum always gives the same answer, no matter how the brackets are distributed:

$$(\mathbf{v}_1 + (\mathbf{v}_2 + (\cdots + (\mathbf{v}_{n-1} + \mathbf{v}_n)\cdots))) \tag{1.2}$$

   gives a result which depends only on $\mathbf{v}_1, \ldots, \mathbf{v}_n$, and so can be written without the brackets as $\mathbf{v}_1 + \mathbf{v}_2 + \cdots + \mathbf{v}_n$.

   The proof is by induction on n: if $n = 3$ the statement is true because of axiom VS1. Let $n \geq 4$, and suppose that the statement is true for every integer k between 3 and $n - 1$. We can write two different sums (1.2) as $\alpha + \beta$ and $\gamma + \delta$ respectively, where α, β are sums in which $\mathbf{v}_1, \ldots, \mathbf{v}_k$ and $\mathbf{v}_{k+1}, \ldots, \mathbf{v}_n$ appear respectively, while γ, δ are such that $\mathbf{v}_1, \ldots, \mathbf{v}_j$ and $\mathbf{v}_{j+1}, \ldots, \mathbf{v}_n$ appear respectively, for some integers j, k less than n.

   If $j = k$, then by the inductive hypothesis, $\alpha = \gamma$ and $\beta = \delta$, whence $\alpha + \beta = \gamma + \delta$. Suppose now that $k < j$. By the inductive hypothesis one has,

$$\begin{aligned} \alpha + (\mathbf{v}_{k+1} + \cdots + \mathbf{v}_j) &= \gamma, \\ (\mathbf{v}_{k+1} + \cdots + \mathbf{v}_j) + \delta &= \gamma + \delta, \end{aligned}$$

   and so $\alpha + \beta = \alpha + (v_{k+1} + \cdots + \mathbf{v}_j) + \delta = \gamma + \delta$.

EXERCISES

1.1 A map $s : \mathbf{N} \to \mathbf{K}$ of the set of natural numbers to $\mathbf{K}$ is called a *sequence of elements of* $\mathbf{K}$. If $s(n) = a_n \in \mathbf{K}$, the sequence s is also denoted by $\{a_n\}_{n\in\mathbf{N}}$.

Let $S_{\mathbf{K}}$ be the set of all sequences of elements of $\mathbf{K}$. In $S_{\mathbf{K}}$ one defines the following operations:

$$\{a_n\} + \{b_n\} = \{a_n + b_n\}, \qquad k\{a_n\} = \{ka_n\}$$

for every $\{a_n\}, \{b_n\} \in S_{\mathbf{K}}$, and $k \in \mathbf{K}$. Show that, with these operations, $S_{\mathbf{K}}$ is a $\mathbf{K}$-vector space (this is a particular example of the vector space $\mathbf{V}$ considered in Example 1.2(3)).

1.2 A sequence $\{a_n\} \in S_{\mathbf{R}}$ is said to be *bounded* if there is a real number $R \in \mathbf{R}$ such that $a_n \leq R$ for every $n \in \mathbf{N}$. Let $B_{\mathbf{R}}$ denote the set of bounded sequences. Show that, with the operations defined in Exercise 1, $B_{\mathbf{R}}$ is a real vector space.

1.3 Let $a, b \in \mathbf{R}, a < b$, and let $C_{(a,b)}$ be the set of all continuous real valued functions defined on the interval (a, b). For every $f, g \in C_{(a,b)}$ the function $f + g : (a, b) \to \mathbf{R}$ is defined by

$$(f+g)(x) = f(x) + g(x) \qquad \text{for every} \qquad x \in (a, b).$$

If $f \in C_{(a,b)}$ and $c \in \mathbf{R}$ then $cf : (a, b) \to \mathbf{R}$ is defined by

$$(cf)\,(x) = cf(x) \qquad \text{for every} \qquad x \in (a, b).$$

Prove that $f+g$ and cf are continuous, so that these define two operations on $C_{(a,b)}$. Prove that with these operations, $C_{(a,b)}$ has the structure of a real vector space.

1.4 Let X be an unknown and let $\mathbf{K}[X]$ be the set of polynomials in X with coefficients in $\mathbf{K}$. For every $f, g \in \mathbf{K}[X]$ and $\alpha \in \mathbf{K}$, let $f + g \in \mathbf{K}[X]$ be the polynomial sum of f and g, and let αf be the polynomial product of α, considered as the constant polynomial, and f. Prove that these operations define on $\mathbf{K}[X]$ the structure of a $\mathbf{K}$-vector space.

2

Matrices

Let m, n be positive integers. An $m \times n$ *matrix* is a rectangular array

$$A = \begin{pmatrix} a_{11} & a_{12} & \dots & a_{1n} \\ a_{21} & a_{22} & \dots & a_{2n} \\ \vdots & \vdots & & \vdots \\ a_{m1} & a_{m2} & \dots & a_{mn} \end{pmatrix}$$

of mn elements of $\mathbf{K}$. We will also write $A = (a_{ij})_{1 \leq i \leq m, 1 \leq j \leq n}$, or just $A = (a_{ij})$. The *i-th row* of the matrix A is the $1 \times n$ matrix

$$A^{(i)} = \begin{pmatrix} a_{i1} & a_{i2} & \dots & a_{in} \end{pmatrix}$$

(for $i = 1, 2, \dots, m$) and the *j-th column* is the $m \times 1$ matrix

$$A_{(j)} = \begin{pmatrix} a_{1j} \\ a_{2j} \\ \vdots \\ a_{mj} \end{pmatrix}$$

(for $j = 1, 2, \dots, n$).

The matrix A has m rows and n columns. Each entry a_{ij} of the matrix has two indices: the first for the row (the *row index*) and the second for the column (the *column index*) to which the entry belongs. a_{ij} is called the *entry in position* i, j.

For example,

$$\begin{pmatrix} 3 & -2 & \frac{4}{5} \\ \sqrt{2} & \pi & 183 \end{pmatrix} \tag{2.1}$$

is a 2×3 matrix with elements in $\mathbf{R}$; its rows are

$$\begin{pmatrix} 3 & -2 & \frac{4}{5} \end{pmatrix}, \qquad \begin{pmatrix} \sqrt{2} & \pi & 183 \end{pmatrix},$$

and its columns are

$$\begin{pmatrix} 3 \\ \sqrt{2} \end{pmatrix}, \quad \begin{pmatrix} -2 \\ \pi \end{pmatrix}, \quad \begin{pmatrix} \frac{4}{5} \\ 183 \end{pmatrix}.$$

The element in position 2, 1 is $\sqrt{2}$, and that in position 1, 3 is $\frac{4}{5}$.

If $m = n$ then A is said to be a *square matrix of order n.*

A $1 \times n$ matrix, i.e. with a single row and n columns, is often called a *row vector* or a *row n-vector*, while an $n \times 1$ matrix, with n rows and one column, is called a *column vector* or a *column n-vector.*

The transpose of an $m \times n$ matrix $A = (a_{ij})$ is the $n \times m$ matrix,

$$A^t = (a_{ji}) = \begin{pmatrix} a_{11} & a_{21} & \dots & a_{m1} \\ a_{12} & a_{22} & \dots & a_{m2} \\ \vdots & \vdots & & \vdots \\ a_{1n} & a_{2n} & \dots & a_{mn} \end{pmatrix},$$

obtained by exchanging the rows and columns. For example, the transpose of matrix (2.1) is

$$\begin{pmatrix} 3 & \sqrt{2} \\ -2 & \pi \\ \frac{4}{5} & 183 \end{pmatrix}.$$

The set of all $m \times n$ matrices with entries in $\mathbf{K}$ is denoted by $M_{m,n}(\mathbf{K})$, and the set of all square matrices of order n by $M_n(\mathbf{K})$. Of course, $M_{n,n}(\mathbf{K}) = M_n(\mathbf{K})$ and $M_1(\mathbf{K}) = \mathbf{K}$ because a 1×1 matrix is just a single element of $\mathbf{K}$.

A useful convention which we follow is to identify $\mathbf{K}^n$ with $M_{1,n}(\mathbf{K})$, the set of column n-vectors. The column vector,

$$\mathbf{x} = \begin{pmatrix} x_1 \\ x_2 \\ \vdots \\ x_n \end{pmatrix}$$

is identified with the element $(x_1, x_2, \dots, x_n) \in \mathbf{K}^n$. We also write

$$\mathbf{x} = \begin{pmatrix} x_1 & x_2 & \cdots & x_n \end{pmatrix}^t.$$

Proposition 2.1
Putting,

$$(a_{ij}) + (b_{ij}) = (a_{ij} + b_{ij})$$
$$k(a_{ij}) = (ka_{ij})$$

for every $(a_{ij}), (b_{ij}) \in M_{m,n}(\mathbf{K})$ *and every* $k \in \mathbf{K}$ *defines on* $M_{m,n}(\mathbf{K})$ *the structure of a* $\mathbf{K}$*-vector space. The zero of the vector space is the* $m \times n$ zero matrix, *which is the* $m \times n$ *matrix with every entry equal to* 0.

The proof is left to the reader.

For $A = (a_{ij})$, $B = (b_{ij})$ and $k \in \mathbf{K}$, the matrices $(a_{ij} + b_{ij})$ and (ka_{ij}) are written $A + B$ and kA respectively.

It is possible to define an operation of multiplication between matrices. This is done as follows.

Given a row n-vector

$$A = \begin{pmatrix} a_1 & a_2 & \cdots & a_n \end{pmatrix}$$

and a column n-vector

$$B = \begin{pmatrix} b_1 \\ b_2 \\ \vdots \\ b_n \end{pmatrix},$$

their product is the element of $\mathbf{K}$ defined by the following identity:

$$\begin{pmatrix} a_1 & a_2 & \cdots & a_n \end{pmatrix} \begin{pmatrix} b_1 \\ b_2 \\ \vdots \\ b_n \end{pmatrix} = a_1 b_1 + a_2 b_2 + \cdots + a_n b_n. \tag{2.2}$$

More generally, given a matrix $A = (a_{ij}) \in M_{m,n}(\mathbf{K})$ and a matrix $B = (b_{jk}) \in M_{n,p}(\mathbf{K})$, their *product* is the matrix $AB \in M_{m,p}(\mathbf{K})$ whose entry in position i, k is the product of the i-th row of A and the k-th column of B. The formula is,

$$AB = (A^{(i)} B_{(k)}) = (a_{i1} b_{1k} + a_{i2} b_{2k} + \cdots + a_{in} b_{nk}).$$

The product of a row vector and a column vector defined in (2.2) is a particular case of the product of two matrices.

Note that the product AB is defined only if the number of columns of A is equal to the number of rows of B; this fact is expressed by saying that A and B *may by multiplied together.*

For example, the matrices

$$\begin{pmatrix} 1 & 2 & 1 \\ 2 & 3 & 0 \end{pmatrix} \quad \text{and} \quad \begin{pmatrix} 2 & 1 & 0 & -1 \\ 0 & 1 & 1 & 0 \\ 1 & 3 & 6 & 2 \end{pmatrix}$$

may be multiplied together, and their product is,

$$\begin{pmatrix} 1 & 2 & 1 \\ 2 & 3 & 0 \end{pmatrix} \begin{pmatrix} 2 & 1 & 0 & -1 \\ 0 & 1 & 1 & 0 \\ 1 & 3 & 6 & 2 \end{pmatrix} = \begin{pmatrix} 3 & 6 & 8 & 1 \\ 4 & 5 & 3 & -2 \end{pmatrix}.$$

On the other hand, the matrices

$$\begin{pmatrix} 2 & 1 & 0 & -1 \\ 0 & 1 & 1 & 0 \\ 1 & 3 & 6 & 2 \end{pmatrix} \quad \text{and} \quad \begin{pmatrix} 1 & 2 & 1 \\ 2 & 3 & 0 \end{pmatrix}$$

cannot be multiplied together.

In particular, any two square matrices of the same order $A, B \in M_n(\mathbf{K})$ may be multiplied. Note, though, that in general $AB \neq BA$. For example,

$$\begin{pmatrix} 1 & 1 \\ 0 & 1 \end{pmatrix} \begin{pmatrix} 1 & 0 \\ 1 & 0 \end{pmatrix} = \begin{pmatrix} 2 & 0 \\ 1 & 0 \end{pmatrix},$$

while,

$$\begin{pmatrix} 1 & 0 \\ 1 & 0 \end{pmatrix} \begin{pmatrix} 1 & 1 \\ 0 & 1 \end{pmatrix} = \begin{pmatrix} 1 & 1 \\ 1 & 1 \end{pmatrix}.$$

If $A = (a_{ij}) \in M_n(\mathbf{K})$, the entries $a_{11}, a_{22}, \ldots, a_{nn}$ constitute the *diagonal* of A. If all entries a_{ij} with $i \neq j$ are zero, then the matrix is called a *diagonal matrix.* A particularly important $n \times n$ diagonal matrix is the *identity matrix* $\mathbf{I}_n = (\delta_{ij})$, where

$$\delta_{ij} = \begin{cases} 1 & \text{if } i = j, \\ 0 & \text{otherwise.} \end{cases} \tag{2.3}$$

The δ_{ij} defined in (2.3) is called the *Kronecker symbol.*

For example,

$$\mathbf{I}_1 = (1), \qquad \mathbf{I}_2 = \begin{pmatrix} 1 & 0 \\ 0 & 1 \end{pmatrix}, \qquad \mathbf{I}_3 = \begin{pmatrix} 1 & 0 & 0 \\ 0 & 1 & 0 \\ 0 & 0 & 1 \end{pmatrix}.$$

A square matrix $A \in M_n(\mathbf{K})$ is said to be *upper triangular* if $a_{ij} = 0$ for every $i > j$ and *lower triangular* if $a_{ij} = 0$ for every $i < j$. Furthermore, if $a_{ij} = 0$ for every $i \geq j$, then A is said to be *strictly upper triangular. Strictly lower triangular* matrices are defined similarly.

For example, of the three square matrices with real entries,

$$\begin{pmatrix} 7 & 0 \\ 0 & 2 \end{pmatrix}, \quad \begin{pmatrix} -1 & \sqrt{3} & -7 \\ 0 & 0 & \frac{1}{2} \\ 0 & 0 & -3 \end{pmatrix}, \quad \begin{pmatrix} 0 & 0 & 0 & 0 \\ 7 & 0 & 0 & 0 \\ 1 & \frac{\pi}{2} & 0 & 0 \\ 0 & 1 & \frac{1}{2} & 0 \end{pmatrix},$$

the first is diagonal, the second upper triangular and the third is strictly lower triangular.

A square matrix $A \in M_n(\mathbf{K})$ satisfying $A^t = A$ is said to be *symmetric*; if on the other hand, $A^t = -A$ then it is said to be *skew symmetric*. For example, the matrix,

$$\begin{pmatrix} -1 & 5 & 3 \\ 5 & 0 & -1 \\ 3 & -1 & \frac{\pi}{2} \end{pmatrix}$$

is symmetric, while

$$\begin{pmatrix} 0 & -5 & \frac{1}{2} \\ 5 & 0 & 3 \\ -\frac{1}{2} & -3 & 0 \end{pmatrix}$$

is skew-symmetric. Notice that any skew-symmetric matrix must have only zeros down the diagonal. On the other hand, the matrix

$$\begin{pmatrix} 1 & 0 \\ 3 & \sqrt{7} \end{pmatrix}$$

is neither symmetric nor skew-symmetric.

Proposition 2.2

1) *If* $A, B \in M_{m,n}(\mathbf{K})$, $C, D \in M_{n,p}(\mathbf{K})$ *and* $k \in \mathbf{K}$, *then*

$$(A + B)C = AC + BC$$

$$A(C + D) = AC + AD$$

$$A(kC) = k(AC) = (kA)C$$

$$A\mathbf{I}_n = A, \quad \mathbf{I}_n C = C.$$

2) *If* $A \in M_{m,n}(\mathbf{K})$, $B \in M_{n,p}(\mathbf{K})$ *and* $C \in M_{p,s}(\mathbf{K})$, *then*

$$(AB)C = A(BC).$$

3) *If* A *and* B *can be multiplied, then* B^t *and* A^t *can be multiplied, and*

$$(AB)^t = B^t A^t.$$

4) *If* $A, B \in M_{m,n}(\mathbf{K})$ *then*

$$A^t + B^t = (A+B)^t.$$

Proof

1) Let $A^{(i)} = (\, a_{i1} \quad a_{i2} \quad \dots \quad a_{in} \,)$, and $B^{(i)} = (\, b_{i1} \quad b_{i2} \quad \dots \quad b_{in} \,)$ be the i-th rows of A and B (for $i = 1, \dots, m$) and

$$C_{(k)} = \begin{pmatrix} c_{1k} \\ c_{2k} \\ \vdots \\ c_{nk} \end{pmatrix}$$

be the k-th column of C (for $k = 1, \dots, p$). The entry of $(A+B)C$ in position i, k is then,

$$(A+B)^{(i)} C_{(k)} = (a_{i1} + b_{i1})c_{1k} + (a_{i2} + b_{i2})c_{2k} + \cdots + (a_{in} + b_{in})c_{nk},$$

while the entry of $AC + BC$ in position i, k is

$$\begin{aligned} A^{(i)} C_{(k)} + B^{(i)} C_{(k)} =& (a_{i1}c_{1k} + a_{i2}c_{2k} + \cdots + a_{in}c_{nk}) \\ &+ (b_{i1}c_{1k} + b_{i2}c_{2k} + \cdots + b_{in}c_{nk}). \end{aligned}$$

These two are clearly equal, which proves the first identity in (1). The second and third identities are proved in a similar manner.

The entry of $A\mathbf{I}_n$ in position i, j is

$$A^{(i)}(\mathbf{I}_n)_{(j)} = a_{i1}0 + \cdots + a_{ij-1}0 + a_{ij}1 + a_{ij+1}0 + \cdots + a_{in}0 = a_{ij},$$

and so $A\mathbf{I}_n = A$. The proof that $\mathbf{I}_n C = C$ is similar.

2) Notice that $AB \in M_{m,p}(\mathbf{K})$ and $BC \in M_{n,s}(\mathbf{K})$, and consequently, not only AB and C but also A and BC can be multiplied. The i-th row of AB is (for $i = 1, \dots, m$)

$$(AB)^{(i)} = (\, A^{(i)}B_{(1)} \quad A^{(i)}B_{(2)} \quad \cdots \quad A^{(i)}B_{(p)} \,),$$

while the h-th column of BC is (for $h = 1, \dots, s$)

$$(BC)_{(h)} = \begin{pmatrix} B^{(1)}C_{(h)} \\ B^{(2)}C_{(h)} \\ \vdots \\ B^{(n)}C_{(h)} \end{pmatrix}.$$

It follows that the entry of $(AB)C$ in position i, h is

$$\begin{aligned}(AB)^{(i)}C_{(h)} =& (A^{(i)}B_{(1)})c_{1h} + (A^{(i)}B_{(2)})c_{2h} + \cdots + (A^{(i)}B_{(p)})c_{ph} \\ =& (a_{i1}b_{11} + a_{i2}b_{21} + \cdots + a_{in}b_{n1})c_{1h} \\ &+ (a_{i1}b_{12} + a_{i2}b_{22} + \cdots + a_{in}b_{n2})c_{2h} \\ &+ \cdots + (a_{i1}b_{1p} + a_{i2}b_{2p} + \cdots + a_{in}b_{np})c_{ph}.\end{aligned} \tag{2.4}$$

The entry of $A(BC)$ in position i, h is, instead,

$$\begin{aligned}A^{(i)}(BC)_{(h)} =& a_{i1}(B^{(1)}C_{(h)}) + a_{i2}(B^{(2)}C_{(h)}) + \cdots + a_{in}(B^{(n)}C_{(h)}) \\ =& a_{i1}(b_{11}c_{1h} + b_{12}c_{2h} + \cdots + b_{1p}c_{ph}) \\ &+ a_{i2}(b_{21}c_{1h} + b_{22}c_{2h} + \cdots + b_{2p}c_{ph}) \\ &+ \cdots + a_{in}(b_{n1}c_{1h} + b_{n2}c_{2h} + \cdots + b_{np}c_{ph}).\end{aligned} \tag{2.5}$$

A comparison of (2.4) and (2.5) shows that they are equal, because each is the sum of all terms of the form $a_{ij}b_{jk}c_{kh}$ as j varies from 1 to n and k varies from 1 to p. Thus the matrices $(AB)C$ and $A(BC)$ coincide entry for entry, and the assertion is proved.

3) Suppose $A \in M_{m,n}(\mathbf{K})$ and $B \in M_{n,p}(\mathbf{K})$. Then $A^t \in M_{n,m}(\mathbf{K})$ and $B^t \in M_{p,n}(\mathbf{K})$ and so B^t and A^t can be multiplied. Furthermore,

$$((AB)^t)_{ij} = (AB)_{ji} = A^{(j)}B_{(i)} = (B^t)^{(i)}(A^t)_{(j)} = (B^tA^t)_{ij}.$$

4) The proof is left to the reader.

□

Proposition 2.2(2) allows us to write ABC for both of the products $(AB)C$ and $A(BC)$, since these two matrices coincide. More generally, if $A_1, A_2, \dots, A_m$ are matrices with entries in $\mathbf{K}$ such that for each $k = 1, 2, \dots, m-1$ the matrices A_k and A_{k+1} can be multiplied, one can easily show that their product does not depend on how the brackets are distributed. Thus for,

$$A_1(A_2(\cdots A_m)\cdots),$$

one need only write $A_1A_2\cdots A_m$. The proof is similar to that given in Observation 1.3(2) and is left to the reader.

A square matrix A of order n is said to be *invertible* if there is a matrix $M \in M_n(\mathbf{K})$ such that $AM = MA = \mathbf{I}_n$. If such an M exists then it is unique: for if N also satisfies $AN = NA = \mathbf{I}_n$, then

$$M = M\mathbf{I}_n = M(AN) = (MA)N = \mathbf{I}_nN = N.$$

This unique M is called the *inverse* of A and is written A^{-1}.

In fact, if $A \in M_n(\mathbf{K})$ is invertible and M satisfies one of the identities $AM = \mathbf{I}_n$ or $MA = \mathbf{I}_n$ then it necessarily satisfies the other too. Indeed, if, for example, $AM = \mathbf{I}_n$, then

$$MA = (A^{-1}A)MA = A^{-1}(AM)A = A^{-1}\mathbf{I}_nA = A^{-1}A = \mathbf{I}_n.$$

In a similar way, one shows that $MA = \mathbf{I}_n$ implies $AM = \mathbf{I}_n$.

The identity matrix $\mathbf{I}_n$ is invertible and is equal to its inverse.

It follows immediately from the definition that $(A^{-1})^{-1} = A$ for any invertible matrix $A \in M_n(K)$.

If $A, B \in M_n(\mathbf{K})$ are invertible then so is their product AB, and $(AB)^{-1} = B^{-1}A^{-1}$. Indeed,

$$(B^{-1}A^{-1})(AB) = B^{-1}(A^{-1}A)B = B^{-1}\mathbf{I}_nB = B^{-1}B = \mathbf{I}_n.$$

More generally, if $A_1, A_2, \ldots, A_k \in M_n(\mathbf{K})$ are all invertible, then their product $A_1A_2\cdots A_k$ is also invertible, and

$$(A_1A_2\cdots A_k)^{-1} = A_k^{-1}\cdots A_2^{-1}A_1^{-1}.$$

The proof is similar to the preceding one.

In Chapter 3 we will describe a procedure for finding the inverse of any invertible matrix.

The subset of $M_n(\mathbf{K})$ of invertible matrices is denoted $\mathrm{GL}_n(\mathbf{K})$.

For any integer $k \geq 1$ the product $AA\cdots A$ of $A \in M_n(\mathbf{K})$ with itself k times is written A^k; by convention, $A^0 = \mathbf{I}_n$. For $A \in \mathrm{GL}_n(\mathbf{K})$ one defines

$$A^{-k} = (A^{-1})^k.$$

A real square matrix $A \in M_n(\mathbf{R})$ is said to be *orthogonal* if $AA^t = \mathbf{I}_n$, that is, if $A^t = A^{-1}$. The set of orthogonal matrices is denoted $\mathrm{O}(n)$. By definition $\mathrm{O}(n) \subset \mathrm{GL}_n(\mathbf{R})$.

The only orthogonal 1×1 matrices are (1) and (-1). A matrix $A \in M_2(\mathbf{R})$ is orthogonal if and only if it has one of the forms,

$$A = \begin{pmatrix} a & -b \\ b & a \end{pmatrix} \tag{2.6}$$

or

$$A = \begin{pmatrix} a & b \\ b & -a \end{pmatrix} \tag{2.7}$$

with $a^2 + b^2 = 1$. Indeed, if $A = (a_{ij})$ then

$$A^t A = \begin{pmatrix} a_{11}^2 + a_{21}^2 & a_{11}a_{12} + a_{21}a_{22} \\ a_{12}a_{11} + a_{22}a_{21} & a_{12}^2 + a_{22}^2 \end{pmatrix},$$

and so $A \in \mathrm{O}(2)$ if and only if

$$a_{11}^2 + a_{21}^2 = 1 = a_{12}^2 + a_{22}^2$$

$$a_{11}a_{12} + a_{21}a_{22} = 0.$$

From these two conditions, it follows that there exists $\rho \neq 0$ such that

$$(a_{11}, a_{21}) = (-\rho a_{22}, \rho a_{12}).$$

From the first two equations, it then follows that $\rho^2 = 1$, that is $\rho = \pm 1$. Thus $a_{12} = \pm a_{21}$ and $a_{22} = \mp a_{11}$, so A is either of the form (2.6) or of the form (2.7).

We will return to a more general discussion of orthogonal matrices in Chapters 20 and 21.

To describe matrices the so-called *block notation* is often useful. This consists of writing a matrix $A \in M_{m,n}(\mathbf{K})$ in the following form:

$$A = \begin{pmatrix} A_{11} & A_{12} & \dots & A_{1k} \\ A_{21} & A_{22} & \dots & A_{2k} \\ \vdots & \vdots & & \vdots \\ A_{h1} & A_{h2} & \dots & A_{hk} \end{pmatrix},$$

where the A_{ij} are themselves matrices of some appropriate sizes. More precisely, $A_{ij} \in M_{m_i,n_j}(\mathbf{K})$ where $m_1 + m_2 + \cdots + m_h = m$ and $n_1 + n_2 + \cdots + n_k = n$.

For example, matrix (2.1) can be written in block notation as,

$$A = (B \quad C),$$

where

$$B = \begin{pmatrix} 3 & -2 \\ \sqrt{2} & \pi \end{pmatrix}, \qquad C = \begin{pmatrix} \frac{4}{5} \\ 183 \end{pmatrix}.$$

Observation 2.3

Matrices can be defined with entries in any domain $\mathbf{D}$ (see Appendix A). The set of all $m \times n$ matrices with entries in $\mathbf{D}$ is denoted by $M_{m,n}(\mathbf{D})$, and the set of square matrices of order n by $M_n(\mathbf{D})$. The most important cases for us will be $\mathbf{D} = \mathbf{Z}$ and $\mathbf{D} = \mathbf{K}[X_1, X_2, \ldots, X_n]$, where $X_1, \ldots, X_n$ are unknowns. The product of two matrices with entries in a domain is defined in exactly the same way as for $\mathbf{D} = \mathbf{K}$. Proposition 2.2 also remains unchanged if the field $\mathbf{K}$ is replaced by a domain $\mathbf{D}$.

EXERCISES

2.1 Calculate:

a) $\begin{pmatrix} 3 & 1 \\ -1 & 1 \end{pmatrix} \begin{pmatrix} 2 & 1 & 4 \\ 0 & 6 & 0 \end{pmatrix} \begin{pmatrix} 1 \\ \sqrt{2} \\ \frac{3}{2} \end{pmatrix}$

b) $\begin{pmatrix} 2 \\ -3 \end{pmatrix} \begin{pmatrix} 1 & 5 & \sqrt{37} & 429\pi & 2 & -2 & 1 \end{pmatrix} \begin{pmatrix} 3 \\ 0 \\ 0 \\ 0 \\ 1 \\ 2 \\ 6 \end{pmatrix}$

c) $\begin{pmatrix} 0 \\ 1 \\ 0 \\ 2 \end{pmatrix} \begin{pmatrix} 5 & 0 & 1 & 0 \end{pmatrix}.$

2.2 Let

$$A = \begin{pmatrix} 1 & 2 \\ 0 & 3 \end{pmatrix} \in M_2(\mathbf{R}).$$

Calculate: a) A^2, b) $3A^3 - \frac{1}{2}A + A^0$, and c) $(A^t)^2 + AA^t + A^tA - 3\mathbf{I}_2$.

2.3 Let

$$A = \begin{pmatrix} 1 & 1 & -1 \\ 0 & 2 & \frac{1}{2} \\ 0 & -2 & -1 \end{pmatrix} \in M_3(\mathbf{R}).$$

Calculate $A^2 - A^t + \mathbf{I}_3$.

2.4 Calculate $\frac{1}{3}\begin{pmatrix} \frac{i}{2} & 1+i \\ -2i & 1 \end{pmatrix}^2 + \mathbf{I}_2$.

2.5 Let $A \in M_n(\mathbf{K})$. Show that $A + A^t$ is symmetric and that $A - A^t$ is skew-symmetric. Deduce that A can be expressed as the sum of a symmetric matrix and a skew-symmetric matrix.

2.6 Express each of the following matrices with entries in $\mathbf{Q}$ as the sum of a symmetric and a skew-symmetric matrix:

a) $\begin{pmatrix} 1 & 2 \\ -1 & 0 \end{pmatrix}$

b) $\begin{pmatrix} 3 & 1 \\ 1 & 0 \end{pmatrix}$ c) $\begin{pmatrix} 1 & 0 & 0 \\ 1 & -1 & -1 \\ 2 & 1 & 0 \end{pmatrix}$.

2.7 Prove that if $A \in M_n(\mathbf{K})$, then A^tA is symmetric.

2.8 A matrix $N \in M_n(\mathbf{K})$ is said to be *nilpotent* if $N^k = \mathbf{0}$ for some integer $k \geq 1$, where $\mathbf{0} \in M_n(\mathbf{K})$ is the zero matrix. Show that, for every $a, b, c \in \mathbf{K}$, the matrices

$$\begin{pmatrix} 0 & a \\ 0 & 0 \end{pmatrix} \quad \text{and} \quad \begin{pmatrix} 0 & a & b \\ 0 & 0 & c \\ 0 & 0 & 0 \end{pmatrix}$$

are nilpotent. Show, more generally, that every strictly upper triangular and every strictly lower triangular matrix is nilpotent.

2.9 Prove that a nilpotent matrix $A \in M_n(\mathbf{K})$ cannot be invertible.

2.10 Determine which of the following matrices are orthogonal:

a) $\begin{pmatrix} \frac{\sqrt{2}}{2} & -\frac{\sqrt{2}}{2} \\ \frac{\sqrt{2}}{2} & \frac{\sqrt{2}}{2} \end{pmatrix}$

b) $\begin{pmatrix} 1 & 0 \\ 0 & -1 \end{pmatrix}$ c) $\begin{pmatrix} 1 & -1 \\ -1 & 1 \end{pmatrix}$

d) $\begin{pmatrix} \frac{\sqrt{3}}{3} & \frac{-2\sqrt{3}}{3} \\ \frac{2\sqrt{3}}{3} & \frac{\sqrt{3}}{3} \end{pmatrix}$ e) $\begin{pmatrix} -\frac{\sqrt{3}}{3} & \frac{\sqrt{6}}{3} \\ \frac{\sqrt{6}}{3} & \frac{\sqrt{3}}{3} \end{pmatrix}$

f) $\begin{pmatrix} -\frac{\sqrt{2}}{2} & 0 & \frac{\sqrt{2}}{2} \\ 0 & 1 & 0 \\ \frac{\sqrt{2}}{2} & 0 & \frac{\sqrt{2}}{2} \end{pmatrix}$ g) $\begin{pmatrix} 0 & -1 & 0 \\ 0 & 0 & -1 \\ -1 & 0 & 0 \end{pmatrix}$

h) $\begin{pmatrix} 0 & 1 & 1 \\ 0 & 0 & 0 \\ 1 & 1 & 0 \end{pmatrix}$ i) $\begin{pmatrix} \frac{1}{2} & \frac{\sqrt{3}}{2} & 0 \\ \frac{\sqrt{3}}{2} & -\frac{1}{2} & 0 \\ 0 & 0 & 1 \end{pmatrix}$

j) $\begin{pmatrix} 1/9 & 8/9 & -4/9 \\ 8/9 & 1/9 & 4/9 \\ -4/9 & 4/9 & 7/9 \end{pmatrix}$ k) $\begin{pmatrix} 1 & 0 & 0 & 0 \\ 0 & 0 & 0 & 1 \\ 0 & \frac{\sqrt{2}}{2} & \frac{\sqrt{2}}{2} & 0 \\ 0 & -\frac{\sqrt{2}}{2} & \frac{\sqrt{2}}{2} & 0 \end{pmatrix}.$

2.11 Let $A, B \in M_n(\mathbf{K})$ be two diagonal matrices of order n:

$$A = \begin{pmatrix} a_1 & 0 & \dots & 0 \\ 0 & a_2 & \dots & 0 \\ \vdots & \vdots & & \vdots \\ 0 & 0 & \dots & a_n \end{pmatrix}, \qquad B = \begin{pmatrix} b_1 & 0 & \dots & 0 \\ 0 & b_2 & \dots & 0 \\ \vdots & \vdots & & \vdots \\ 0 & 0 & \dots & b_n \end{pmatrix}.$$

Prove that

$$AB = BA = \begin{pmatrix} a_1 b_1 & 0 & \dots & 0 \\ 0 & a_2 b_2 & \dots & 0 \\ \vdots & \vdots & & \vdots \\ 0 & 0 & \dots & a_n b_n \end{pmatrix}.$$

3

Systems of linear equations

Matrices arise in a natural way when studying 'systems of linear equations', or 'simultaneous equations'.

Let $X_1, X_2, \ldots, X_n$ be unknowns (or 'indeterminates'). A *linear equation* in the unknowns $X_1, X_2, \ldots, X_n$ with coefficients in $\mathbf{K}$ is an equation of the form,

$$a_1X_1 + \cdots + a_nX_n = b \tag{3.1}$$

where $a_1, \ldots, a_n, b \in \mathbf{K}$. It is sometimes written in the equivalent form,

$$a_1X_1 + \cdots + a_nX_n - b = 0.$$

Equation (3.1) must be understood as a relation between unknowns or between variable quantities which are represented by the unknowns $X_1, X_2, \ldots, X_n$.

A *solution of equation* (3.1) is an element $(x_1, \ldots, x_n)$ of $\mathbf{K}^n$ which gives rise to an identity when substituted into (3.1) in place of the n-tuple $(X_1, \ldots, X_n)$.

Equation (3.1) is said to be *homogeneous* if $b = 0$, otherwise it is said to be *non-homogeneous*.

If we consider simultaneously $m \geq 1$ linear equations in the unknowns $X_1, \ldots, X_n$:

$$\begin{array}{rcl} a_{11}X_1 + a_{12}X_2 + \cdots + a_{1n}X_n & = & b_1 \\ a_{21}X_1 + a_{22}X_2 + \cdots + a_{2n}X_n & = & b_2 \\ \vdots & & \vdots \\ a_{m1}X_1 + a_{m2}X_2 + \cdots + a_{mn}X_n & = & b_m, \end{array} \tag{3.2}$$

we obtain a *system of m equations in n unknowns* $X_1, \ldots, X_n$. The system is said to be *homogeneous* if all the $b_i = 0$, otherwise, if at least one $b_i \neq 0$, it is said to be *non-homogeneous.*

A *solution of system* (3.2) is an element $(x_1, \ldots, x_n) \in \mathbf{K}^n$ which is simultaneously a solution of all m equations. The system is said to be *compatible* if it has at least 1 solution, otherwise it is said to be *incompatible.* Every homogeneous system is compatible as the element $(0, \ldots, 0)$ is a solution, called the *trivial solution*; any other solution is called a *non-trivial* solution.

Notice that, conversely, if the system (3.2) admits $(0, \ldots, 0)$ as a solution then it is homogeneous.

For example, the system of linear equations with real coefficients

$$\begin{aligned} X_1 + 2X_2 &= 1 \\ X_1 + 2X_2 &= 0 \end{aligned}$$

is incompatible, because the left-hand sides of the equations are identical, but the right-hand sides are not, and so there cannot be any element $(x_1, x_2) \in \mathbf{R}^2$ which satisfies both equations. On the other hand, the system

$$\begin{aligned} X_1 + X_2 &= 1 \\ X_1 - X_2 &= 3 \end{aligned}$$

is compatible, and has solution $(2, -1)$, which can be found in the following manner. Adding the two equations term by term gives the new equation $2X_1 = 4$, which is satisfied by the unique value $X_1 = 2$; substituting this value in the first equation gives the unique solution $X_2 = -1$. Moreover, the pair $(2, -1)$ is also a solution of the second equation, and so it is the unique solution of the system.

The system

$$\begin{aligned} X_1 + 3X_2 &= -1 \\ 2X_1 + 6X_2 &= -2 \end{aligned}$$

is compatible, and admits an infinity of solutions $(-1 - 3t, t)$ for every value of $t \in \mathbf{R}$. Indeed, the two equations are proportional and so they have the same solutions: solving the first, for example, gives the solutions above.

The system

$$\begin{aligned} a_{11}X_1 + a_{12}X_2 + \cdots + a_{1n}X_n &= 0 \\ a_{21}X_1 + a_{22}X_2 + \cdots + a_{2n}X_n &= 0 \\ &\vdots \\ a_{m1}X_1 + a_{m2}X_2 + \cdots + a_{mn}X_n &= 0, \end{aligned} \tag{3.3}$$

is called the *homogeneous system associated to* (3.2).

Proposition 3.1
If system (3.2) *is compatible, its solutions are precisely those obtained from any given solution by adding to it any solution of the associated homogeneous system* (3.3).

Proof
Let Σ and Σ_0 be the two subsets of $\mathbf{K}^n$ whose elements are the solutions of (3.2) and (3.3) respectively. If $(y_1, \dots, y_n) \in \Sigma$ and $(x_1, \dots, x_n) \in \Sigma_0$, then

$$(y_1, \dots, y_n) + (x_1, \dots, x_n) = (y_1 + x_1, \dots, y_n + x_n) \in \Sigma.$$

Indeed, for each $j = 1, \dots, m$ one has

$$\begin{aligned} a_{j1}(y_1 + x_1) &+ a_{j2}(y_2 + x_2) + \cdots + a_{jn}(y_n + x_n) \\ &= (a_{j1}y_1 + a_{j2}y_2 + \cdots + a_{jn}y_n) \\ &\quad + (a_{j1}x_1 + a_{j2}x_2 + \cdots + a_{jn}x_n) \\ &= b_j + 0 = b_j. \end{aligned}$$

Conversely, fix $(y_1, \dots, y_n) \in \Sigma$. For every $(z_1, \dots, z_n) \in \Sigma$ one has

$$(z_1 - y_1, z_2 - y_2, \dots, z_n - y_n) \in \Sigma_0$$

because

$$\begin{aligned} a_{j1}(z_1 - y_1) &+ \cdots + a_{jn}(z_n - y_n) \\ &= (a_{j1}z_1 + \cdots + a_{jn}z_n) - (a_{j1}y_1 + \cdots + a_{jn}y_n) \\ &= b_j - b_j = 0. \end{aligned}$$

The propostition now follows from the fact that

$$(z_1, \dots, z_n) = (y_1, \dots, y_n) + (z_1 - y_1, \dots, z_n - y_n). \qquad \square$$

We can associate to system (3.2) an $m \times n$ matrix $A = (a_{ij}) \in M_{m,n}(\mathbf{K})$ formed from the coefficients of the unknowns in the m equations, and called the *matrix of coefficients* of system (3.2). The matrix A can be augmented by adding to it the column vector

$$\mathbf{b} = \begin{pmatrix} b_1 \\ b_2 \\ \vdots \\ b_n \end{pmatrix}$$

as the $(n+1)$-st column (the b_i being the constant terms on the right-hand side of the equations in (3.2)). This gives a new matrix which has m rows and $n+1$ columns:

$$(A\mathbf{b}) = \begin{pmatrix} a_{11} & a_{12} & \dots & a_{1n} & b_1 \\ a_{21} & a_{22} & \dots & a_{2n} & b_2 \\ \vdots & \vdots & & \vdots & \vdots \\ a_{m1} & a_{m2} & \dots & a_{mn} & b_m \end{pmatrix},$$

which is called the *augmented matrix* for system (3.2).

We can interpret the m right-hand sides of the equations in (3.2) as the components of a column vector, and rewrite (3.2) as an equation between column vectors:

$$\begin{pmatrix} a_{11}X_1 + a_{12}X_2 + \cdots + a_{1n}X_n \\ a_{21}X_1 + a_{22}X_2 + \cdots + a_{2n}X_n \\ \vdots \\ a_{m1}X_1 + a_{m2}X_2 + \cdots + a_{mn}X_n \end{pmatrix} = \begin{pmatrix} b_1 \\ b_2 \\ \vdots \\ b_n \end{pmatrix}. \tag{3.4}$$

Putting

$$\mathbf{X} = \begin{pmatrix} X_1 \\ X_2 \\ \vdots \\ X_n \end{pmatrix},$$

we can consider $\mathbf{X}$ as a column vector, so the left-hand side of (3.4) is the matrix product $A\mathbf{X}$. System (3.2) can thus also be written in the much more concise form

$$A\mathbf{X} = \mathbf{b}. \tag{3.5}$$

Conversely, it is clear that for every $m \times (n+1)$ matrix there is a system of m linear equations in the n unknowns $X_1, \dots, X_n$ for which the given matrix is the augmented matrix. We will often make use of this one-to-one correspondence between matrices and systems of linear equations in order to simplify the study of systems of equations by replacing them with matrices.

A system of m equations in the n unknowns $X_1, \ldots, X_n$ is said to be in *echelon form* if it is in the following form:

$$\begin{aligned} a_{11}X_1 + a_{12}X_2 + \cdots \qquad\qquad \cdots + a_{1n}X_n &= b_1 \\ a_{22}X_2 + \cdots \qquad\qquad \cdots + a_{2n}X_n &= b_2 \\ &\vdots \\ a_{mm}X_m + \cdots + a_{mn}X_n &= b_n \end{aligned} \tag{3.6}$$

with $a_{11}, a_{22}, \ldots, a_{mm} \neq 0$. The matrix of coefficients of (3.6) is

$$\begin{pmatrix} a_{11} & a_{12} & & \cdots & & & a_{1n} \\ 0 & a_{22} & & \cdots & & & a_{2n} \\ 0 & 0 & & \cdots & & & a_{3n} \\ \vdots & \vdots & & & & & \vdots \\ 0 & 0 & \cdots & 0 & a_{mm} & \cdots & a_{mn} \end{pmatrix}.$$

In particular, to be in echelon form one must have $m \leq n$.

Suppose $m = n$. Then the last equation in (3.6) has a unique solution $x_n = b_n a_{nn}^{-1}$. When this value x_n is substituted for X_n in the penultimate equation, that equation has a unique solution x_{n-1}. The values x_n and x_{n-1} are then substituted in turn in the third to last equation, giving rise to a unique solution x_{n-2}. Proceeding in this way, one arrives at a unique solution of the whole system (3.6). Thus, *a system of n equations in n unknowns in ecehelon form has a unique solution.*

If $m < n$, then (3.6) can be rewritten as

$$\begin{aligned} a_{11}X_1 + a_{12}X_2 + \cdots + a_{1m}X_m &= b_1 - (a_{1m+1}X_{m+1} + \cdots + a_{1n}X_n) \\ a_{22}X_2 + \cdots + a_{2m}X_m &= b_2 - (a_{2m+1}X_{m+1} + \cdots + a_{2n}X_n) \\ &\vdots \\ a_{mm}X_m &= b_m - (a_{mm+1}X_{m+1} + \cdots + a_{mn}X_n). \end{aligned}$$

Assigning any values $t_{m+1}, \ldots, t_n$ to the unknowns $X_{m+1}, \ldots, X_n$ gives a system of m equations in the m unknowns $X_1, \ldots, X_m$ which is in echelon form:

$$\begin{aligned} a_{11}X_1 + a_{12}X_2 + \cdots + a_{1m}X_m &= b_1 - (a_{1m+1}t_{m+1} + \cdots + a_{1n}t_n) \\ a_{22}X_2 + \cdots + a_{2m}X_m &= b_2 - (a_{2m+1}t_{m+1} + \cdots + a_{2n}t_n) \\ \vdots \qquad & \qquad \qquad \vdots \\ a_{mm}X_m &= b_m - (a_{mm+1}t_{m+1} + \cdots + a_{mn}t_n), \end{aligned} \tag{3.7}$$

which has a unique solution. We can therefore deduce that system (3.6) has an infinity of solutions obtained from (3.7) by varying the values of $t_{m+1}, \ldots, t_n$. From the way in which the solutions are calculated, it follows that every solution of (3.7) can be expressed as an n-tuple,

$$(S_1(t_{m+1}, \ldots, t_n), S_2(t_{m+1}, \ldots, t_n), \cdots, S_n(t_{m+1}, \ldots, t_n)) \tag{3.8}$$

in which each $S_i(t_{m+1}, \ldots, t_n)$ is a polynomial of degree 1 in the parameters $t_{m+1}, \ldots, t_n$. Expression (3.8) is called the *general solution* of system (3.6).

Each S_i has a constant term c_i obtained by putting $t_{m+1} = \cdots = t_n = 0$; together these form an n-tuple $(c_1, \ldots, c_n)$. From this and Proposition 3.1 it follows that the n-tuple of homogeneous polynomials in $t_{m+1}, \ldots, t_n$

$$(S_1(t_{m+1}, \ldots, t_n) - c_1, S_2(t_{m+1}, \ldots, t_n) - c_2, \cdots, S_n(t_{m+1}, \ldots, t_n) - c_n)$$

is the general solution of the homogeneous system associated to (3.6).

A particular consequence of all this is that *a system in echelon form is always compatible.* We will express the fact that the solutions of (3.6) arise as functions of $n - m$ parameters which can be given arbitrary values by saying that *the system in echelon form* (3.6) *has an* $(n - m)$*-parameter family of solutions.* In the case that $n = m$ this will mean that the system has precisely one solution.

A linear equation like (3.1) in which $(a_1, \ldots, a_n) \neq (0, \ldots, 0)$ can be considered as a system in echelon form (provided $a_1 \neq 0$, which can be ensured by exchanging two of the variables if necessary). Therefore system (3.1) has an $(n - 1)$-parameter family of solutions.

Two systems of linear equations in the same unknowns $X_1, \ldots, X_n$ are said to be *equivalent* if they have the same set of solutions. It is not necessary for two systems to have the same number of equations in order to be equivalent.

We are now going to study a procedure, called *Gaussian elimination* (or *Gauss-Jordan elimination*), which allows one to establish whether or not a given system is compatible, and if it is to find all the solutions systematically. This procedure consists of changing the given system into an equivalent one, but in echelon form. This change is effected by a sequence of what are called, 'elementary operations on the equations of the system'. These correspond to 'elementary row operations' on the associated augmented matrix.

There are three types of *elementary row operation on a matrix*:

I) interchange two rows of the matrix;
II) multiply a row of the matrix by a non-zero scalar;
III) add to one row of the matrix any multiple of another row.

The corresponding *elementary operations on a system of equations* are the following:

I) interchange two equations of the system;
II) multiply (left and right hand sides of) an equation by a non-zero scalar;
III) add to one equation of the system any multiple of another equation.

If an elementary operation of type (I) is performed on a system of equations, the new system is equivalent to the old one because the set of solutions of a system does not depend on the order in which the equations are written down. Similarly, an operation of type (II) does not change the set of solutions of the system because two proportional equations have the same solutions. Furthermore, an operation of type (III) does not change the set of solutions: indeed, if an n-tuple $(x_1, \dots, x_n) \in \mathbf{K}^n$ satisfies two equations

$$\begin{aligned} a_{i1}X_1 + a_{i2}X_2 + \cdots + a_{in}X_n &= b_i \\ a_{j1}X_1 + a_{j2}X_2 + \cdots + a_{jn}X_n &= b_j, \end{aligned} \tag{3.9}$$

then for any $c \in \mathbf{K}$ it is also a solution of the equations

$$\begin{aligned} a_{i1}X_1 + a_{i2}X_2 + \cdots + a_{in}X_n &= b_i \\ a_{j1}X_1 + a_{j2}X_2 + \cdots + a_{jn}X_n & \\ + c(a_{i1}X_1 + a_{i2}X_2 + \cdots + a_{in}X_n) &= b_j + cb_i. \end{aligned} \tag{3.10}$$

One can show in the same way that every solution of (3.10) is also a solution of (3.9).

Thus, *performing any elementary operation on a system of equations gives rise to a new system which is equivalent to the old one.*

We are now in a position to describe the method of Gaussian elimination. Suppose now that we are given a system of equations like (3.2). First note that if the left-hand side of any of the equations, say the i-th one, is identically zero, that is, if the i-th equation reads

$$0 = b_i,$$

then it is satisfied identically if $b_i = 0$, while it is incompatible if $b_i \neq 0$. In the first case we can remove this equation and so obtain a system equivalent to the given one, but with one less equation, while in the second, the system is incompatible. We can therefore assume that none of the left-hand sides in (3.2) are identically zero.

Furthermore, we can suppose that $a_{i1} \neq 0$, for some i: this can be achieved, if necessary, by interchanging two of the unknowns. Via an operation of type (I) we can then make $a_{11} \neq 0$, and multiplying the first equation by a_{11}^{-1} (an operation of type (II)) we can reduce to the case where $a_{11} = 1$. Next we add to each of the other equations the first equation multiplied by $-a_{21}, -a_{31}, \ldots$, and $-a_{m1}$ respectively (a sequence of operations of type (III)) we obtain a new system of the form

$$\begin{aligned} X_1 + a'_{12}X_2 + \cdots + a'_{1n}X_n &= b'_1 \\ a'_{22}X_2 + \cdots + a'_{2n}X_n &= b'_2 \\ \vdots \qquad\qquad & \quad \vdots \\ a'_{m2}X_2 + \cdots + a'_{mn}X_n &= b'_m. \end{aligned} \tag{3.11}$$

If any of the equations of system (3.11) are of the form $0 = 0$, we can omit them without changing the set of solutions. If, on the other hand, there appears an equation of the form $0 = b'_i$, with $b'_i \neq 0$, then the system is incompatible, and therefore so is system (3.2), and we can stop there. We can therefore suppose again that none of the left-hand sides of (3.11) is identically zero.

We now continue with system (3.11) ignoring the first equation, and following the same line of reasoning as before on the remaining equations. Changing the order of the variables and using elementary operations of types (I) and (II) if necessary, we can suppose that $a'_{22} = 1$. Adding to each of the successive equations the second multiplied by

$-a'_{32}, -a'_{42}, \ldots, -a'_{m2}$ (elementary operations of type (III)), we obtain a new system of the form

$$\begin{aligned} X_1 + a'_{12}X_2 + a'_{13}X_3 + \cdots + a'_{1n}X_n &= b'_1 \\ X_2 + a''_{23}X_3 + \cdots + a''_{2n}X_n &= b''_2 \\ a''_{33}X_3 + \cdots + a''_{3n}X_n &= b''_3 \\ &\vdots \\ a''_{s3}X_3 + \cdots + a''_{sn}X_n &= b''_s. \end{aligned} \tag{3.12}$$

After having eliminated from (3.12) all equations of the form $0 = 0$, we check that there are no equations of the form $0 = b''_i$, with $b''_i \neq 0$. If there is, then (3.12), and hence (3.2), is incompatible and we stop there. Otherwise, we repeat the same procedure, but now to (3.12) ignoring the first two equations.

This procedure can be iterated until the system is either incompatible or in echelon form. In the first case, we can conclude that the original system was incompatible; in the second we can calculate the solution set of the system in echelon form, which is also the solution set of (3.2), and the Gaussian elimination procedure has finished. The general solution of the system in echelon form is also the *general solution of system* (3.2).

We will now give a few examples of the process of Gaussian elimination. In practice, it is more straightforward to do elementary row operations on the augmented matrix, rather than operate on the system of equations. Furthermore, it is sensible to do any reordering of the variables, corresponding to permuting columns of the augmented matrix, only after having done all the row operations on the matrix.

Observations and examples 3.2

1. $\mathbf{K} = \mathbf{R}$

$$\begin{aligned} X_1 + 2X_2 + 3X_3 &= 1 \\ 2X_1 + X_2 + 4X_3 &= 2 \\ 3X_1 - 3X_2 + X_3 &= 1. \end{aligned}$$

We now perform the row operations on the augmented matrix associated to this system of equations:

$$\begin{pmatrix} 1 & 2 & 3 & 1 \\ 2 & 1 & 4 & 2 \\ 3 & -3 & 1 & 1 \end{pmatrix} \to \begin{pmatrix} 1 & 2 & 3 & 1 \\ 0 & -3 & -2 & 0 \\ 0 & -9 & -8 & -2 \end{pmatrix} \to$$

$$\rightarrow \begin{pmatrix} 1 & 2 & 3 & 1 \\ 0 & 1 & \frac{2}{3} & 0 \\ 0 & 0 & -2 & -2 \end{pmatrix} \rightarrow \begin{pmatrix} 1 & 2 & 3 & 1 \\ 0 & 1 & \frac{2}{3} & 0 \\ 0 & 0 & 1 & 1 \end{pmatrix}.$$

The system reduced to echelon form, then, is

$$\begin{aligned} X_1 + 2X_2 + 3X_3 &= 1 \\ X_2 + \tfrac{2}{3}X_3 &= 0 \\ X_3 &= 1, \end{aligned}$$

which has the unique solution $(-\frac{2}{3}, -\frac{2}{3}, 1)$.

2. $\mathbf{K} = \mathbf{R}$

$$\begin{aligned} X_3 + 2X_4 &= 3 \\ 2X_1 + 4X_2 - 2X_3 \qquad &= 4 \\ 2X_1 + 4X_2 - X_3 + 2X_4 &= 7. \end{aligned} \tag{3.13}$$

The elementary row operations on the augmented matrix are:

$$\begin{pmatrix} 0 & 0 & 1 & 2 & 3 \\ 2 & 4 & -2 & 0 & 4 \\ 2 & 4 & -1 & 2 & 7 \end{pmatrix} \rightarrow \begin{pmatrix} 1 & 2 & -1 & 0 & 2 \\ 2 & 4 & -1 & 2 & 7 \\ 0 & 0 & 1 & 2 & 3 \end{pmatrix} \rightarrow$$

$$\rightarrow \begin{pmatrix} 1 & 2 & -1 & 0 & 2 \\ 0 & 0 & 1 & 2 & 3 \\ 0 & 0 & 1 & 2 & 3 \end{pmatrix} \rightarrow \begin{pmatrix} 1 & 2 & -1 & 0 & 2 \\ 0 & 0 & 1 & 2 & 3 \end{pmatrix}.$$

The corresponding system is,

$$\begin{aligned} X_1 + 2X_2 - X_3 \qquad &= 2 \\ X_3 + 2X_4 &= 3. \end{aligned}$$

Taking the variables in the order X_1, X_3, X_2, X_4 these equations can be rewritten as,

$$\begin{aligned} X_1 - X_3 + 2X_2 \qquad &= 2 \\ X_3 + \qquad 2X_4 &= 3. \end{aligned}$$

We have thus found a system in echelon form which is equivalent to system (3.13). The solution set of both is therefore,

$$(x_1, x_2, x_3, x_4) = (5 - 2t - 2u, t, 3 - 2u, u), \quad t, u \in \mathbf{R},$$

and so (3.13) has a 2-parameter family of solutions.

3. **K = R**

$$\begin{aligned} X_2 - X_3 &= -1 \\ X_1 \qquad + X_3 &= 1 \\ 2X_1 + X_2 + X_3 &= 2. \end{aligned}$$

The elementary row operations on the augmented matrix are,

$$\begin{pmatrix} 0 & 1 & -1 & -1 \\ 1 & 0 & 1 & 1 \\ 2 & 1 & 1 & 2 \end{pmatrix} \to \begin{pmatrix} 1 & 0 & 1 & 1 \\ 0 & 1 & -1 & -1 \\ 2 & 1 & 1 & 2 \end{pmatrix} \to$$

$$\to \begin{pmatrix} 1 & 0 & 1 & 1 \\ 0 & 1 & -1 & -1 \\ 0 & 1 & -1 & 0 \end{pmatrix} \to \begin{pmatrix} 1 & 0 & 1 & 1 \\ 0 & 1 & -1 & -1 \\ 0 & 0 & 0 & 1 \end{pmatrix}.$$

The third row corresponds to the incompatible equation $0 = 1$ and the system is therefore incompatible.

4. Every homogeneous system of m equations in n unknowns, with $n \geq m$, has an N-parameter family of solutions for some $N \geq n-m$. Indeed, the system is compatible because it is homogeneous, and the process of Gaussian elimination reduces it to a system of p equations in echelon form, with $p \leq m$. Thus the original system has an $(n-p)$-parameter family of solutions, and $n - p \geq m - p$. Let us see an example with $\mathbf{K} = \mathbf{R}$.

$$\begin{aligned} X_1 + X_2 + 2X_3 + \ \ X_4 \qquad\quad &= 0 \\ X_1 + X_2 + \ \ X_3 + 2X_4 - \ \ X_5 &= 0 \\ X_1 + X_2 \qquad\quad + 3X_4 - 2X_5 &= 0 \\ X_1 + X_2 + 3X_3 \qquad\quad + \ \ X_5 &= 0. \end{aligned} \tag{3.14}$$

This system is homogeneous. In this case it is sufficient to consider the matrix of coefficients, rather than the associated augmented matrix. We will perform the elementary row operations on this matrix.

$$\begin{pmatrix} 1 & 1 & 2 & 1 & 0 \\ 1 & 1 & 1 & 2 & -1 \\ 1 & 1 & 0 & 3 & -2 \\ 1 & 1 & 3 & 0 & 1 \end{pmatrix} \to \begin{pmatrix} 1 & 1 & 2 & 1 & 0 \\ 0 & 0 & -1 & 1 & -1 \\ 0 & 0 & -2 & 2 & -2 \\ 0 & 0 & 1 & -1 & 1 \end{pmatrix} \to$$

$$\to \begin{pmatrix} 1 & 1 & 2 & 1 & 0 \\ 0 & 0 & 1 & -1 & 1 \\ 0 & 0 & 0 & 0 & 0 \\ 0 & 0 & 0 & 0 & 0 \end{pmatrix} \to \begin{pmatrix} 1 & 1 & 2 & 1 & 0 \\ 0 & 0 & 1 & -1 & 1 \end{pmatrix}$$

and exchanging the second and third columns gives

$$\begin{pmatrix} 1 & 1 & 2 & 1 & 0 \\ 0 & 0 & 1 & -1 & 1 \end{pmatrix} \to \begin{pmatrix} 1 & 2 & 1 & 1 & 0 \\ 0 & 1 & 0 & -1 & 1 \end{pmatrix}.$$

This gives the system in echelon form,

$$\begin{aligned} X_1 + 2X_3 + X_2 + X_4 \qquad &= 0 \\ X_3 \qquad - X_4 + X_5 &= 0 \end{aligned}$$

which has a 3-parameter family of solutions. Thus the general solution of system (3.14) is

$$(x_1, x_2, x_3, x_4, x_5) = (-t - 3u + 2v, t, u - v, u, v). \quad t, u, v \in \mathbf{R}.$$

5. Suppose that

$$A\mathbf{X} = \mathbf{b}, \tag{3.15}$$

with $A \in M_n(\mathbf{K})$ and $\mathbf{b} \in \mathbf{K}^n$, is a system of n equations in n unknowns and that A is invertible. Then this is a compatible system and it has a unique solution $\mathbf{x} = (x_1, \ldots, x_n)^t \in \mathbf{K}^n$ given by the expression,

$$\mathbf{x} = A^{-1}\mathbf{b}. \tag{3.16}$$

In fact, if the value of $\mathbf{x}$ given in (3.16) is substituted for $\mathbf{X}$ in (3.15) one obtains the identity $AA^{-1}\mathbf{b} = \mathbf{b}$ and so $\mathbf{x}$ is indeed a solution. Conversely, every solution $\mathbf{y} \in \mathbf{K}^n$ of (3.15) satisfies $A\mathbf{y} = \mathbf{b}$; multiplying through by A^{-1} we get $\mathbf{y} = A^{-1}\mathbf{b} = \mathbf{x}$.

Equation (3.16) provides a method for finding the solution of a system of n equations in n unknowns, such as (3.15), in which A is invertible, provided one knows how to calculate A^{-1}. This method is called the method of inversion.

We will see shortly a method for finding A^{-1} by means of elementary row operations on A.

6. An *elementary matrix of order n* is a matrix $R \in M_n(\mathbf{K})$ which can be obtained from the identity matrix by elementary row operations. For example, each of the following matrices is elementary:

$$\begin{pmatrix} 1 & 0 & 0 & 0 \\ 0 & 0 & 0 & 1 \\ 0 & 0 & 1 & 0 \\ 0 & 1 & 0 & 0 \end{pmatrix}, \quad \begin{pmatrix} 1 & 0 & 0 & 0 \\ 0 & 1 & 0 & 0 \\ 0 & 0 & 2 & 0 \\ 0 & 0 & 0 & 1 \end{pmatrix}, \quad \begin{pmatrix} 1 & 0 & 0 & 0 \\ 0 & 1 & 5 & 0 \\ 0 & 0 & 1 & 0 \\ 0 & 0 & 0 & 1 \end{pmatrix}.$$

Let us introduce some notation for elementary matrices of order n:

R^n_{ij} is the matrix obtained by interchanging the i-th and j-th rows of $\mathbf{I}_n$;

$R^n_i(c)$ is the matrix obtained by multiplying the i-th row of $\mathbf{I}_n$ by $c \in \mathbf{K}^*$;

$R^n_{ij}(c)$ is the matrix obtained by adding to the i-th row of $\mathbf{I}_n$ the j-th row multiplied by $c \in \mathbf{K}$.

We will often use the simpler notation R_{ij}, $R_i(c)$, and $R_{ij}(c)$ to denote these elementary matrices.

The importance of these elementary matrices lies in the fact that, if $A \in M_{n,m}(\mathbf{K})$, then *every elementary row operation on A can be produced by multiplying A on the left by a suitable elementary matrix.* The proof of this is left to the reader.

The following identities can be verified easily:

$$\begin{aligned} R_{ij}^{-1} &= R_{ij} \\ R_i(c)^{-1} &= R_i(c^{-1}) \\ R_{ij}(c)^{-1} &= R_{ij}(-c). \end{aligned}$$

In particular we see that the elementary matrices are invertible, and that their inverses are elementary matrices of the same type.

7. Let $A \in M_n(\mathbf{K})$. The following conditions are equivalent:
a) A is invertible;
b) A can be expressed as a product of elementary matrices.
(b) $\Rightarrow$ (a): If A is a product of elementary matrices then it is invertible because each of its factors is.
(a) $\Rightarrow$ (b): As we saw in Example 5, the system $A\mathbf{X} = \mathbf{0}$ has the unique solution $\mathbf{x} = \mathbf{0}$. Using Gaussian elimination, this system can therefore be reduced to a system in echelon form:

$$\begin{aligned} X_1 + a'_{12}X_2 + \cdots + a'_{1n}X_n &= 0 \\ X_2 + \cdots + a'_{2n}X_n &= 0 \\ \vdots \qquad\qquad & \vdots \\ X_{n-1} + a'_{n-1n}X_n &= 0 \\ X_n &= 0. \end{aligned}$$

With further operations of type (III), it is possible to reduce this to a system in the form $\mathbf{X} = \mathbf{0}$, that is,

$$\begin{aligned} X_1 \qquad\qquad &= 0 \\ X_2 \qquad &= 0 \\ &\vdots \\ X_n &= 0. \end{aligned}$$

Since this sequence of operations corresponds to multiplying the left hand side of $A\mathbf{X} = \mathbf{0}$ by the product of a finite number of elementary matrices, we have that

$$R^{(1)}R^{(2)}\cdots R^{(s)}A\mathbf{X} = \mathbf{X} = \mathbf{I}_n\mathbf{X}$$

for certain elementary matrices $R^{(1)}, \ldots, R^{(s)}$. Thus

$$R^{(1)}R^{(2)}\cdots R^{(s)}A = \mathbf{I}_n$$

and hence, by the uniqueness of A^{-1}, one finds that

$$R^{(1)}R^{(2)}\cdots R^{(s)} = A^{-1},$$

and so

$$A = (R^{(1)}R^{(2)}\cdots R^{(s)})^{-1} = (R^{(s)})^{-1}\cdots(R^{(2)})^{-1}(R^{(1)})^{-1}$$

is a product of elementary matrices.

8. Let $A, B \in M_n(\mathbf{K})$. If $M \in M_n(\mathbf{K})$ then the matrices M and $(A \quad B) \in M_{n,2n}(\mathbf{K})$ can be multiplied, and the definition of product implies that

$$M(A \quad B) = (MA \quad MB).$$

Suppose that the matrix $A \in M_n(\mathbf{K})$ is invertible, and consider the matrix $(A \quad \mathbf{I}_n) \in M_{n,2n}(\mathbf{K})$. Multiplying on the left by A^{-1} gives,

$$A^{-1}(A \quad \mathbf{I}_n) = (\mathbf{I}_n \quad A^{-1}).$$

It follows from Example 7 that A^{-1} can be expressed as a product of elementary matrices, and so we conclude that the matrix $(\mathbf{I}_n \quad A^{-1})$ can be obtained from $(A \quad \mathbf{I}_n)$ by elementary row operations.

As an example, consider the matrix, $\begin{pmatrix} 1 & 1 \\ 2 & -1 \end{pmatrix}$. We have,

$$\begin{pmatrix} 1 & 1 & 1 & 0 \\ 2 & -1 & 0 & 1 \end{pmatrix} \to \begin{pmatrix} 1 & 1 & 1 & 0 \\ 0 & -3 & -2 & 1 \end{pmatrix} \to$$

$$\to \begin{pmatrix} 1 & 1 & 1 & 0 \\ 0 & 1 & 2/3 & -1/3 \end{pmatrix} \to \begin{pmatrix} 1 & 0 & 1/3 & 1/3 \\ 0 & 1 & 2/3 & -1/3 \end{pmatrix}.$$

Thus A is invertible, and

$$A^{-1} = \begin{pmatrix} 1/3 & 1/3 \\ 2/3 & -1/3 \end{pmatrix}.$$

On the other hand the matrix $\begin{pmatrix} 2 & 1 \\ 1 & 1/2 \end{pmatrix}$ is not invertible, and in fact the matrix

$$\begin{pmatrix} 2 & 1 & 1 & 0 \\ 1 & 1/2 & 0 & 1 \end{pmatrix}$$

cannot be transformed to one of the form $(\mathbf{I}_2 \quad B)$ by elementary operations. The proof of this fact is left to the reader.

Consider now the matrix

$$A = \begin{pmatrix} 1 & 1 & 2 \\ 0 & 1 & 0 \\ -1 & 0 & 1 \end{pmatrix}.$$

One has,

$$\begin{pmatrix} 1 & 1 & 2 & 1 & 0 & 0 \\ 0 & 1 & 0 & 0 & 1 & 0 \\ -1 & 0 & 1 & 0 & 0 & 1 \end{pmatrix} \to \begin{pmatrix} 1 & 1 & 2 & 1 & 0 & 0 \\ 0 & 1 & 0 & 0 & 1 & 0 \\ 0 & 1 & 3 & 1 & 0 & 1 \end{pmatrix} \to$$

$$\to \begin{pmatrix} 1 & 1 & 2 & 1 & 0 & 0 \\ 0 & 1 & 0 & 0 & 1 & 0 \\ 0 & 0 & 3 & 1 & -1 & 1 \end{pmatrix} \to \begin{pmatrix} 1 & 0 & 2 & 1 & -1 & 0 \\ 0 & 1 & 0 & 0 & 1 & 0 \\ 0 & 0 & 3 & 1 & -1 & 1 \end{pmatrix} \to$$

$$\to \begin{pmatrix} 1 & 0 & 0 & \frac{1}{3} & -\frac{1}{3} & -\frac{2}{3} \\ 0 & 1 & 0 & 0 & 1 & 0 \\ 0 & 0 & 3 & 1 & -1 & 1 \end{pmatrix} \to \begin{pmatrix} 1 & 0 & 0 & \frac{1}{3} & -\frac{1}{3} & -\frac{2}{3} \\ 0 & 1 & 0 & 0 & 1 & 0 \\ 0 & 0 & 1 & \frac{1}{3} & -\frac{1}{3} & \frac{1}{3} \end{pmatrix}.$$

Thus A is invertible and

$$A^{-1} = \begin{pmatrix} \frac{1}{3} & -\frac{1}{3} & -\frac{2}{3} \\ 0 & 1 & 0 \\ \frac{1}{3} & -\frac{1}{3} & \frac{1}{3} \end{pmatrix}.$$

In Chapter 6 we will describe another method for calculating the inverse of a matrix.

EXERCISES

3.1 Solve the following systems of equations by Gaussian elimination:

a) $(\mathbf{K} = \mathbf{Q})$

$$\begin{aligned} X - 3Y + 5Z &= 0 \\ 2X - 4Y + 2Z &= 0 \\ 5X - 11Y + 9Z &= 0 \end{aligned}$$

b) $(\mathbf{K} = \mathbf{Q})$

$$\begin{aligned} X_1 - 2X_2 + 3X_3 + 4X_4 + 5X_5 &= 0 \\ X_1 + 4X_2 + 7X_4 + 2X_5 &= 0 \\ X_1 + 4X_2 + 7X_4 + 2X_5 &= 0 \\ 2X_1 + 2X_2 + 3X_3 + 11X_4 + 7X_5 &= 0 \\ 3X_1 + 6X_2 + 3X_3 + 18X_4 + 9X_5 &= 0 \end{aligned}$$

c) $(\mathbf{K} = \mathbf{R})$

$$\begin{aligned} X_1 + 2X_2 - \sqrt{2}X_3 &= 0 \\ 3X_1 - (\sqrt{2} + 6)X_3 &= 0 \\ -X_1 + X_2 + 3X_3 &= -1 \end{aligned}$$

d) $(\mathbf{K} = \mathbf{C})$

$$\begin{aligned} 2X_2 + X_4 + 5X_5 &= i \\ 2X_1 + 2X_3 + X_4 - 3X_5 &= i \\ X_1 + X_2 + X_3 + X_4 + X_5 &= 0 \end{aligned}$$

e) $(\mathbf{K} = \mathbf{R})$

$$\begin{aligned} X_3 + 2X_4 &= 3 \\ 2X_1 + 4X_2 - 2X_3 &= 4 \\ 2X_1 + 4X_2 - X_3 + 2X_4 &= 7. \end{aligned}$$

3.2 Prove that the diagonal matrix

$$A = \begin{pmatrix} a_1 & 0 & \cdots & 0 \\ 0 & a_2 & \cdots & 0 \\ \vdots & \vdots & & \vdots \\ 0 & 0 & \cdots & a_n \end{pmatrix} \in M_n(\mathbf{K})$$

is invertible if and only if all $a_1, \ldots, a_n \neq 0$, and in this case its inverse is

$$A^{-1} = \begin{pmatrix} a_1^{-1} & 0 & \cdots & 0 \\ 0 & a_2^{-1} & \cdots & 0 \\ \vdots & \vdots & & \vdots \\ 0 & 0 & \cdots & a_n^{-1} \end{pmatrix} \in M_n(\mathbf{K}).$$

3.3 Calculate $3A^{-1} - AB^{-2}$, where $A = \begin{pmatrix} 1 & -1 \\ 1 & 1 \end{pmatrix}$ and $B = \begin{pmatrix} 0 & 2 \\ -1 & 1 \end{pmatrix}$.

3.4 Calculate the inverse, if it exists, of each of the following matrices.

a) $(\mathbf{K} = \mathbf{Q})$ $\begin{pmatrix} 1/2 & 1/3 \\ 5 & 2 \end{pmatrix}$ b) $(\mathbf{K} = \mathbf{R})$ $\begin{pmatrix} 3 & \sqrt{3} \\ 1 & 1/\sqrt{3} \end{pmatrix}$

c) $(\mathbf{K} = \mathbf{R})$ $\begin{pmatrix} 0 & -1 & \sqrt{2} \\ 1 & 0 & 3 \end{pmatrix} \begin{pmatrix} -3 & 2 \\ -\sqrt{2} & 4 \\ -1 & 0 \end{pmatrix}$

d) $(\mathbf{K} = \mathbf{C})$ $\begin{pmatrix} 1 & i \\ 2i & -1 \end{pmatrix}$ e) $(\mathbf{K} = \mathbf{C})$ $\begin{pmatrix} 2 & 2-i \\ 2+i & -2 \end{pmatrix}$

f) $(\mathbf{K} = \mathbf{C})$ $\begin{pmatrix} 1 & 1/2 \\ 2i & 1 \end{pmatrix}$ g) $(\mathbf{K} = \mathbf{Q})$ $\begin{pmatrix} 6 & -3 & -2 \\ 5 & -2 & -2 \\ 5 & -3 & -1 \end{pmatrix}$

h) $(\mathbf{K} = \mathbf{Q})$ $\begin{pmatrix} -1 & 2 & -1 \\ -5 & 13 & -10 \\ 2 & -5 & 4 \end{pmatrix}$ i) $(\mathbf{K} = \mathbf{Q})$ $\begin{pmatrix} 1 & 0 & 1 \\ 1 & 1 & 0 \\ -1 & 1 & 0 \end{pmatrix}$

j) $(\mathbf{K} = \mathbf{C})$ $\begin{pmatrix} 2i & 0 & 1 \\ 0 & 0 & i \\ 1 & 1 & 1 \end{pmatrix}$ k) $(\mathbf{K} = \mathbf{Q})$ $\begin{pmatrix} -1 & 0 & 0 & 2 \\ 0 & 1 & 0 & 0 \\ 1 & 0 & 0 & 1 \\ 1 & 0 & 1 & 1 \end{pmatrix}$

l) $(\mathbf{K} = \mathbf{C})$ $\begin{pmatrix} 0 & 0 & 0 & i \\ 0 & 0 & 1 & 0 \\ 0 & 1 & 0 & 0 \\ i & 0 & 0 & 0 \end{pmatrix}$.

3.5 Solve the following systems of equations by the method of inversion.

a) $(\mathbf{K} = \mathbf{Q})$

$$\begin{aligned} X + Y - \tfrac{1}{2}Z &= 1 \\ 12Y - Z &= 12 \\ X + 3Y &= 3 \end{aligned}$$

b) $(\mathbf{K} = \mathbf{C})$

$$\begin{aligned} iX - Y &= 2i \\ 3X - 2iY &= 1 \end{aligned}$$

c) $(\mathbf{K} = \mathbf{R})$

$$\begin{aligned} X \phantom{+ \sqrt{2}Y} + \phantom{\tfrac{1}{2}} Z &= \sqrt{2} \\ X + \sqrt{2}Y + \tfrac{1}{2}Z &= 2\sqrt{2} \\ \tfrac{\sqrt{2}}{2}X + 2Y + \tfrac{\sqrt{2}}{2}Z &= 3. \end{aligned}$$

3.6 Express each of the following square matrices with real entries as a product of elementary matrices.

a) $\begin{pmatrix} 0 & 2 \\ 1/2 & 0 \end{pmatrix}$ b) $\begin{pmatrix} 1 & -1 \\ 2 & 0 \end{pmatrix}$ c) $\begin{pmatrix} 3 & 5 \\ 1 & 2 \end{pmatrix}$

d) $\begin{pmatrix} 2 & 1 & 0 \\ 1 & 1 & 0 \\ 0 & 0 & 2 \end{pmatrix}$ e) $\begin{pmatrix} 1 & 3 & 0 \\ 2 & 1 & 1 \\ 2 & -1 & 0 \end{pmatrix}$.

4

Some linear algebra

Let $\mathbf{V}$ be a vector space over $\mathbf{K}$.

Definition 4.1
A non-empty subset $\mathbf{W}$ of $\mathbf{V}$ is called a *vector subspace* of $\mathbf{V}$ if:
1) for each $\mathbf{w_1}, \mathbf{w_2} \in \mathbf{W}$, the sum $\mathbf{w_1} + \mathbf{w_2} \in \mathbf{W}$, and
2) for each $\mathbf{w} \in \mathbf{W}$ and each $k \in \mathbf{K}$, the product $k\mathbf{w} \in \mathbf{W}$.

Conditions (1) and (2) of Definition (4.1) imply that the operations of sum and scalar product on $\mathbf{V}$ define similar operations on $\mathbf{W}$. Moreover, condition (2) implies that $\mathbf{0} = 0\mathbf{w} \in \mathbf{W}$ and $-\mathbf{w} \in \mathbf{W}$, by using the scalars 0 and -1 respectively. It follows that $\mathbf{W}$ satisfies axioms VS2 and VS3. Now, the axioms VS4, ... , VS8 are satisfied by $\mathbf{V}$ and therefore also by $\mathbf{W}$. Consequently, $\mathbf{W}$ is a vector space in its own right.

It is clear that if $\mathbf{U}$ is a subspace of $\mathbf{W}$, and $\mathbf{W}$ is a subspace of $\mathbf{V}$ then $\mathbf{U}$ is a subspace of $\mathbf{V}$. Also, if $\mathbf{U}$ and $\mathbf{W}$ are subspaces of $\mathbf{V}$ and $\mathbf{U} \subset \mathbf{W}$ then $\mathbf{U}$ is a subspace of $\mathbf{W}$.

Examples 4.2

1. Two immediate examples of subspaces of an arbitrary vector space $\mathbf{V}$ are $\mathbf{V}$ itself and the subset consisting only of $\mathbf{0}$, which is denoted $\langle \mathbf{0} \rangle$. These two subspaces are called the *trivial* or *improper subspaces.*

   Let $\mathbf{v} \in \mathbf{V}$ be any element. The set

$$\langle \mathbf{v} \rangle = \{c\mathbf{v} \,|\, c \in \mathbf{K}\}$$

consisting of all multiples of $\mathbf{v}$ is another example of vector subspace of $\mathbf{V}$; the proof is left to the reader. In the ordinary plane and space, the subspaces of the form $\langle \mathbf{v} \rangle$ are those obtained by fixing a straight line and considering all vectors parallel to it.

2. Let $\mathbf{V} = \mathbf{K}^n$, with $n \geq 2$. The subset H_1 of $\mathbf{V}$ consisting of n-tuples of the form $(0, x_2, \ldots, x_n)$ is a subspace of $\mathbf{K}^n$. Indeed, H_1 is non-empty and moreover, for each $x_2, \ldots, x_n, y_2, \ldots, y_n$ and $c \in \mathbf{K}$, one has

$$(0, x_2, \ldots, x_n) + (0, y_2, \ldots, y_n) = (0, x_2 + y_2, \ldots, x_n + y_n) \in H_1$$

and

$$c(0, x_2, \ldots, x_n) = (0, cx_2, \ldots, cx_n) \in H_1.$$

In a similar way one proves that for every index i between 1 and n, the subset H_i of $\mathbf{K}^n$ consisting of those n-tuples whose i-th entry is 0 is a subspace of $\mathbf{K}^n$.

Given any $a_1, \ldots, a_n \in \mathbf{K}^n$ not all zero, let

$$H = \{(x_1, \ldots, x_n) \in \mathbf{K}^n \mid a_1x_1 + \cdots + a_nx_n = 0\}.$$

If $(x_1, \ldots, x_n)$, $(y_1, \ldots, y_n) \in H$ and $c \in \mathbf{K}$ then

$$\begin{aligned} a_1(x_1 + y_1) + \cdots + a_n(x_n + y_n) &= a_1x_1 + \cdots + a_nx_n \\ &\quad + a_1y_1 + \cdots + a_ny_n \\ &= 0 + 0 = 0 \end{aligned}$$

and

$$a_1(cx_1) + \cdots + a_n(cx_n) = c(a_1x_1 + \cdots + a_nx_n) = c0 = 0,$$

that is, $(x_1, \ldots, x_n) + (y_1, \ldots, y_n) \in H$ and $c(x_1, \ldots, x_n) \in H$; therefore H is a vector subspace of $\mathbf{K}^n$. The subspaces H_i are particular cases of H which are obtained by putting $a_i = 1$ and $a_j = 0$ if $j \neq i$.

In a similar way, the reader can check that *the set $\Sigma \in \mathbf{K}^n$ consisting of the solutions of any given system of linear homogeneous equations in n unknowns is a vector subspace of $\mathbf{K}^n$.*

3. Let $\mathbf{V}$ be the real vector space of geometric vectors of ordinary space, let π be a plane and P a point of π. *The set $\mathbf{W}$ of vectors of $\mathbf{V}$ of the form $\overrightarrow{PQ}$ for some $Q \in \pi$ is a vector subspace of $\mathbf{V}$.*

Indeed, for every $Q, R \in \pi$ we have,

$$\overrightarrow{PQ} + \overrightarrow{PR} \in \mathbf{W}$$

because the fourth vertex of a parallelogram with vertices P, Q, R is also contined in π, while for every $c \in \mathbf{R}$ and $Q \in \pi$ one has

$$c\overrightarrow{PQ} \in \mathbf{W}$$

because the straight line containing P and Q is contained in π.

Note that $\mathbf{W}$ depends only on π and not on the point $P \in \pi$ used to define $\mathbf{W}$.

It follows immediately from Definition 4.1 that if $\mathbf{U}$ and $\mathbf{W}$ are vector subspaces of $\mathbf{V}$, then so is their intersection $\mathbf{U} \cap \mathbf{W}$.

More generally, given any collection $\{\mathbf{W}_i\}_{i \in I}$ of vector subspaces of $\mathbf{V}$, the intersection $\cap_{i \in I} \mathbf{W}_i$ is a vector subspace of $\mathbf{V}$. The proof is straightforward and is left to the reader.

The union $\mathbf{U} \cup \mathbf{W}$ of two subspaces is not in general a subspace of $\mathbf{V}$. For example, if $\mathbf{u}$ and $\mathbf{w}$ are two vectors in $\mathbf{V}$ which are not proportional, then $\langle \mathbf{u} \rangle \cup \langle \mathbf{w} \rangle$ does not satisfy Condition (1) of Definition 4.1, because it contains $\mathbf{u}$ and $\mathbf{w}$ but not $\mathbf{u} + \mathbf{w}$, as this is neither a multiple of $\mathbf{u}$ nor a multiple of $\mathbf{w}$.

The subset of $\mathbf{V}$ consisting of all vectors of the form $\mathbf{u} + \mathbf{w}$, as $\mathbf{u}$ varies in $\mathbf{U}$ and $\mathbf{w}$ varies in $\mathbf{W}$ is denoted $\mathbf{U} + \mathbf{W}$.

If $\mathbf{u}_1, \mathbf{u}_2 \in \mathbf{U}$, $\mathbf{w}_1, \mathbf{w}_2 \in \mathbf{W}$ and $c \in \mathbf{K}$ then

$$(\mathbf{u}_1 + \mathbf{w}_1) + (\mathbf{u}_2 + \mathbf{w}_2) = (\mathbf{u}_1 + \mathbf{u}_2) + (\mathbf{w}_1 + \mathbf{w}_2) \in \mathbf{U} + \mathbf{W}$$

and

$$c(\mathbf{u}_1 + \mathbf{w}_1) = (c\mathbf{u}_1 + c\mathbf{w}_1) \in \mathbf{U} + \mathbf{W}$$

and so $\mathbf{U} + \mathbf{W}$ is a vector subspace of $\mathbf{V}$. It is called the *sum* of the subspaces $\mathbf{U}$ and $\mathbf{W}$.

Notice that $\mathbf{U} + \mathbf{W}$ contains $\mathbf{U} \cup \mathbf{W}$ because it contains all vectors of the form $\mathbf{u} = \mathbf{u} + \mathbf{0}$ and $\mathbf{w} = \mathbf{0} + \mathbf{w}$.

If $\mathbf{U} \cap \mathbf{W} = \langle \mathbf{0} \rangle$ then $\mathbf{U} + \mathbf{W}$ is called the *direct sum* of $\mathbf{U}$ and $\mathbf{W}$, and is denoted by $\mathbf{U} \oplus \mathbf{W}$.

Every vector in $\mathbf{U} \oplus \mathbf{W}$ *can be expressed uniquely as a sum of the form* $\mathbf{u} + \mathbf{w}$. Indeed, if

$$\mathbf{u} + \mathbf{w} = \mathbf{u}' + \mathbf{w}'$$

for some $\mathbf{u}, \mathbf{u}' \in \mathbf{U}$ and $\mathbf{w}, \mathbf{w}' \in \mathbf{W}$, then $\mathbf{u} - \mathbf{u}' = \mathbf{w}' - \mathbf{w} \in \mathbf{U} \cap \mathbf{W}$, and so

$$\mathbf{u} - \mathbf{u}' = \mathbf{w}' - \mathbf{w} = \mathbf{0},$$

that is, $\mathbf{u} = \mathbf{u}'$ and $\mathbf{w} = \mathbf{w}'$.

If $\mathbf{V} = \mathbf{U} \oplus \mathbf{W}$ then the subspaces $\mathbf{U}$ and $\mathbf{W}$ are said to be *supplementary* in $\mathbf{V}$.

In the case that $\mathbf{U} = \langle \mathbf{u} \rangle$, $\mathbf{W} = \langle \mathbf{w} \rangle$ with $\mathbf{u}, \mathbf{w} \in \mathbf{V}$ not proportional, one has that $\langle \mathbf{u} \rangle \cap \langle \mathbf{w} \rangle = \langle \mathbf{0} \rangle$. The direct sum $\langle \mathbf{u} \rangle \oplus \langle \mathbf{w} \rangle$ consists of all vectors of the form $a\mathbf{u} + b\mathbf{w}$, as a, b vary in $\mathbf{K}$. Fig. 4.1 depicts this situation in ordinary space.

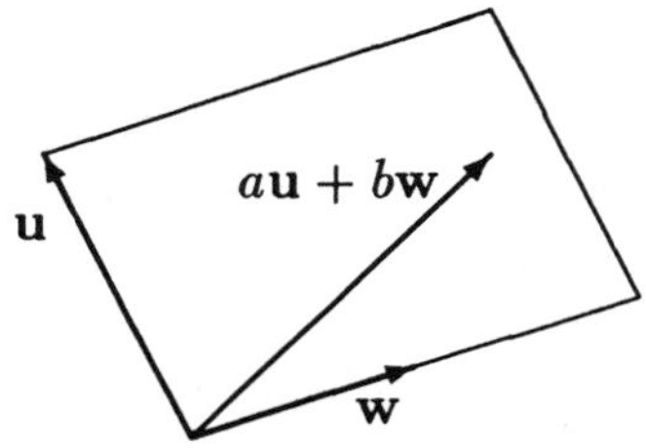

Fig. 4.1

If $\mathbf{U}$ *and* $\mathbf{W}$ *are* $\mathbf{K}$*-vector spaces, their Cartesian product* $\mathbf{U} \times \mathbf{W}$ *is a* $\mathbf{K}$*-vector space with addition and scalar multiplication defined in the following manner:*

$$\begin{aligned}(\mathbf{u}, \mathbf{w}) + (\mathbf{u}', \mathbf{w}') &= (\mathbf{u} + \mathbf{u}', \mathbf{w} + \mathbf{w}') \\ a(\mathbf{u}, \mathbf{w}) &= (a\mathbf{u}, a\mathbf{w})\end{aligned}$$

for every $(\mathbf{u}, \mathbf{w}), (\mathbf{u}', \mathbf{w}') \in \mathbf{U} \times \mathbf{W}$, *and* $a \in \mathbf{K}$. *The zero vector of* $\mathbf{U} \times \mathbf{W}$ *is* $(\mathbf{0}, \mathbf{0})$.

The confirmation of these statements is left to the reader.

The subsets

$$\begin{aligned}\mathbf{U}' &= \{(\mathbf{u}, \mathbf{0}) \mid \mathbf{u} \in \mathbf{U}\} \\ \mathbf{W}' &= \{(\mathbf{0}, \mathbf{w}) \mid \mathbf{w} \in \mathbf{W}\}\end{aligned}$$

are two subspaces of $\mathbf{U} \times \mathbf{W}$. Clearly $\mathbf{U}' \cap \mathbf{W}' = (\mathbf{0}, \mathbf{0})$, and moreover,

$$(\mathbf{u}, \mathbf{w}) = (\mathbf{u}, \mathbf{0}) + (\mathbf{0}, \mathbf{w})$$

for every $(\mathbf{u}, \mathbf{w}) \in \mathbf{U} \times \mathbf{W}$. Consequently,

$$\mathbf{U} \times \mathbf{W} = \mathbf{U}' \oplus \mathbf{W}'.$$

For example, the vector space $\mathbf{K}^{m+n}$ can be identified with the space $\mathbf{K}^m \times \mathbf{K}^n$, and is the direct sum of the two subspaces,

$$\begin{aligned} \mathbf{K}^{m\prime} &= \{(x_1, x_2, \ldots, x_m, 0, \ldots, 0) \,|\, x_1, \ldots, x_m \in \mathbf{K}\} \\ \mathbf{K}^{n\prime} &= \{(0, \ldots, 0, y_1, y_2, \ldots, y_n) \,|\, y_1, \ldots, y_n \in \mathbf{K}\}. \end{aligned}$$

A procedure for constructing vector subspaces is given by the notion of 'linear combination'.

Let $\mathbf{v}_1, \ldots, \mathbf{v}_n \in \mathbf{V}$ and $a_1, \ldots, a_n \in \mathbf{K}$. The vector

$$a_1\mathbf{v}_1 + \cdots + a_n\mathbf{v}_n \tag{4.1}$$

is called a *linear combination* of the vectors $\mathbf{v}_1, \ldots, \mathbf{v}_n$, and $a_1, \ldots, a_n$ are called the *coefficients of the linear combination.*

If all the coefficients are zero, then (4.1) is equal to the vector $\mathbf{0}$, and this is called the *trivial linear combination of* $\mathbf{v}_1, \ldots, \mathbf{v}_n$. Every linear combination of $\mathbf{v}_1, \ldots, \mathbf{v}_n$ in which some of the coefficients are non-zero is said to be *non-trivial.*

(*Warning:* A linear combination may be non-trivial but still be equal to $\mathbf{0}$. This is the case, for example, with $0\mathbf{v} + a\mathbf{0}$, for every $\mathbf{v} \in \mathbf{V}$ and every $a \in \mathbf{K}^*$.)

The linear combinations of a single vector $\mathbf{v} \in \mathbf{V}$ are just its multiples.

Note that

$$\mathbf{v}_i = 0\mathbf{v}_1 + \cdots + 0\mathbf{v}_{i-1} + 1\mathbf{v}_i + 0\mathbf{v}_{i+1} + \cdots + 0\mathbf{v}_n$$

is a linear combination of $\mathbf{v_1}, \ldots, \mathbf{v_n}$, for each $i = 1, \ldots, n$.

It follows immediately from Definition 4.1 that if $\mathbf{W}$ is a vector subspace of $\mathbf{V}$, and $\mathbf{v}_1, \ldots, \mathbf{v}_n$ are elements of $\mathbf{W}$, then every linear combination of $\mathbf{v}_1, \ldots, \mathbf{v}_n$ belongs to $\mathbf{W}$.

Proposition 4.3
Let $\mathbf{v}_1, \ldots, \mathbf{v}_n$ be a finite subset of $\mathbf{V}$. The set $\langle \mathbf{v}_1, \ldots, \mathbf{v}_n \rangle$ consisting of all linear combinations of $\mathbf{v}_1, \ldots, \mathbf{v}_n$ is a vector subspace of $\mathbf{V}$. It is equal to the intersection of all the subspaces containing $\{\mathbf{v}_1, \ldots, \mathbf{v}_n\}$.

Proof

If $a_1\mathbf{v}_1 + \cdots + a_n\mathbf{v}_n$ and $b_1\mathbf{v}_1 + \cdots + b_n\mathbf{v}_n$ are arbitrary elements of $\langle \mathbf{v}_1, \ldots, \mathbf{v}_n \rangle$ and $c \in \mathbf{K}$ then,

$$(a_1\mathbf{v}_1 + \cdots + a_n\mathbf{v}_n) + (b_1\mathbf{v}_1 + \cdots + b_n\mathbf{v}_n) = \\ (a_1 + b_1)\mathbf{v}_1 + (a_2 + b_2)\mathbf{v}_2 + \cdots + (a_n + b_n)\mathbf{v}_\mathbf{n}$$

and

$$c(a_1\mathbf{v}_1 + \cdots + a_n\mathbf{v}_n) = ca_1\mathbf{v}_1 + \cdots + ca_n\mathbf{v}_n$$

are elements of $\langle \mathbf{v}_1, \ldots, \mathbf{v}_n \rangle$, and so $\langle \mathbf{v}_1, \ldots, \mathbf{v}_n \rangle$ is a vector subspace of $\mathbf{V}$.

Let $\mathbf{W}$ be the intersection of all subspaces of $\mathbf{V}$ that contain $\{\mathbf{v}_1, \ldots, \mathbf{v}_n\}$. Since $\langle \mathbf{v}_1, \ldots, \mathbf{v}_n \rangle$ is a vector subspace of $\mathbf{V}$ containing $\{\mathbf{v}_1, \ldots, \mathbf{v}_n\}$, it follows that $\mathbf{W} \subset \langle \mathbf{v}_1, \ldots, \mathbf{v}_n \rangle$. Conversely, $\mathbf{W}$, being a subspace, contains all linear combinations of its elements, and so in particular it contains those of $\mathbf{v}_1, \ldots, \mathbf{v}_n$; that is $\mathbf{W} \supset \langle \mathbf{v}_1, \ldots, \mathbf{v}_n \rangle$. Combining these, we see that $\mathbf{W} = \langle \mathbf{v}_1, \ldots, \mathbf{v}_n \rangle$. □

We will call $\langle \mathbf{v}_1, \ldots, \mathbf{v}_n \rangle$ the *subspace generated by* $\mathbf{v}_1, \ldots, \mathbf{v}_n$.

Notice that, if $1 \le m < n$, the subspace $\langle \mathbf{v}_1, \ldots, \mathbf{v}_m \rangle$ is contained in the subspace $\langle \mathbf{v}_1, \ldots, \mathbf{v}_n \rangle$, because every linear combination of $\langle \mathbf{v}_1, \ldots, \mathbf{v}_m \rangle$ is also a linear combination of $\langle \mathbf{v}_1, \ldots, \mathbf{v}_n \rangle$:

$$a_1\mathbf{v}_1 + \cdots + a_m\mathbf{v}_m = a_1\mathbf{v}_1 + \cdots + a_m\mathbf{v}_m + 0\mathbf{v}_{m+1} + \cdots + 0\mathbf{v}_n.$$

We will say that $\mathbf{v}_1, \ldots, \mathbf{v}_n$ *generate* $\mathbf{V}$, or that $\{\mathbf{v}_1, \ldots, \mathbf{v}_n\}$ is a *system of generators* for $\mathbf{V}$, if $\langle \mathbf{v}_1, \ldots, \mathbf{v}_n \rangle = \mathbf{V}$. Thus $\mathbf{v}_1, \ldots, \mathbf{v}_n$ generate $\mathbf{V}$ if and only if for each $\mathbf{v} \in \mathbf{V}$ there exist $a_1, \ldots, a_n \in \mathbf{K}$ such that

$$\mathbf{v} = a_1\mathbf{v}_1 + \cdots + a_n\mathbf{v}_n.$$

Definition 4.4

The vectors $\mathbf{v}_1, \ldots, \mathbf{v}_n \in \mathbf{V}$ are *linearly dependent* if there are scalars $a_1, \ldots, a_n \in \mathbf{K}$ not all zero such that,

$$a_1\mathbf{v}_1 + \cdots + a_n\mathbf{v}_n = \mathbf{0},$$

that is, if the vector $\mathbf{0}$ can be expressed as a non-trivial linear combination of $\mathbf{v}_1, \ldots, \mathbf{v}_n$. Otherwise $\mathbf{v}_1, \ldots, \mathbf{v}_n$ are said to be *linearly independent*.

There are some simple consequences of this definition.

Proposition 4.5
A single vector $\mathbf{v}$ *is linearly dependent if and only if* $\mathbf{v} = \mathbf{0}$.

Proposition 4.6
If $\mathbf{v}_1$ *and* $\mathbf{v}_2$ *are two vectors with* $\mathbf{v}_2$ *proportional to* $\mathbf{v}_1$, *that is* $\mathbf{v}_2 = a\mathbf{v}_1$ *for some* $a \in \mathbf{K}$, *then* $\mathbf{v}_1$ *and* $\mathbf{v}_2$ *are linearly dependent.*

Indeed, we have that $a\mathbf{v}_1 - \mathbf{v}_2 = \mathbf{0}$ is a linear combination of $\mathbf{v}_1$ and $\mathbf{v}_2$ with coefficients a and -1, which are not both zero.

Conversely, *if* $\mathbf{v}_1$ *and* $\mathbf{v}_2$ *are two linearly dependent vectors, then one of them is a multiple of the other.* Indeed, $a_1\mathbf{v}_1 + a_2\mathbf{v}_2 = 0$ for some $a_1, a_2 \in \mathbf{K}$ (not both zero), and thus $a_1\mathbf{v}_1 = -a_2\mathbf{v}_2$. Supposing that $a_2 \neq 0$, for example, then $\mathbf{v}_2 = a\mathbf{v}_1$ where $a = -a_1/a_2$.

Proposition 4.7
The vectors $\mathbf{v}_1, \ldots, \mathbf{v}_n \in \mathbf{V}$ *with* $n \geq 2$, *are linearly dependent if and only if at least one of them can be expressed as a linear combination of the others.*

Indeed, if $\mathbf{v}_1, \ldots, \mathbf{v}_n$ are linearly dependent, then

$$\mathbf{0} = a_1\mathbf{v}_1 + \cdots + a_n\mathbf{v}_n,$$

with $a_i \neq 0$ for some i; then

$$a_i\mathbf{v}_i = -(a_1\mathbf{v}_1 + \cdots + a_{i-1}\mathbf{v}_{i-1} + a_{i+1}\mathbf{v}_{i+1} + \cdots + a_n\mathbf{v}_n),$$

that is

$$\begin{aligned} \mathbf{v}_i &= -a_i^{-1}(a_1\mathbf{v}_1 + \cdots + a_{i-1}\mathbf{v}_{i-1} + a_{i+1}\mathbf{v}_{i+1} + \cdots + a_n\mathbf{v}_n) \\ &= -a_i^{-1}a_1\mathbf{v}_1 - \cdots - a_i^{-1}a_{i-1}\mathbf{v}_{i-1} \\ &\quad -a_i^{-1}a_{i+1}\mathbf{v}_{i+1} - \cdots - a_i^{-1}a_n\mathbf{v}_n. \end{aligned}$$

Conversely, if for some i

$$\mathbf{v}_i = b_1\mathbf{v}_1 + \cdots + b_{i-1}\mathbf{v}_{i-1} + b_{i+1}\mathbf{v}_{i+1} + \cdots + b_n\mathbf{v}_n,$$

then

$$\mathbf{0} = b_1\mathbf{v}_1 + \cdots + b_{i-1}\mathbf{v}_{i-1} - \mathbf{v}_i + b_{i+1}\mathbf{v}_{i+1} + \cdots + b_n\mathbf{v}_n$$

and so the $\mathbf{v}_1, \ldots, \mathbf{v}_n$ are linearly dependent. □

Proposition 4.8
If the set $\{\mathbf{v}_1, \ldots, \mathbf{v}_n\}$ contains the vector $\mathbf{0}$ then it is linearly dependent.

Indeed, suppose that $\mathbf{v}_i = \mathbf{0}$ for some i. Then,

$$0\mathbf{v}_1 + \cdots + 0\mathbf{v}_{i-1} + 1\mathbf{v}_i + 0\mathbf{v}_{i+1} + \cdots + 0\mathbf{v}_n = \mathbf{v}_i = \mathbf{0},$$

and so $\mathbf{0}$ is a non-trivial linear combination of $\mathbf{v}_1, \ldots, \mathbf{v}_n$.

One can also deduce Proposition 4.8 from Proposition 4.7 as follows: $\mathbf{v}_i = \mathbf{0}$ can be expressed as a linear combination of $\mathbf{v}_1, \ldots, \mathbf{v}_{i-1}$, $\mathbf{v}_{i+1}, \ldots, \mathbf{v}_n$, namely as their trivial linear combination, and so by Proposition 4.7 the vectors $\mathbf{v}_1, \ldots, \mathbf{v}_n$ are linearly dependent.

Proposition 4.9
If $\{\mathbf{v}_1, \ldots, \mathbf{v}_n\}$ are linearly independent and $1 \leq m \leq n$, then $\{\mathbf{v}_1, \ldots, \mathbf{v}_m\}$ are linearly independent. Equivalently, if $\{\mathbf{v}_1, \ldots, \mathbf{v}_m\}$ are linearly dependent, then so are $\{\mathbf{v}_1, \ldots, \mathbf{v}_n\}$.

We prove the second version. If $\{\mathbf{v}_1, \ldots, \mathbf{v}_m\}$ are linearly dependent then,

$$\mathbf{0} = a_1\mathbf{v}_1 + \cdots + a_m\mathbf{v}_m = a_1\mathbf{v}_1 + \cdots + a_m\mathbf{v}_m + 0\mathbf{v}_{m+1} + \cdots + 0\mathbf{v}_n$$

for some $a_1, \ldots, a_m \in \mathbf{K}$ not all zero, and so the right hand side is a non-trivial linear combination of the $\mathbf{v}_1, \ldots, \mathbf{v}_n$ which is equal to $\mathbf{0}$, that is $\mathbf{v}_1, \ldots, \mathbf{v}_n$ are linearly dependent.

Proposition 4.10
If $\mathbf{v}_1, \ldots, \mathbf{v}_n$ are linearly independent, and $a_1, \ldots, a_n, b_1, \ldots, b_n \in \mathbf{K}$ are such that

$$a_1\mathbf{v}_1 + \cdots + a_n\mathbf{v}_n = b_1\mathbf{v}_1 + \cdots + b_n\mathbf{v}_n,$$

then $a_1 = b_1$, $a_2 = b_2$, ..., $a_n = b_n$.

Indeed,

$$\begin{aligned} \mathbf{0} &= (a_1\mathbf{v}_1 + \cdots + a_n\mathbf{v}_n) - (b_1\mathbf{v}_1 + \cdots + b_n\mathbf{v}_n) \\ &= (a_1 - b_1)\mathbf{v}_1 + \cdots (a_n - b_n)\mathbf{v}_n \end{aligned}$$

so we must have that $a_1 - b_1 = a_2 - b_2 = \cdots = a_n - b_n = 0$.

Definition 4.11
A finite subset $\{\mathbf{v}_1, \ldots, \mathbf{v}_n\}$ of $\mathbf{V}$ is said to be a *finite basis*, or simply a *basis*, of $\mathbf{V}$ if $\mathbf{v}_1, \ldots, \mathbf{v}_n$ are linearly independent and generate $\mathbf{V}$.

If $\{\mathbf{v}_1, \ldots, \mathbf{v}_n\}$ is a basis, then, since $\mathbf{v}_1, \ldots, \mathbf{v}_n$ generate $\mathbf{V}$, for each $\mathbf{v} \in \mathbf{V}$ there exist $a_1, \ldots, a_n \in \mathbf{K}$ such that

$$\mathbf{v} = a_1\mathbf{v}_1 + \cdots + a_n\mathbf{v}_n. \tag{4.2}$$

Furthermore, by Proposition 4.10, the $a_1, \ldots, a_n$ are uniquely determined by $\mathbf{v}$.

The coefficients $a_1, \ldots, a_n$ of the linear combination (4.2) are called the *coordinates of* $\mathbf{v}$ *with respect to the basis* $\{\mathbf{v}_1, \ldots, \mathbf{v}_n\}$, and $(a_1, \ldots, a_n)$ is called the *n-tuple of coordinates of* $\mathbf{v}$. Thus, once we are given a basis $\{\mathbf{v}_1, \ldots, \mathbf{v}_n\}$ of $\mathbf{V}$, then to every vector there is associated a unique n-tuple of coordinates. Conversely, every n-tuple $(a_1, \ldots, a_n) \in \mathbf{K}^n$ determines, by (4.2), a unique vector for which it is the n-tuple of coordinates.

The expression $\mathbf{v}(a_1, \ldots, a_n)$ denotes the vector with coordinates $a_1, \ldots, a_n$.

The vector space $\{\mathbf{0}\}$ consisting of the single vector $\mathbf{0}$ does not have a finite basis because its only vector is linearly dependent. There are other vector spaces, different from $\{\mathbf{0}\}$, which do not possess a finite basis — see Example 4.15(5).

We will prove soon that if a vector space $\mathbf{V}$ has a basis consisting of n vectors, then every other basis of $\mathbf{V}$ also consists of n vectors. This fundamental result follows from the following theorem.

Theorem 4.12
Let $\{\mathbf{v}_1, \ldots, \mathbf{v}_n\}$ be a system of generators of $\mathbf{V}$ and let $\mathbf{w}_1, \ldots, \mathbf{w}_m$ be elements of $\mathbf{V}$. If $m > n$ then $\mathbf{w}_1, \ldots, \mathbf{w}_m$ are linearly dependent.

Proof
If $\mathbf{w}_1, \ldots, \mathbf{w}_n$ are linearly dependent then so are $\mathbf{w}_1, \ldots, \mathbf{w}_m$, by Proposition 4.9. We can therefore restrict our attention to the case where $\mathbf{w}_1, \ldots, \mathbf{w}_n$ are linearly independent. It will be sufficient to show that $\mathbf{w}_1, \ldots, \mathbf{w}_n$ generate $\mathbf{V}$ because then $\mathbf{w}_m$ can be expressed as a linear

combination of $\mathbf{w}_1, \ldots, \mathbf{w}_n$ and so by Proposition 4.7 it would follow that $\mathbf{w}_1, \ldots, \mathbf{w}_m$ are indeed linearly dependent.

From the hypothesis that $\mathbf{v}_1, \ldots, \mathbf{v}_n$ generate $\mathbf{V}$ it follows that there are scalars $a_1, \ldots, a_n$ such that

$$\mathbf{w}_1 = a_1\mathbf{v}_1 + \cdots + a_n\mathbf{v}_n.$$

Now, since $\mathbf{w}_1, \ldots, \mathbf{w}_n$ are linearly independent, $\mathbf{w}_1 \neq \mathbf{0}$ and so the coefficients $a_1, \ldots, a_n$ are not all zero. After possibly reordering $\mathbf{v}_1, \ldots,$ $\mathbf{v}_n$ we can suppose that $a_1 \neq 0$. Then,

$$\mathbf{v}_1 = a_1^{-1}\mathbf{w}_1 - a_1^{-1}a_2\mathbf{v}_2 - \cdots - a_1^{-1}a_n\mathbf{v}_n,$$

that is, $\mathbf{v}_1 \in \langle \mathbf{w}_1, \mathbf{v}_2, \ldots, \mathbf{v}_n \rangle$. Since $\mathbf{v}_2, \ldots, \mathbf{v}_n \in \langle \mathbf{w}_1, \mathbf{v}_2, \ldots, \mathbf{v}_n \rangle$ we have

$$\langle \mathbf{w}_1, \mathbf{v}_2, \ldots, \mathbf{v}_n \rangle \supset \langle \mathbf{v}_1, \ldots, \mathbf{v}_n \rangle = \mathbf{V},$$

that is, $\mathbf{w}_1, \mathbf{v}_2, \ldots, \mathbf{v}_n$ generate $\mathbf{V}$.

Suppose now that for some $1 \leq s \leq n-1$ we have that

$$\langle \mathbf{w}_1, \ldots, \mathbf{w}_s, \mathbf{v}_{s+1}, \ldots, \mathbf{v}_n \rangle = \mathbf{V}. \tag{4.3}$$

It follows that there are scalars $b_1, \ldots, b_s, c_{s+1}, \ldots, c_n$ such that

$$\mathbf{w}_{s+1} = b_1\mathbf{w}_1 + \cdots + b_s\mathbf{w}_s + c_{s+1}\mathbf{v}_{s+1} + \cdots + c_n\mathbf{v}_n.$$

Since $\mathbf{w}_1, \ldots, \mathbf{w}_s, \mathbf{w}_{s+1}$ are linearly independent, at least one of the coefficients $c_{s+1}, \ldots, c_n$ must be non-zero. After reordering $\mathbf{v}_{s+1}, \ldots,$ $\mathbf{v}_n$ if necessary, we can suppose that $c_{s+1} \neq 0$. It follows that

$$\mathbf{v}_{s+1} = -c_{s+1}^{-1}b_1\mathbf{w}_1 - \cdots - c_{s+1}^{-1}b_s\mathbf{w}_s + c_{s+1}^{-1}\mathbf{w}_{s+1} - \cdots - c_{s+1}^{-1}c_n\mathbf{v}_n,$$

and so $\mathbf{v}_{s+1} \in \langle \mathbf{w}_1, \ldots, \mathbf{w}_s, \mathbf{w}_{s+1}, \mathbf{v}_{s+2}, \ldots, \mathbf{v}_n \rangle$ (or $\mathbf{v}_n \in \langle \mathbf{w}_1, \ldots, \mathbf{w}_n \rangle$ if $s = n-1$). Thus we have that,

$$\begin{aligned} \langle \mathbf{w}_1, \ldots, \mathbf{w}_s, \mathbf{w}_{s+1}, \mathbf{v}_{s+2}, \ldots, \mathbf{v}_n \rangle &\supset \langle \mathbf{w}_1, \ldots, \mathbf{w}_s, \mathbf{v}_{s+1}, \mathbf{v}_{s+2}, \ldots, \mathbf{v}_n \rangle \\ &= \mathbf{V}, \end{aligned}$$

that is $\mathbf{w}_1, \ldots, \mathbf{w}_{s+1}, \mathbf{v}_{s+2}, \ldots, \mathbf{v}_n$ (respectively $\mathbf{w}_1, \ldots, \mathbf{w}_n$) generate $\mathbf{V}$.

We have already established that (4.3) is true if $s = 1$; the statement of the theorem then follows by induction. □

Corollary 4.13
Let $\{\mathbf{v}_1, \ldots, \mathbf{v}_n\}$ *and* $\{\mathbf{w}_1, \ldots, \mathbf{w}_m\}$ *be two bases of the vector space* $\mathbf{V}$*. Then* $m = n$*.*

Proof
Since $\mathbf{v}_1, \ldots, \mathbf{v}_n$ generate $\mathbf{V}$, and $\mathbf{w}_1, \ldots, \mathbf{w}_m$ are linearly independent, it follows from Theorem 4.12 that $m \leq n$. On the other hand, since $\mathbf{w}_1, \ldots, \mathbf{w}_m$ generate $\mathbf{V}$, and $\mathbf{v}_1, \ldots, \mathbf{v}_n$ are linearly independent, it also follows that $n \leq m$. Thus $m = n$. □

Definition 4.14
If the $\mathbf{K}$-vector space $\mathbf{V}$ has a finite basis $\{\mathbf{v}_1, \ldots, \mathbf{v}_n\}$, the number n is called the *dimension of* $\mathbf{V}$, and is denoted $\dim_{\mathbf{K}}(\mathbf{V})$, or simply $\dim(\mathbf{V})$. If $\mathbf{V} = \{\mathbf{0}\}$ consists only of the zero vector we put $\dim(\mathbf{V}) = 0$.

If $\mathbf{V}$ has a finite basis or if $\mathbf{V} = \{\mathbf{0}\}$ we say that $\mathbf{V}$ is finite dimensional.

It follows from Corollary 4.13 that the dimension of a vector space $\mathbf{V}$ is well-defined because it depends only on $\mathbf{V}$ and not on the particular choice of basis.

Examples 4.15

1. Let $\mathbf{v} = \overrightarrow{OP}$ be a non-zero geometric vector in the ordinary line. Since every other vector is a multiple of $\mathbf{v}$, the set $\{\mathbf{v}\}$ is a basis for the space of geometric vectors in the ordinary line, which therefore has dimension 1.

   Let $\mathbf{V}$ be the real vector space of geometric vectors in the ordinary plane, and let $\mathbf{v}_1$ and $\mathbf{v}_2$ be two elements of $\mathbf{V}$ which are not proportional, and which we will think of as being based at the same point O:

   $$\mathbf{v}_1 = \overrightarrow{OP_1}, \quad \mathbf{v}_2 = \overrightarrow{OP_2}$$

   for some points P_1 and P_2 (see Fig. 4.2). By Proposition 4.6 it follows that $\mathbf{v}_1$ and $\mathbf{v}_2$ are linearly independent. Denote by ℓ_1 and ℓ_2 the lines through O and P_1 and through O and P_2 respectively.

   Let $\mathbf{v} = \overrightarrow{OP}$ be any vector in the plane. Denote by Q_1 the point of intersection of ℓ_1 with the line parallel to ℓ_2 passing through P,

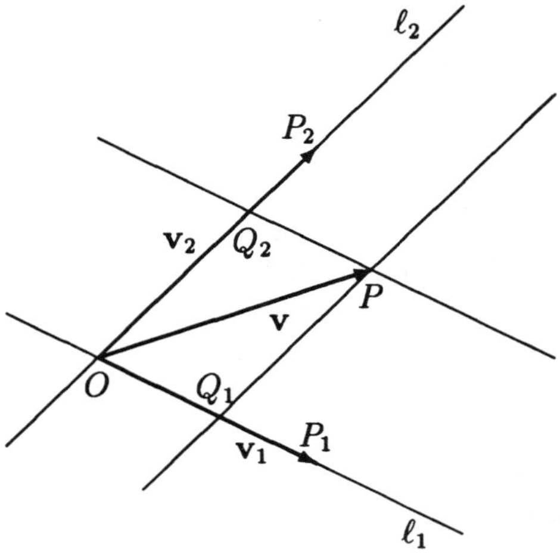

Fig. 4.2

and by Q_2 the point of intersection of ℓ_2 with the line parallel to ℓ_1 passing through P. Let $a_1, a_2 \in \mathbf{R}$ be such that $\overrightarrow{OQ}_1 = a_1\overrightarrow{OP}_1$ and $\overrightarrow{OQ}_2 = a_2\overrightarrow{OP}_2$.

Since

$$\mathbf{v} = \overrightarrow{OQ}_1 + \overrightarrow{OQ}_2 = a_1\overrightarrow{OP}_1 + a_2\overrightarrow{OP}_2 = a_1\mathbf{v}_1 + a_2\mathbf{v}_2$$

we see that $\mathbf{v}$ is a linear combination of $\mathbf{v}_1$ and $\mathbf{v}_2$, and so $\mathbf{v}_1$ and $\mathbf{v}_2$ generate $\mathbf{V}$. Therefore $\{\mathbf{v}_1, \mathbf{v}_2\}$ is a basis for $\mathbf{V}$, and $\mathbf{V}$ has dimension 2.

In ordinary space, three geometric vectors are linearly independent if and only if the points O, P_1, P_2 and P_3 do not all lie in a single plane. This follows immediately from the fact that linear dependance of $\mathbf{v}_1, \mathbf{v}_2, \mathbf{v}_3$ is equivalent to one of them being a linear combination of the other two. If $\mathbf{v}_1, \mathbf{v}_2, \mathbf{v}_3$ are linearly independent then using a similar consruction to the one used in the plane (Fig 4.2), one can check that every geometric vector can be expressed as a linear combination of $\mathbf{v}_1, \mathbf{v}_2, \mathbf{v}_3$. Thus $\{\mathbf{v}_1, \mathbf{v}_2, \mathbf{v}_3\}$ constitutes a basis, and the geometric vectors of ordinary space form a real vector space of dimension 3.

2. Let $n \geq 1$. Consider the numerical vector space $\mathbf{K}^n$, and let

$$\mathbf{E}_1 = (1, 0, \ldots, 0), \ \ \mathbf{E}_2 = (0, 1, \ldots, 0), \ \ \ldots, \ \ \mathbf{E}_n = (0, 0, \ldots, 1).$$

The vectors $\mathbf{E}_1, \ldots, \mathbf{E}_n$ generate $\mathbf{K}^n$. Indeed, for any $(x_1, \ldots, x_n) \in \mathbf{K}^n$ one has,

$$(x_1, \ldots, x_n) = x_1\mathbf{E}_1 + x_2\mathbf{E}_2 + \cdots + x_n\mathbf{E}_n. \tag{4.4}$$

Moreover, since $(x_1, \ldots, x_n) = \mathbf{0}$ if and only if $x_1 = \cdots = x_n = 0$, it follows from (4.4) that $\mathbf{E}_1, \ldots, \mathbf{E}_n$ are linearly independent. Thus $\{\mathbf{E}_1, \ldots, \mathbf{E}_n\}$ is a basis of $\mathbf{K}^n$, and so $\mathbf{K}^n$ has dimension n. The basis $\{\mathbf{E}_1, \ldots, \mathbf{E}_n\}$ is called the *canonical basis* of $\mathbf{K}^n$.

For $n = 1$ we have that the field $\mathbf{K}$, considered as a vector space over itself, has dimension 1, with its canonical basis consisting only of $1 \in \mathbf{K}$.

It follows from (4.4) that the coordinates $(x_1, \ldots, x_n)$ of a vector with respect to the canonical basis coincide with its components $x_1, \ldots, x_n$.

3. Let $A \in M_{m,n}(\mathbf{K})$, $\mathbf{X} = (X_1\, X_2 \ldots X_n)^t$ be unknowns, and let

$$A\mathbf{X} = \mathbf{0} \tag{4.5}$$

be the homogeneous system of m equations in the unknowns $\mathbf{X}$, whose matrix of coefficients is A. Denote by Σ_0 the set of solutions of the system. As we already noticed in Example 4.2(2), Σ_0 is a vector subspace of $\mathbf{K}^n$. The m equations (4.5) are called *Cartesian equations for the subspace* Σ_0 *of* $\mathbf{K}^n$. Thus, by definition, two different systems of homogeneous linear equations in the unknowns $\mathbf{X}$ are equivalent if and only if they are Cartesian equations for the same subspace of $\mathbf{K}^n$.

The elements of Σ_0 can be viewed as linear dependence relations between the columns $A_{(1)}, A_{(2)}, \ldots, A_{(n)}$ of A, considered as vectors in $\mathbf{K}^m$. Indeed, $\mathbf{x} = (x_1, \ldots, x_n)^t \in \mathbf{K}^n$ is a solution of (4.5) if and only if $A\mathbf{x} = \mathbf{0}$, or, equivalently, if and only if

$$A_{(1)}x_1 + A_{(2)}x_2 + \cdots + A_{(n)}x_n = \mathbf{0}.$$

4. Consider a system of m homogeneous linear equations in echelon form:

$$\begin{array}{rr} a_{11}X_1 + a_{12}X_2 + \cdots & \cdots + a_{1n}X_n = 0 \\ a_{22}X_2 + \cdots & \cdots + a_{2n}X_n = 0 \\ & \vdots \\ & a_{mm}X_m + \cdots + a_{mn}X_n = 0 \end{array} \tag{4.6}$$

with $a_{11}, a_{22}, \ldots, a_{mm} \neq 0$. If $m = n$, system (4.6) has only the trivial solution. If $n > m$ the $(n-m)$-parameter family of solutions is obtained by giving arbitrary values to the unknowns $X_{m+1}, \ldots, X_n$, and then solving the resulting m equations for the remaining unknowns $X_1, \ldots, X_m$. Consider in particular the $n - m$ solutions

$$\begin{aligned} s_{m+1} &= (s_{m+11}, s_{m+12}, \ldots, s_{m+1m}, 1, 0, 0, \ldots, 0) \\ s_{m+2} &= (s_{m+21}, s_{m+22}, \ldots, s_{m+2m}, 0, 1, 0, \ldots, 0) \\ &\vdots \\ s_n &= (s_{n1}, s_{n2}, \ldots, s_{nm}, 0, 0, 0, \ldots, 1). \end{aligned}$$

obtained from the $(n-m)$-tuples

$$(t_{m+1}, t_{m+2}, \ldots, t_n) = \begin{cases} (1, 0, 0, \ldots, 0) \\ (0, 1, 0, \ldots, 0) \\ \vdots \\ (0, 0, 0, \ldots, 1) \end{cases}$$

respectively.

For every $t_{m+1}, t_{m+2}, \ldots, t_n \in \mathbf{K}$ one has

$$\begin{aligned} &t_{m+1}s_{m+1} + t_{m+2}s_{m+2} + \cdots + t_n s_n = \\ &(t_{m+1}s_{m+11} + \cdots + t_n s_{n1}, t_{m+1}s_{m+12} + \cdots + t_n s_{n2}, \\ &\quad \ldots, t_{m+1}s_{m+1m} + \cdots + t_n s_{nm}, t_{m+1}, t_{m+2}, \ldots, t_n). \end{aligned} \tag{4.7}$$

The right hand side is equal to $(0, 0, \ldots, 0)$ if and only if $t_{m+1} = t_{m+2} = \cdots = t_n = 0$. Thus $s_{m+1}, s_{m+2}, \ldots, s_n$ are linearly independent.

The right hand side of (4.7) is the solution of (4.6) corresponding to choosing the values $t_{m+1}, t_{m+2}, \ldots, t_n$ for the unknowns $X_{m+1}, X_{m+2}, \ldots, X_n$. Equation (4.7) shows that any such solution is a linear combination of the $s_{m+1}, s_{m+2}, \ldots, s_n$. Thus s_{m+1}, $s_{m+2}, \ldots, s_n$ generate the space Σ_0 of solutions of (4.6). It follows that $\{s_{m+1}, s_{m+2}, \ldots, s_n\}$ is a basis for Σ_0. In particular, $\dim(\Sigma_0) = n - m$. Recall from Chapter 3 that system (4.6) has an $(n-m)$-parameter family of solutions, and thus the number of parameters is the same as the dimension of the space of solutions.

More generally, since any system of linear homogeneous equations (4.5) can be reduced to one in echelon form, we have that *a*

system of homogeneous linear equations has an N-parameter family of solutions if and only if the space of its solutions has dimension N.

5. Let X be an unknown. The $\mathbf{K}$ vector space $\mathbf{K}[X]$ of polynomials in X with coefficients in $\mathbf{K}$ does not have finite dimension.

   To see this, suppose that $\mathbf{K}[X]$ does have a finite basis $\{f_1(X), \ldots, f_n(X)\}$. Let $d_1, \ldots, d_n$ be the degrees of $f_1(X), \ldots, f_n(X)$ respectively, and put $D = \max\{d_1, \ldots, d_n\}$. Let $f(X)$ be a polynomial of degree $d > D$. Since $\{f_1(X), \ldots, f_n(X)\}$ is a finite basis, there exist $a_1, \ldots, a_n \in \mathbf{K}$ such that

   $$f(X) = a_1 f_1(X) + \cdots + a_n f_n(X).$$

   However, the polynomial on the right hand side has degree at most D, and so cannot be equal to $f(X)$. This contradiction implies that a finite basis of $\mathbf{K}[X]$ cannot exist.

   In a similar fashion one can show that the $\mathbf{K}$ vector space $\mathbf{K}[X_1, X_2, \ldots, X_n]$ of polynomials in the unknowns $X_1, \ldots, X_n$ with coefficients in $\mathbf{K}$ is not finite dimensional.

6. For any positive integer d, the vector space $\mathbf{K}[X_1, \ldots, X_n]_d$ consists of all homogeneous polynomials of degree d in $X_1, \ldots, X_n$ with coefficients in $\mathbf{K}$ together with the polynomial 0. This vector space has a finite basis: one can take the set of all monomials of degree d. Thus $\mathbf{K}[X_1, \ldots, X_n]_d$ has dimension $\binom{n+d-1}{d}$ by Lemma A.11. In particular the vector space $\mathbf{K}[X_1, \ldots, X_n]_1$ of all homogeneous polynomials of degree 1 in $X_1, \ldots, X_n$ has dimension n.

7. The vector space $M_{m,n}(\mathbf{K})$ of all $m \times n$ matrices with entries in $\mathbf{K}$ has dimension mn. To see this consider, for each $1 \leq i \leq m$, and each $1 \leq j \leq n$, the matrix $\mathbf{1}_{ij}$ which has a 1 in the ij-th position and 0s elsewhere. We obtain a set of mn matrices $\{\mathbf{1}_{11}, \mathbf{1}_{12}, \ldots, \mathbf{1}_{mn}\}$ which forms a basis for $M_{m,n}(\mathbf{K})$. Indeed, if $A = (a_{ij}) \in M_{m,n}(\mathbf{K})$, then there is a unique expression for A as

   $$A = a_{11}\mathbf{1}_{11} + a_{12}\mathbf{1}_{12} + \cdots + a_{mn}\mathbf{1}_{mn}.$$

Proposition 4.16

Suppose that $\dim(\mathbf{V}) = n$.

1) *If* $\mathbf{v}_1, \ldots, \mathbf{v}_n \in \mathbf{V}$ *are linearly independent then* $\{\mathbf{v}_1, \ldots, \mathbf{v}_n\}$ *is a basis for* $\mathbf{V}$.

2) *If* $\mathbf{v}_1, \dots, \mathbf{v}_k \in \mathbf{V}$ *are linearly independent then there are vectors* $\mathbf{v}_{k+1}, \dots, \mathbf{v}_n \in \mathbf{V}$ *such that* $\{\mathbf{v}_1, \dots, \mathbf{v}_n\}$ *is a basis for* $\mathbf{V}$.

Proof
1) It is enough to show that $\langle \mathbf{v}_1, \dots, \mathbf{v}_n \rangle = \mathbf{V}$. By hypothesis there is a basis $\{\mathbf{b}_1, \dots, \mathbf{b}_n\}$ of $\mathbf{V}$, and this implies, by Theorem 4.12, that for every $\mathbf{v} \in \mathbf{V}$, the vectors $\mathbf{v}_1, \dots, \mathbf{v}_n, \mathbf{v}$ are linearly dependent. Therefore there exist $a_1, \dots, a_n, a \in \mathbf{K}$ not all zero, such that

$$a_1\mathbf{v}_1 + \cdots + a_n\mathbf{v}_n + a\mathbf{v} = \mathbf{0}.$$

Since the $\mathbf{v}_1, \dots, \mathbf{v}_n$ are linearly independent, a must be non-zero. Thus we have that

$$\mathbf{v} = -a^{-1}a_1\mathbf{v}_1 - \cdots - a^{-1}a_n\mathbf{v}_n,$$

and so $\mathbf{v} \in \langle \mathbf{v}_1, \dots, \mathbf{v}_n \rangle$. As $\mathbf{v}$ was arbitrary, we have proved the assertion.
2) By Theorem 4.12, we have $k \leq n$. If $k = n$ the result follows from part (1), so suppose now that $k < n$.

Now, $\mathbf{v}_1, \dots, \mathbf{v}_k$ cannot generate $\mathbf{V}$, as otherwise there would be a basis with $k \neq n$ elements, contradicting Corollary 4.13. Thus we can find a vector $\mathbf{v}_{k+1} \in \mathbf{V} \setminus \langle \mathbf{v}_1, \dots, \mathbf{v}_k \rangle$.

Suppose that $a_1, \dots, a_k, a_{k+1} \in \mathbf{K}$ are such that

$$a_1\mathbf{v}_1 + \cdots + a_k\mathbf{v}_k + a_{k+1}\mathbf{v}_{k+1} = \mathbf{0}.$$

Then $a_{k+1} = 0$, for otherwise

$$\mathbf{v}_{k+1} = -a_{k+1}^{-1}a_1\mathbf{v}_1 - \cdots - a_{k+1}^{-1}a_n\mathbf{v}_n,$$

so then $\mathbf{v}_{k+1} \in \langle \mathbf{v}_1, \dots, \mathbf{v}_k \rangle$, contrary to the hypothesis. Moreover, since $a_{k+1} = 0$, and $\mathbf{v}_1, \dots, \mathbf{v}_k$ are linearly independent, it follows that $a_1 = \ldots = a_k = 0$. Thus the vectors $\mathbf{v}_1, \dots, \mathbf{v}_k, \mathbf{v}_{k+1}$ are linearly independent.

If $k + 1 = n$ then from part (1) we can conclude that $\{\mathbf{v}_1, \dots, \mathbf{v}_k, \mathbf{v}_{k+1}\}$ is a basis, so the assertion is proved. If instead $k + 1 < n$ we can repeat the argument above to find $\mathbf{v}_{k+2} \in \mathbf{V}$ such that $\{\mathbf{v}_1, \dots, \mathbf{v}_k, \mathbf{v}_{k+1}, \mathbf{v}_{k+2}\}$ are linearly independent. Iterating this procedure $n - k$ times we can find vectors $\mathbf{v}_{k+1}, \mathbf{v}_{k+2}, \dots, \mathbf{v}_n \in \mathbf{V}$ such that $\{\mathbf{v}_1, \dots, \mathbf{v}_n\}$ are linearly independent, and so $\{\mathbf{v}_1, \dots, \mathbf{v}_n\}$ is a basis, by part (1). □

Corollary 4.17
If $\dim(\mathbf{V}) = n$*, then every vector subspace* $\mathbf{W}$ *of* $\mathbf{V}$ *has dimension at most* n.

Proof
If $\{\mathbf{w}_1, \ldots, \mathbf{w}_m\}$ is a set of linearly independent vectors in $\mathbf{W}$ then $m \leq n$, by Theorem 4.12. Thus, if $\mathbf{W}$ is finite dimensional, every basis of $\mathbf{W}$ must have no more than n elements, and so $\dim(\mathbf{W}) \leq n$. On the other hand, if $\mathbf{W}$ were not finite dimensional, then $\langle \mathbf{w}_1, \ldots, \mathbf{w}_m \rangle \neq \mathbf{W}$ for every subset $\{\mathbf{w}_1, \ldots, \mathbf{w}_m\}$ of linearly independent vectors in $\mathbf{W}$ so there would be another vector $\mathbf{w}_{m+1} \in \mathbf{W}$ such that $\{\mathbf{w}_1, \ldots, \mathbf{w}_m, \mathbf{w}_{m+1}\}$ are linearly independent (cf. proof of Proposition 4.16(2)). This implies the existence of arbitrarily large sets of linearly independent vectors in $\mathbf{V}$, which is impossible, again by Theorem 4.12. □

In particular, if $\dim(\mathbf{W}) = 0$ then $\mathbf{W} = \mathbf{0}$ and if $\dim(\mathbf{W}) = n$ then $\mathbf{W} = \mathbf{V}$.

For any subspace $\mathbf{W}$ of $\mathbf{V}$, the number

$$\dim(\mathbf{V}) - \dim(\mathbf{W})$$

is called the *codimension of* $\mathbf{W}$ *in* $\mathbf{V}$.

We now prove an important result relating the dimensions of two subspaces, the dimensions of their intersection and of their sum. It is called *Grassman's formula.*

Theorem 4.18
Let $\mathbf{U}$ *and* $\mathbf{W}$ *be two finite dimensional subspaces of a vector space* $\mathbf{V}$. *Then* $\mathbf{U} \cap \mathbf{W}$ *and* $\mathbf{U} + \mathbf{W}$ *are also finite dimensional, and*

$$\dim(\mathbf{U}) + \dim(\mathbf{W}) = \dim(\mathbf{U} + \mathbf{W}) + \dim(\mathbf{U} \cap \mathbf{W}). \qquad (4.8)$$

In particular, $\mathbf{U} + \mathbf{W}$ *is the direct sum of* $\mathbf{U}$ *and* $\mathbf{W}$ *if and only if* $\dim(\mathbf{U}) + \dim(\mathbf{W}) = \dim(\mathbf{U} + \mathbf{W})$.

Proof
Since $\mathbf{U} \cap \mathbf{W}$ is a subspace of $\mathbf{U}$ it is finite dimensional, and so has a basis. Let $\{\mathbf{z}_1, \ldots, \mathbf{z}_q\}$ be such a basis. By Proposition 4.12(2), there exist $\mathbf{u}_1, \ldots, \mathbf{u}_t \in \mathbf{U}$ and $\mathbf{w}_1, \ldots, \mathbf{w}_s \in \mathbf{W}$ such that

$$\{\mathbf{z}_1, \ldots, \mathbf{z}_q, \mathbf{u}_1, \ldots, \mathbf{u}_t\} \quad \text{and} \quad \{\mathbf{z}_1, \ldots, \mathbf{z}_q, \mathbf{w}_1, \ldots, \mathbf{w}_s\}$$

are bases of $\mathbf{U}$ and $\mathbf{W}$ respectively. Since

$$\dim(\mathbf{U}) + \dim(\mathbf{W}) - \dim(\mathbf{U} \cap \mathbf{W}) = q + t + s$$

it is enough for us to show that $\{\mathbf{z}_1, \dots, \mathbf{z}_q, \mathbf{u}_1, \dots, \mathbf{u}_t, \mathbf{w}_1, \dots, \mathbf{w}_s\}$ is a basis for $\mathbf{U} + \mathbf{W}$.

Let $\mathbf{u} + \mathbf{w} \in \mathbf{U} + \mathbf{W}$. There are scalars $a_1, \dots, a_q$, $a'_1, \dots, a'_q$, $b_1, \dots, b_t$, $c_1, \dots, c_s$ such that

$$\begin{aligned} \mathbf{u} &= a_1\mathbf{z}_1 + \cdots + a_q\mathbf{z}_q + b_1\mathbf{u}_1 + \cdots + b_t\mathbf{u}_t \\ \mathbf{w} &= a'_1\mathbf{z}_1 + \cdots + a'_q\mathbf{z}_q + c_1\mathbf{w}_1 + \cdots + c_s\mathbf{w}_s. \end{aligned}$$

Thus,

$$\begin{aligned} \mathbf{u} + \mathbf{w} &= (a_1 + a'_1)\mathbf{z}_1 + \cdots + (a_q + a'_q)\mathbf{z}_q + \\ &\quad b_1\mathbf{u}_1 + \cdots + b_t\mathbf{u}_t + c_1\mathbf{w}_1 + \cdots + c_s\mathbf{w}_s, \end{aligned}$$

and $\mathbf{z}_1, \dots, \mathbf{z}_q, \mathbf{u}_1, \dots, \mathbf{u}_t, \mathbf{w}_1, \dots, \mathbf{w}_s$ generate $\mathbf{U} + \mathbf{W}$.

Suppose now that there are $a_1, \dots, a_q$, $b_1, \dots, b_t$, $c_1, \dots, c_s \in \mathbf{K}$ such that

$$a_1\mathbf{z}_1 + \cdots + a_q\mathbf{z}_q + b_1\mathbf{u}_1 + \cdots + b_t\mathbf{u}_t + c_1\mathbf{w}_1 + \cdots + c_s\mathbf{w}_s = \mathbf{0}. \quad (4.9)$$

Rearranging this equation gives

$$c_1\mathbf{w}_1 + \cdots + c_s\mathbf{w}_s = -(a_1\mathbf{z}_1 + \cdots + a_q\mathbf{z}_q + b_1\mathbf{u}_1 + \cdots + b_t\mathbf{u}_t). \quad (4.10)$$

Since the left hand side of (4.10) belongs to $\mathbf{W}$, the right hand side lies in $\mathbf{U} \cap \mathbf{W}$. As $\{\mathbf{z}_1, \dots, \mathbf{z}_q\}$ is a basis of $\mathbf{U} \cap \mathbf{W}$, we have

$$a_1\mathbf{z}_1 + \cdots + a_q\mathbf{z}_q + b_1\mathbf{u}_1 + \cdots + b_t\mathbf{u}_t = e_1\mathbf{z}_1 + \cdots + e_q\mathbf{z}_q$$

for some $e_1, \dots, e_q \in \mathbf{K}$. Equivalently,

$$(a_1 - e_1)\mathbf{z}_1 + \cdots (a_q - e_q)\mathbf{z}_q + b_1\mathbf{u}_1 + \cdots + b_t\mathbf{u}_t = \mathbf{0}.$$

By the linear independence of $\mathbf{z}_1, \dots, \mathbf{z}_q$, $\mathbf{u}_1, \dots, \mathbf{u}_t$, all the coefficients are zero, and in particular $b_1 = \cdots = b_t = 0$. From (4.9) it follows that

$$a_1\mathbf{z}_1 + \cdots + a_q\mathbf{z}_q + c_1\mathbf{w}_1 + \cdots + c_s\mathbf{w}_s = \mathbf{0}.$$

The linear independence of $\{\mathbf{z}_1, \dots, \mathbf{z}_q, \mathbf{w}_1, \dots, \mathbf{w}_s\}$ now implies that

$$a_1 = \cdots = a_q = c_1 = \cdots = c_s = 0.$$

Therefore $\mathbf{z}_1, \ldots, \mathbf{z}_q, \mathbf{u}_1, \ldots, \mathbf{u}_t, \mathbf{w}_1, \ldots, \mathbf{w}_s$ are linearly independent.

The final assertion of the theorem follows immediately from (4.8) and the definition of direct sum. □

EXERCISES

4.1 Establish which of the following sets of vectors are linearly independent, which generate the space and which form a basis for the space:

In $\mathbf{R}^2$:
a) $\{(1, 123), (-\pi, -\pi)\}$,
b) $\{(2, -1/3), (-1, 1/6)\}$,
c) $\{(4/5, 5/4), (4, 5)\}$,
d) $\{(1, 2), (11, -7\sqrt{2}), (-1, 1)\}$.

In $\mathbf{R}^3$:
e) $\{(1, 1, 3), (2, 2, 0), (3, 3, -3)\}$,
f) $\{(1, -1, -\sqrt{5}), (1, 1, \sqrt{5}), (0, 1, 2\sqrt{5})\}$,
g) $\{(1, 0, 0), (1, 1, 1), (0, 1, 2), (-1, -2, -3)\}$.

In $\mathbf{C}^4$:
h) $\{(1, 0, i, 0), (i, 0, i, 0), (0, 1, 1, 0), (0, i, 0, i)\}$,
i) $\{(0, 1, 1, 0), (0, -i, -2i, 1), (0, i, 0, 1), (1, 0, 0, 0)\}$.

4.2 Show that the matrices,

$$\begin{pmatrix} 1 & 1 & 1 \\ 2 & 0 & 1 \end{pmatrix}, \quad \begin{pmatrix} 1 & 1 & 0 \\ -1 & 0 & 1 \end{pmatrix}, \quad \begin{pmatrix} 2 & -2 & 1 \\ 1 & 0 & 0 \end{pmatrix} \quad \in M_{2,3}(\mathbf{Q})$$

are linearly independent.

4.3 Which of the following subsets of $\mathbf{R}^3$ are vector subspaces?
a) $\{(0, 0, 0)\}$,
b) $\{(x, 0, 0) \mid x \in \mathbf{R}, x \neq 0\}$,
c) $\{(x, y, z) \mid x - 2y + z = 1\}$,
d) $\{(t, t, t) \mid 0 \leq t \leq 1\}$,
e) $\{(t, t, t) \mid 0 < t < 1\}$,
f) $\mathbf{R}^3 \setminus \{(0, 0, 1)\}$,
g) $H_1 \cup H_2 \cup H_3$, where $H_i = \{(x_1, x_2, x_3) \mid x_i = 0\}$,
h) $\{(x, y, z) \mid x^2 + y^2 + z^2 = 1\}$,

i) $\{(x,y,z) \mid x+y-5z=0,\ 2(x+y)=0\}$,
j) $\{(t,1,t) \mid t \in \mathbf{R}\}$.

4.4 Let $\mathbf{V}$ be a real vector space of dimension 3, and let $\{\mathbf{i},\mathbf{j},\mathbf{k}\}$ be a basis of $\mathbf{V}$. Let $\mathbf{U} = \langle \mathbf{i}+\mathbf{j}, \mathbf{i}-\mathbf{j}\rangle$, $\mathbf{W} = \langle \mathbf{j}+\mathbf{k}, \mathbf{j}-\mathbf{k}\rangle$. Show that $\mathbf{V} = \mathbf{U}+\mathbf{W}$, and that the sum is not a direct sum.

4.5 Prove that $\mathbf{R}^4 = \mathbf{U} \oplus \mathbf{V}$, where

$$\begin{aligned}\mathbf{U} &= \langle (1,0,-\sqrt{5},0),\ (\sqrt{5},0,-1,0)\rangle \\ \mathbf{W} &= \langle (0,-2,0,3),\ (0,1,0,1)\rangle\end{aligned}$$

4.6 Show that $\mathbf{R}^3 = \mathbf{U} \oplus \mathbf{W}$, where

$$\begin{aligned}\mathbf{U} &= \{(x,y,z) \mid x-y=0\} \\ \mathbf{W} &= \langle (1,0,1)\rangle\end{aligned}$$

4.7 Using only elementary operations on the vectors, find a basis for the subspace of $\mathbf{Q}^4$ generated by the following vectors:

$$\mathbf{v}_1 = (1,1,2,3), \quad \mathbf{v}_2 = (3,2,1,0),$$

$$\mathbf{v}_3 = (-1,0,3,6), \quad \mathbf{v}_4 = (2,2,2,2).$$

4.8 Prove that the n vectors

$$(1,1,\ldots,1),\ (0,1,1,\ldots,1),\ (0,0,1,\ldots,1),\ \ldots,\ (0,\ldots,0,1)$$

form a basis of $\mathbf{K}^n$.

4.9 Let $\mathbf{V}$ be a vector space over $\mathbf{K}$, and suppose that $\mathbf{v}_1,\ldots,\mathbf{v}_n$ are linearly independent. Show that for any $\lambda_1,\ldots,\lambda_n \in \mathbf{K}^*$, the vectors $\lambda_1\mathbf{v}_1,\ldots,\lambda_n\mathbf{v}_n$ are linearly independent.

4.10 Let $1 \leq i \leq n$. Find a basis for the subspace H_i of $\mathbf{K}^n$ (as defined in Example 4.2(2)).

4.11 Prove that $\mathrm{GL}_n(\mathbf{K})$ is not a vector subspace of $M_n(\mathbf{K})$.

4.12 Let $A = (a_{ij}) \in M_n(\mathbf{K})$. The *trace* of A is

$$\mathrm{tr}(A) = a_{11} + a_{22} + \cdots + a_{nn}.$$

Prove that the subset $\mathcal{T}_0$ of $M_n(\mathbf{K})$ consisting of matrices of trace 0, is a vector subspace, and find its dimension.

4.13 The matrices in $M_2(\mathbf{C})$,

$$\sigma_1 = \begin{pmatrix} 0 & 1 \\ 1 & 0 \end{pmatrix} \qquad \sigma_2 = \begin{pmatrix} 0 & -i \\ i & 0 \end{pmatrix} \qquad \sigma_3 = \begin{pmatrix} 1 & 0 \\ 0 & -1 \end{pmatrix}$$

are called the *Pauli matrices.* Prove that:

a) $\sigma_1^2 = \sigma_2^2 = \sigma_3^2 = \mathbf{I}_2$,

$$\begin{aligned} \sigma_1\sigma_2 &= i\sigma_3, & \sigma_2\sigma_1 &= -i\sigma_3, \\ \sigma_2\sigma_3 &= i\sigma_1, & \sigma_3\sigma_2 &= -i\sigma_1, \\ \sigma_3\sigma_1 &= i\sigma_2, & \sigma_1\sigma_3 &= -i\sigma_2. \end{aligned}$$

b) $\{\mathbf{I}_2, \sigma_1, \sigma_2, \sigma_3\}$ is a basis of $M_2(\mathbf{C})$

Calculate the coordinates with respect to this basis of the matrices:

$$\mathbf{1}_{11} = \begin{pmatrix} 1 & 0 \\ 0 & 0 \end{pmatrix}, \quad \mathbf{1}_{12} = \begin{pmatrix} 0 & 1 \\ 0 & 0 \end{pmatrix},$$

$$\mathbf{1}_{21} = \begin{pmatrix} 0 & 0 \\ 1 & 0 \end{pmatrix}, \quad \mathbf{1}_{22} = \begin{pmatrix} 0 & 0 \\ 0 & 1 \end{pmatrix}.$$

4.14 Prove that the vector space $S_{\mathbf{K}}$ of sequences of elements of $\mathbf{K}$ is not finite dimensional.

4.15 Prove that the set $B_{\mathbf{R}}$ of bounded sequences of real numbers is a vector subspace of $S_{\mathbf{R}}$. Prove also that $B_{\mathbf{R}}$ is not finite dimensional.

4.16 Let $S_{\mathbf{K}}$ be the vector space of sequences of elements of $\mathbf{K}$, and let $P_{\mathbf{K}}$ be the subset of elements of $S_{\mathbf{K}}$ consisting of those sequences $\{a_n\}$ such that $a_n = 0$ for all n sufficiently large. (That is, $\{a_n\} \in P_{\mathbf{K}}$ if there is an N such that $a_n = 0$ for all $n > N$.) Prove that $P_{\mathbf{K}}$ is a vector subspace of $S_{\mathbf{K}}$.

4.17 Let $a, b \in \mathbf{R}$ with $a < b$. Prove that the vector space $C_{(a,b)}$ of all continuous functions on the open interval (a, b) is not finite dimensional.

4.18 Let X be an unknown, and $d \geq 1$ an integer. Prove that the subset $\mathbf{K}[X]_{\leq d}$ of $\mathbf{K}[X]$ consisting of polynomials of degree at most d, together with the polynomial 0, is a vector subspace of dimension $d + 1$.

4.19 Prove that the homogeneous polynomials of degree 2,

$$X_2^2, \quad X_1^2 + X_2^2, \quad X_0^2 + X_1^2 + X_2^2, \quad X_0X_1, \quad X_1X_2, \quad X_0X_2$$

form a basis of $\mathbf{K}[X_0, X_1, X_2]$.

4.20 Let $\mathbf{W}_0$, $\mathbf{W}_{01}$, and $\mathbf{W}_{012}$ be subsets of $\mathbf{K}[X_0, X_1, X_2]_2$ as follows:

$$\begin{aligned}
\mathbf{W}_0 &= \{F \mid F \text{ is divisible by } X_0\} \\
\mathbf{W}_{01} &= \{F \mid F = X_0 L_0 + X_1 L_1 \\
&\qquad \text{for some } L_0, L_1 \in \mathbf{K}[X_0, X_1, X_2]_1\} \\
\mathbf{W}_{012} &= \{F \mid F = X_0 L_0 + X_1 L_1 + X_2 L_2 \\
&\qquad \text{for some } L_0, L_1, L_2 \in \mathbf{K}[X_0, X_1, X_2]_1\}.
\end{aligned}$$

Prove that $\mathbf{W}_0$, $\mathbf{W}_{01}$ and $\mathbf{W}_{012}$ are vector subspaces of $\mathbf{K}[X_0, X_1, X_2]_2$ and find their dimensions.

4.21 A sequence $\{a_n\}$ of elements of $\mathbf{K}$ is a *Fibonacci sequence* if for every $n \geq 0$ it satisfies $a_{n+2} = a_{n+1} + a_n$. Prove that the set of Fibonacci sequences constitute a vector subspace $F_{\mathbf{K}}$ of $S_{\mathbf{K}}$, and that $\dim(F_{\mathbf{K}}) = 2$.

5

Rank

As we have seen, the method of Gaussian elimination is very useful in practice for solving systems of linear equations. However, it does not lend itself so effectively to theoretical questions, for which it is preferable to have general criteria for the solubility of such systems of equations in terms of the matrix associated to the system. Criteria of this type can be obtained using the notion of 'rank'.

If $\mathbf{V}$ is a vector space, and $\{\mathbf{v}_1, \dots, \mathbf{v}_m\}$ is a finite subset of $\mathbf{V}$, the *rank* of $\{\mathbf{v}_1, \dots, \mathbf{v}_m\}$ is the maximum number of linearly independent vectors in the set $\{\mathbf{v}_1, \dots, \mathbf{v}_m\}$. Equivalently, the rank of $\{\mathbf{v}_1, \dots, \mathbf{v}_m\}$ is the dimension of the vector subspace $\langle \mathbf{v}_1, \dots, \mathbf{v}_m \rangle$.

If $A \in M_{m,n}(\mathbf{K})$, the *row rank* of A is the rank of the set of its rows, that is, the maximum number of linearly independent rows of A, considered as vectors in $\mathbf{K}^n$. Similarly, one defines the *column rank* of A to be the rank of the set of its columns, considered as vectors in $\mathbf{K}^m$.

The importance of this notion is based on the following result.

Theorem 5.1
The row rank and the column rank of a matrix coincide.

Proof
Let r be the row rank, and c the column rank, of the matrix $A \in M_{m,n}(\mathbf{K})$. If $r = 0$ then all the entries of A are zero, and so $c = 0$. We can therefore assume that $r > 0$.

A linear dependence relation between the columns of A is a non-trivial solution of the homogeneous system

$$A\mathbf{X} = \mathbf{0} \tag{5.1}$$

where $\mathbf{X} = (X_1 \ \dots \ X_n)^t$. Thus the column rank of A can be found from the set of solutions of (5.1). If the order of the rows of A is changed, then c is unchanged because the set of solutions of a system of equations is unchanged by a change in the order of the equations (i.e. a sequence of elementary operations of type (I)). Moreover r does not change if the rows are reordered, because the rank of a set of vectors does not depend on their order.

Therefore we can suppose that the first r rows of A are linearly independent. The matrix

$$A^* = \begin{pmatrix} a_{11} & a_{12} & \dots & a_{1n} \\ a_{21} & a_{22} & \dots & a_{2n} \\ \vdots & \vdots & & \vdots \\ a_{r1} & a_{r2} & \dots & a_{rn} \end{pmatrix}$$

has row rank r.

Consider the system

$$A^*\mathbf{X} = \mathbf{0} \tag{5.2}$$

and let $(x_1, \dots, x_n)$ be a solution. If $r < m$, then for $i = r+1, \dots, m$, the ith row of A is a linear combination of the rows of A^*, and so

$$\begin{aligned} & a_{i1}x_1 + \cdots + a_{in}x_n \\ & = \alpha_1(a_{11}x_1 + \cdots a_{1n}x_n) + \cdots + \alpha_r(a_{r1}x_1 + \cdots a_{rn}x_n) \\ & = 0 + \cdots + 0 = 0 \end{aligned}$$

for some $\alpha_1, \dots, \alpha_r \in \mathbf{K}$. Therefore $(x_1, \dots, x_n)$ is also a solution of (5.1). Conversely, it is obvious that any solution of (5.1) is also a solution of (5.2). Thus (5.1) and (5.2) are equivalent, and so the column rank of A^* is equal to c. Since the columns of A^* are vectors in $\mathbf{K}^r$, we conclude that $c \leq r$.

Applying the same reasoning to A^t, we can also conclude that $r \leq c$, and the assertion is proved. □

The common value of the row rank and the column rank of A is called simply the *rank* of A, and is denoted by $\text{rk}(A)$. It is clear that for $A \in M_{m,n}(\mathbf{K})$, $\text{rk}(A) \leq \min\{m, n\}$.

The following proposition ensures that performing elementary row operations on a matrix does not affect its rank.

Proposition 5.2
Let $A \in M_{m,n}(\mathbf{K})$*. If* $B \in M_{m,n}(\mathbf{K})$ *is obtained from A by a succession of elementary row operations then* $\mathrm{rk}(B) = \mathrm{rk}(A)$.

Proof
It suffices to consider the cases where B is obtained from A by a single elementary row operation.

If B is obtained from A by an operation of type (I) then the rank remains unchanged because the rank of a set of vectors does not depend upon their order.

If the operation is of type (II), then the rows of B are

$$A^{(1)}, A^{(2)}, \ldots, cA^{(t)}, \ldots, A^{(m)},$$

for some $c \in \mathbf{K}^*$, and $1 \le t \le m$. Clearly,

$$\langle A^{(1)}, \ldots, cA^{(t)}, \ldots, A^{(m)} \rangle = \langle A^{(1)}, \ldots, A^{(t)}, \ldots, A^{(m)} \rangle,$$

so again $\mathrm{rk}(A) = \mathrm{rk}(B)$.

Finally, if the operation is of type (III), then the rows of B are $A^{(1)}, A^{(2)}, \ldots, A^{(s)} + cA^{(t)}, \ldots, A^{(m)}$ for some $c \in \mathbf{K}$, and $1 \le s, t \le m$, $s \ne t$. Since these rows are linear combinations of those of A one has

$$\langle A^{(1)}, \ldots, A^{(s)} + cA^{(t)}, \ldots, A^{(m)} \rangle \subset \langle A^{(1)}, \ldots, A^{(m)} \rangle.$$

On the other hand, $A^{(s)} = (A^{(s)} + cA^{(t)}) - cA^{(t)}$, and so the reverse inclusion also holds:

$$\langle A^{(1)}, \ldots, A^{(s)} + cA^{(t)}, \ldots, A^{(m)} \rangle \supset \langle A^{(1)}, \ldots, A^{(m)} \rangle,$$

so again, $\mathrm{rk}(A) = \mathrm{rk}(B)$. □

Proposition 5.3

1) *If A and B are matrices that can be multiplied, then*

$$\mathrm{rk}(AB) \le \min\{\mathrm{rk}(A), \mathrm{rk}(B)\}.$$

2) *If* $A \in \mathrm{GL}_m(\mathbf{K})$, $B \in M_{m,n}(\mathbf{K})$ *and* $C \in \mathrm{GL}_n(\mathbf{K})$*, then*

$$\mathrm{rk}(AB) = \mathrm{rk}(B) = \mathrm{rk}(BC) \tag{5.3}$$

Proof

1) Let $A = (a_{ij}) \in M_{m,n}(\mathbf{K})$ and $B = (b_{jh}) \in M_{n,s}(\mathbf{K})$. For each $i = 1, \dots, m$ the i-th row of AB is given by

$$\begin{aligned} A^{(i)} &= (a_{i1}b_{11} + \cdots + a_{in}b_{n1}, a_{i1}b_{12} + \cdots + a_{in}b_{n2}, \dots, \\ &\qquad a_{i1}b_{1s} + \cdots + a_{in}b_{ns}) \\ &= (a_{i1}b_{11}, a_{i1}b_{12}, \dots, a_{i1}b_{1s}) + (a_{i2}b_{21}, a_{i2}b_{22}, \dots, a_{i2}b_{2s}) \\ &\qquad + \cdots + (a_{in}b_{n1}, a_{in}b_{n2}, \dots, a_{in}b_{ns}) \\ &= a_{i1}B^{(1)} + a_{i2}B^{(2)} + \cdots + a_{in}B^{(n)}. \end{aligned}$$

Thus, since every $A^{(i)}B$ is a linear combination of the rows of B, one has

$$\langle A^{(1)}B, A^{(2)}B, \dots, A^{(m)}B \rangle \subset \langle B^{(1)}, B^{(2)}, \dots, B^{(n)} \rangle$$

and so $\mathrm{rk}(AB) \leq \mathrm{rk}(B)$.

On the other hand,

$$\mathrm{rk}(AB) = \mathrm{rk}((AB)^t) = \mathrm{rk}(B^t A^t) \leq \mathrm{rk}(A^t) = \mathrm{rk}(A).$$

2) By (1) we have that

$$\mathrm{rk}(AB) \leq \mathrm{rk}(B) = \mathrm{rk}(A^{-1}(AB)) \leq \mathrm{rk}(AB)$$

and so $\mathrm{rk}(AB) = \mathrm{rk}(B)$. The second identity in (5.3) is proved similarly. □

For square matrices there is the following important result.

Theorem 5.4

A square matrix of order n is invertible if and only if it has rank n.

Proof

By Proposition 5.3(2), an invertible matrix A has the same rank as $\mathbf{I}_n = A^{-1}A$. The rank of $\mathbf{I}_n$ is n because its rows form the canonical basis $\{E_1, \dots, E_n\}$ of $\mathbf{K}^n$. Conversely, if A has rank n its rows $A^{(1)}, \dots, A^{(n)}$ form a basis of $\mathbf{K}^n$. Thus, for each $i = 1, \dots, n$, there exist $b_{i1}, \dots, b_{in} \in \mathbf{K}$ such that

$$E_i = b_{i1}A^{(1)} + b_{i2}A^{(2)} + \cdots + b_{in}A^{(n)} \qquad (5.4)$$

Let B be the square matrix $(b_{ij}) \in M_n(\mathbf{K})$. Then (5.4) is equivalent to the identity

$$\mathbf{I}_n = BA,$$

and so A is invertible. □

A $p \times q$ *submatrix* of a matrix $A \in M_{m,n}(\mathbf{K})$ is a matrix obtained from A by choosing p of its rows and q of its columns, and taking the pq entries at the intersections of these rows and columns. Choosing indices

$$1 \leq i_1 < i_2 < \cdots < i_p \leq m$$
$$1 \leq j_1 < j_2 < \cdots < j_q \leq n,$$

the corresponding $p \times q$ submatrix of A is denoted by

$$A(i_1\, i_2\, \ldots\, i_p \,|\, j_1\, j_2\, \ldots\, j_q).$$

For example, if

$$A = \begin{pmatrix} 5 & -2 & 1 & 0 & \sqrt{3} \\ 8 & 0 & -1 & 3 & 1/10 \\ 2 & -5 & \pi & \sqrt{2} & 6 \\ 1/2 & 1 & 1 & 0 & -5 \end{pmatrix} \in M_{4,5}(\mathbf{R})$$

then

$$A(2\,4 \,|\, 3\,4\,5) = \begin{pmatrix} -1 & 3 & 1/10 \\ 1 & 0 & -5 \end{pmatrix} \in M_{2,3}(\mathbf{R}).$$

Proposition 5.5
If B is a submatrix of A then $\mathrm{rk}(B) \leq \mathrm{rk}(A)$.

Proof
Let $A \in M_{m,n}(\mathbf{K})$ and let $B = A(i_1\, i_2\, \ldots\, i_p \,|\, j_1\, j_2\, \ldots\, j_q)$. Consider the submatrix of A:

$$C = A(i_1\, i_2\, \ldots\, i_p \,|\, 1\, 2\, \ldots\, n) \in M_{p,n}(\mathbf{K}),$$

formed from the i_1-th, i_2-th, ..., i_p-th rows of A. The inequality

$$\mathrm{rk}(C) \leq \mathrm{rk}(A) \tag{5.5}$$

is obvious when interpreted as a relation between the row ranks of the two matrices. On the other hand B is a submatrix of C:

$$B = C(1\, 2\, \ldots\, p \,|\, j_1\, j_2\, \ldots\, j_q),$$

formed from the j_1-th, j_2-th, ..., j_q-th columns of C. In this case the inequality

$$\mathrm{rk}(B) \leq \mathrm{rk}(C) \tag{5.6}$$

is obvious when seen as a relation between the column ranks.

Together, (5.5) and (5.6) imply that $\mathrm{rk}(B) \leq \mathrm{rk}(A)$. □

As a consequence of the preceding results we obtain the following theorem.

Theorem 5.6
The rank of a matrix A is the maximum order of its invertible square submatrices.

Proof
Let ρ be the maximum order of the invertible square submatrices of A, and let $r = \mathrm{rk}(A)$. From Propositions 5.4 and 5.5 it follows that $\rho \leq r$. On the other hand, there are r linearly independent rows of A, say $A^{(i_1)}, A^{(i_2)}, \ldots, A^{(i_r)}$. It follows that the matrix $B = A(i_1\, i_2\, \ldots\, i_r \,|\, 1\, 2\, \ldots\, n)$ has rank r, and so it has r linearly independent columns, say $B_{(j_1)}, B_{(j_2)}, \ldots, B_{(j_r)}$. The square submatrix

$$B(1\, \ldots\, r \,|\, j_1\, \ldots\, j_r)$$

of B has rank r, that is, it is invertible. Since

$$B(1\, \ldots\, r \,|\, j_1\, \ldots\, j_r) = A(i_1\, \ldots\, i_r \,|\, j_1\, \ldots\, j_r)$$

is a submatrix af A one also has that $\rho \geq r$. □

The notion of 'determinant', which will be introduced in the next chapter, and Theorem 5.6 together provide a practical method for computing the rank af a matrix (cf. Corollary 6.6).

From the proof of Theorem 5.6 it follows that if a square submatrix of A,

$$B = A(i_1\, \ldots\, i_r \,|\, j_1\, \ldots\, j_r),$$

is invertible then the rows $A^{(i_1)}, A^{(i_2)}, \ldots, A^{(i_r)}$ of A are linearly independent and the columns $A_{(j_1)}, A_{(j_2)}, \ldots, A_{(j_r)}$ are linearly independent.

We can now apply these considerations to systems of linear equations. The notion of rank permits the following simple criterion for the compatiblity of a system of equations.

Theorem 5.7 (Kronecker-Rouché-Capelli)
A system of m equations in n unknowns

$$AX = b, \tag{5.7}$$

where $A \in M_{m,n}(\mathbf{K})$, $\mathbf{b} \in M_{m,1}(\mathbf{K})$ and $\mathbf{X} = (X_1\, X_2\, \ldots\, X_n)^t$, is compatible if and only if

$$\mathrm{rk}(A) = \mathrm{rk}(A\mathbf{b}).$$

In this case the space of solutions of system (5.7) *has dimension $n-r$, where $r = \mathrm{rk}(A)$.*

Proof
Let

$$A = \begin{pmatrix} a_{11} & a_{12} & \ldots & a_{1n} \\ a_{21} & a_{22} & \ldots & a_{2n} \\ \vdots & \vdots & & \vdots \\ a_{m1} & a_{m2} & \ldots & a_{mn} \end{pmatrix}.$$

An n-tuple $(x_1, \ldots, x_n) \in \mathbf{K}^n$ is a solution of (5.7) if and only if

$$x_1 \begin{pmatrix} a_{11} \\ a_{21} \\ \vdots \\ a_{m1} \end{pmatrix} + x_2 \begin{pmatrix} a_{12} \\ a_{22} \\ \vdots \\ a_{m2} \end{pmatrix} + \ldots + x_n \begin{pmatrix} a_{1n} \\ a_{2n} \\ \vdots \\ a_{mn} \end{pmatrix} = \begin{pmatrix} b_1 \\ b_2 \\ \vdots \\ b_m \end{pmatrix}. \tag{5.8}$$

Equation (5.8) expresses the fact that $\mathbf{b}$ is a linear combination of the columns of A. This is satisfied if and only if the augmented matrix $(A\mathbf{b})$ has the same column rank as A, that is, if and only if $\mathrm{rk}(A\mathbf{b}) = \mathrm{rk}(A)$. The first part of theorem is thus proved.

If system (5.7) is compatible, and $r = \mathrm{rk}(A)$, then we can suppose that its first r equations are linearly independent, and substitute (5.7) with the equivalent system,

$$\begin{aligned} a_{11}X_1 + a_{12}X_2 + \cdots + a_{1n}X_n &= b_1 \\ a_{21}X_1 + a_{22}X_2 + \cdots + a_{2n}X_n &= b_2 \\ \vdots \qquad & \quad \vdots \\ a_{r1}X_1 + a_{r2}X_2 + \cdots + a_{rn}X_n &= b_r. \end{aligned} \tag{5.9}$$

Applying Gaussian elimination to system (5.9), one can see that none of the equations become $0 = 0$, because this would imply that that equation would be a linear combination of the others. Thus system (5.9) can be transformed into an equivalent system in echelon form with r equations. Therefore (5.9), and so (5.7), has an $(n-r)$-parameter family of solutions, and consequently the space of solutions is $(n-r)$-dimensional — see Example 4.15(4) □

EXERCISES

5.1 Calculate the rank of each of the following matrices with rational entries:

a) $\begin{pmatrix} 1/2 & 3 & 1 & -1 \\ 1 & 4 & 2 & 0 \\ -1/2 & -2 & -1 & 0 \end{pmatrix}$ b) $\begin{pmatrix} 1 & 1 & -1 \\ 0 & 1 & 1 \\ 1 & -1 & -1 \\ 0 & 0 & 1 \\ 0 & 0 & 0 \end{pmatrix}$

c) $\begin{pmatrix} 1 & -2 & 3 & 4 & 5 \\ 1 & 4 & 0 & 7 & 2 \\ 2 & 2 & 3 & 11 & 7 \\ 3 & 6 & 3 & 18 & 9 \end{pmatrix}.$

5.2 Prove that every $n \times m$ matrix with entries in $\mathbf{K}$ with rank at most 1 are of the form

$$\begin{pmatrix} a_1 \\ \vdots \\ a_n \end{pmatrix} \begin{pmatrix} b_1 & \dots & b_m \end{pmatrix}$$

for suitable $a_1, \dots, a_n, b_1, \dots, b_m \in \mathbf{K}$.

6

Determinants

In this chapter we describe a way of associating an element of $\mathbf{K}$ to any square matrix A with entries in $\mathbf{K}$. This is called the 'determinant' of A. The determinant, as we shall see, is an instrument of fundamental practical importance in linear algebra.

We will make use of the *summation symbol* Σ to denote the sum of any finite number of terms indexed by one or more indices; the sets of values of the indices will be indicated under and/or over the symbol Σ.

Similarly, the symbol Π will be used to denote products.

Definition 6.1
Let $n \geq 1$ and

$$A = \begin{pmatrix} a_{11} & a_{12} & \dots & a_{1n} \\ a_{21} & a_{22} & \dots & a_{2n} \\ \vdots & \vdots & & \vdots \\ a_{n1} & a_{n2} & \dots & a_{nn} \end{pmatrix} \in M_n(\mathbf{K}).$$

The *determinant* of A is the element of $\mathbf{K}$ given by

$$\det(A) = \sum_{p \in S_n} \varepsilon(p) a_{1p(1)} a_{2p(2)} \dots a_{np(n)}, \tag{6.1}$$

where S_n denotes the set of all permutations of $\{1, 2, \dots, n\}$, and $\varepsilon(p)$ is the sign of the permutation $p \in S_n$ (see Appendix B). $\det(A)$ is also denoted $\det(a_{ij})$ or $|A|$ or again $|a_{ij}|$.

Expression (6.1) is the sum of $n!$ terms. Ignoring the signs, these terms are all possible products of n entries of A belonging to different rows and different columns.

If $n = 1$ and $A = (a)$ then $\det(A) = a$.

If $n = 2$ and $A = \begin{pmatrix} a_{11} & a_{12} \\ a_{21} & a_{22} \end{pmatrix}$, then

$$\begin{vmatrix} a_{11} & a_{12} \\ a_{21} & a_{22} \end{vmatrix} = a_{11}a_{22} - a_{12}a_{21}.$$

If $n = 3$ then

$$\begin{vmatrix} a_{11} & a_{12} & a_{13} \\ a_{21} & a_{22} & a_{23} \\ a_{31} & a_{32} & a_{33} \end{vmatrix} = a_{11}a_{22}a_{33} - a_{11}a_{23}a_{32} + a_{12}a_{23}a_{31}$$
$$-a_{12}a_{21}a_{33} + a_{13}a_{21}a_{32} - a_{13}a_{22}a_{31}.$$

As n increases it becomes difficult to compute the determinant of a general $n \times n$ matrix directly from the definition (6.1). We will see shortly some simpler ways of doing it, without using (6.1)

Let $A \in M_n(\mathbf{K})$. As usual, denote the rows of A by $A^{(1)}, \ldots, A^{(n)}$ and the columns by $A_{(1)}, \ldots, A_{(n)}$. Using the block notation, we can write either

$$A = \begin{pmatrix} A^{(1)} & A^{(2)} & \ldots & A^{(n)} \end{pmatrix}$$

or

$$A = \begin{pmatrix} A_{(1)} \\ A_{(2)} \\ \vdots \\ A_{(n)} \end{pmatrix}.$$

We can therefore write

$$\det(A) = \det \begin{pmatrix} A^{(1)} & A^{(2)} & \ldots & A^{(n)} \end{pmatrix} = \det \begin{pmatrix} A_{(1)} \\ A_{(2)} \\ \vdots \\ A_{(n)} \end{pmatrix}.$$

With this notation, the determinant will be considered as a function of n row vectors, or of n column vectors.

Theorem 6.2
Let $n \geq 1$ be an integer, and let

$$A = (a_{ij}) = \begin{pmatrix} A^{(1)} & A^{(2)} & \dots & A^{(n)} \end{pmatrix} = \begin{pmatrix} A_{(1)} \\ A_{(2)} \\ \vdots \\ A_{(n)} \end{pmatrix} \in M_n(\mathbf{K}).$$

Then:

1) $\det(A^t) = \det(A)$.
2) *If $A^{(i)} = cV + c'V'$, for some $1 \leq i \leq n$, $c, c' \in \mathbf{K}$, that is, if the i-th row of A is a linear combination of two row n-vectors, then*

$$\det(A) = c \det \begin{pmatrix} A^{(1)} \\ \vdots \\ V \\ \vdots \\ A^{(n)} \end{pmatrix} + c' \det \begin{pmatrix} A^{(1)} \\ \vdots \\ V' \\ \vdots \\ A^{(n)} \end{pmatrix}.$$

   Similarly, if $A_{(i)} = cW + c'W'$, for some $1 \leq i \leq n$, $c, c' \in \mathbf{K}$, where W and W' are two row n-vectors, then

$$\begin{aligned} \det(A) &= c \det \begin{pmatrix} A^{(1)} & \dots & W & \dots & A^{(n)} \end{pmatrix} \\ &\quad + c' \det \begin{pmatrix} A^{(1)} & \dots & W' & \dots & A^{(n)} \end{pmatrix}. \end{aligned}$$

3) *If the matrix $B \in M_n(\mathbf{K})$ is obtained from A by swapping a pair of rows or a pair of columns, then* $\det(B) = -\det(A)$.
4) *If A has two identical rows or columns then* $\det(A) = 0$.
5) $\det(\mathbf{I}_n) = 1$.

Proof
1) By definition,

$$\det(A^t) = \sum_{p \in S_n} \varepsilon(p) a_{p(1)1} a_{p(2)2} \dots a_{p(n)n}. \tag{6.2}$$

Ignoring the signs, the terms in (6.2) are the same as in (6.1). Indeed, the term

$$a_{p(1)1} a_{p(2)2} \dots a_{p(n)n} \tag{6.3}$$

can also be written as

$$a_{1q(1)} a_{2q(2)} \dots a_{nq(n)}, \tag{6.4}$$

where $q = p^{-1} \in S_n$. Now, to see that the signs of the summands in $\det(A)$ and $\det(A^t)$ are the same, note that $\varepsilon(p^{-1}) = \varepsilon(p)$. Thus $\det(A) = \det(A^t)$.

2) By part (1), the two statements are equivalent, therefore we need only prove the first one. Let

$$V = (\, v_1 \quad \dots \quad v_n \,) \quad \text{and} \quad V' = (\, v_1' \quad \dots \quad v_n' \,).$$

Then,

$$(\, a_{i1} \quad a_{i2} \quad \dots \quad a_{in} \,) = (\, cv_1 + c'v_1' \quad cv_2 + c'v_2' \quad \dots \quad cv_n + c'v_n' \,)$$

and

$$\begin{aligned}
\det(A) &= \sum_{p\in S_n} \varepsilon(p) a_{1p(1)} a_{2p(2)} \dots a_{np(n)} \\
&= \sum_{p\in S_n} \varepsilon(p) a_{1p(1)} \dots (cv_{p(i)} + c'v'_{p(i)}) \dots a_{np(n)} \\
&= c \sum_{p\in S_n} \varepsilon(p) a_{1p(1)} \dots v_{p(i)} \dots a_{np(n)} \\
&\quad + c' \sum_{p\in S_n} \varepsilon(p) a_{1p(1)} \dots v'_{p(i)} \dots a_{np(n)} \\
&= c \det \begin{pmatrix} A^{(1)} \\ \vdots \\ V \\ \vdots \\ A^{(n)} \end{pmatrix} + c' \det \begin{pmatrix} A^{(1)} \\ \vdots \\ V' \\ \vdots \\ A^{(n)} \end{pmatrix}.
\end{aligned}$$

3) By part (1) it is sufficient to consider the case where B is obtained from A by swapping two rows. Suppose the i-th and j-th rows are swapped, where $1 \le i < j \le n$. Putting $B = (b_{hk})$, we have

$$\begin{aligned}
\det(B) &= \sum_{p\in S_n} \varepsilon(p) b_{1p(1)} \dots b_{ip(i)} \dots b_{jp(j)} \dots b_{np(n)} \\
&= \sum_{p\in S_n} \varepsilon(p) a_{1p(1)} \dots a_{jp(i)} \dots a_{ip(j)} \dots a_{np(n)} \\
&= \sum_{p\in S_n} \varepsilon(p) a_{1pot(1)} \dots a_{ipot(i)} \dots a_{jpot(j)} \dots a_{npot(n)}
\end{aligned}$$

where t denotes the transposition which exchanges i and j. Since $\varepsilon(p \circ t) = -\varepsilon(p)$, and because as p varies over all permutations so does

$p \circ t$, we obtain

$$\det(B) = \sum_{q \in S_n} -\varepsilon(q) a_{1q(1)} \dots a_{iq(i)} \dots a_{jq(j)} \dots a_{nq(n)} = -det(A).$$

4) Suppose that A has two identical rows. Swapping these two rows gives the matrix A again, so by part (3) we get $\det(A) = -\det(A)$, whence $\det(A) = 0$.
5) The only summand in (6.1) which is non-zero for $A = \mathbf{I}_n$ is

$$a_{11}a_{22}\dots a_{nn} = 1.$$

□

The following fact is easily deduced from Theorem 6.2.

Corollary 6.3
If $A, B \in M_n(\mathbf{K})$ then $\det(AB) = \det(A)\det(B)$. In particular if A is invertible then $\det(A^{-1}) = \det(A)^{-1}$.

Proof
Let $A = (a_{ij})$ and

$$B = \begin{pmatrix} B^{(1)} \\ B^{(2)} \\ \vdots \\ B^{(n)} \end{pmatrix}.$$

By the theorem, we have

$$\det(AB) = \det \begin{pmatrix} a_{11}B^{(1)} + \dots + a_{1n}B^{(n)} \\ a_{21}B^{(1)} + \dots + a_{2n}B^{(n)} \\ \vdots \\ a_{n1}B^{(1)} + \dots + a_{nn}B^{(n)} \end{pmatrix}$$

$$= \sum_{p \in S_n} \det \begin{pmatrix} a_{1p(1)}B^{p(1)} \\ a_{2p(2)}B^{p(2)} \\ \vdots \\ a_{np(n)}B^{p(n)} \end{pmatrix} = \sum_{p \in S_n} a_{1p(1)}a_{2p(2)} \dots a_{np(n)} \begin{pmatrix} B^{p(1)} \\ B^{p(2)} \\ \vdots \\ B^{p(n)} \end{pmatrix}$$

$$= \sum_{p \in S_n} \varepsilon(p) a_{1p(1)}a_{2p(2)} \dots a_{np(n)} \det \begin{pmatrix} B^1 \\ B^2 \\ \vdots \\ B^n \end{pmatrix} = \det(A)\det(B).$$

The second statement follows immediately from the first using $AA^{-1} = \mathbf{I}_n$ and Theorem 6.2(5). □

A fundamental property of the determinant is its close relationship with the rank, which is expressed in the following theorem.

Theorem 6.4
Let $A \in M_n(\mathbf{K})$. *Then* $\det(A) \neq 0$ *if and only if* $\mathrm{rk}(A) = n$.

Proof
If $\mathrm{rk}(A) = n$ then A is invertible by Theorem 5.4. Corollary 6.3 then implies that $\det(A^{-1}) = \det(A)^{-1}$, and so $\det(A) \neq 0$.

Suppose that $\mathrm{rk}(A) < n$, that is that the rows are linearly dependent. After possibly swapping two of the rows (which by Theorem 6.2(3) will not affect the result) we can suppose that the first row is a linear combination of the others:

$$A^{(1)} = c_2 A^{(2)} + \cdots + c_n A^{(n)}, \qquad c_2, \ldots, c_n \in \mathbf{K}.$$

By Theorem 6.2(2), we have

$$\det(A) = c_2 \det \begin{pmatrix} A^{(2)} \\ A^{(2)} \\ \vdots \\ A^{(n)} \end{pmatrix} + c_3 \det \begin{pmatrix} A^{(3)} \\ A^{(2)} \\ \vdots \\ A^{(n)} \end{pmatrix} + \cdots + c_n \det \begin{pmatrix} A^{(n)} \\ A^{(2)} \\ \vdots \\ A^{(n)} \end{pmatrix}.$$

Since each of the matrices that appear on the right hand side have two identical rows, all the determinants are zero by Theorem 6.2(4). Thus

$$\det(A) = c_2 0 + c_3 0 + \cdots + c_n 0 = 0.$$

□

Definition 6.5
Let $M \in M_{m,n}(\mathbf{K})$. A *minor* of M is the determinant of a square submatrix. The *order* of the minor is the order of the corresponding square submatrix.

The following corollary is an immediate consequence of Theorems 5.6 and 6.4.

Corollary 6.6
Let $M \in M_{m,n}(\mathbf{K})$. The rank of M is equal to the largest order of its non-zero minors.

The following Proposition follows easily from the preceding theorems.

Proposition 6.7
Let $A = (a_{ij}) \in M_n(\mathbf{K})$.
1) *If A has a row or a column which is zero, then* $\det(A) = 0$.
2) *If $B \in M_n(\mathbf{K})$ is obtained from A by adding to one of its rows (or columns) a scalar multiple of another row (or, respectively, column), then* $\det(B) = \det(A)$.

Proof
1) This follows from Theorem 6.4, since if A has a row or column which is 0 then $\mathrm{rk}(A) < n$.
2) Suppose that $B = (\, B_{(1)} \quad \ldots \quad B_{(n)} \,)$ with $B_{(i)} = A_{(i)} + cA_{(j)}$, for some $j \neq i$, and $B_{(k)} = A_{(k)}$ for every $k \neq i$. Then,

$$\begin{aligned} \det(B) &= \det(A) + c\det(\, A_{(1)} \quad \ldots \quad A_{(j)} \quad \ldots \quad A_{(j)} \quad \ldots \quad A_{(n)} \,) \\ &= \det(A) + c\,0 = \det(A) \end{aligned}$$

The second determinant on the right hand side is zero because the matrix has two identical columns. The case where $B^{(i)} = A^{(i)} + cA^{(j)}$ is proved in the same way. □

We now give a definition of considerable importance for calculating determinants.

Definition 6.8
Let $A = (a_{ij}) \in M_n(\mathbf{K})$. For each $1 \leq i, j \leq n$ let

$$A(1 \ldots \hat{\imath} \ldots n \,|\, 1 \ldots \hat{\jmath} \ldots n)$$

be the square submatrix of A of order $n-1$ obtained by removing the i- th row and the j-th column.

The *cofactor* of the entry a_{ij} of A is

$$A_{ij} = (-1)^{i+j} \det(A(1 \ldots \hat{\imath} \ldots n \,|\, 1 \ldots \hat{\jmath} \ldots n)).$$

The *matrix of cofactors* of A is

$$\operatorname{cof}(A) = (A_{ij}) \in M_n(\mathbf{K}).$$

The following result provides an inductive procedure for calculating the determinant of a matrix.

Proposition 6.9
Let $A = (a_{ij}) \in M_n(\mathbf{K})$. For each $1 \leq i \leq n$ one has

$$\det(A) = a_{i1}A_{i1} + a_{i2}A_{i2} + \cdots + a_{in}A_{in} \tag{6.5}$$

and for each $1 \leq j \leq n$ one has

$$\det(A) = a_{1j}A_{1j} + a_{2j}A_{2j} + \cdots + a_{nj}A_{nj}. \tag{6.6}$$

Identities (6.5) and (6.6) are called the expansion of $\det(A)$ by the i-th row and by the j-th column respectively.

Proof
Replacing A with A^t shows that (6.5) and (6.6) are equivalent, so we only need prove (6.5). After $i-1$ transpositions of adjacent rows of A the i-th row can be brought to the top, leaving the other rows in the same order. Call this new matrix A', so

$$A' = \begin{pmatrix} A^{(i)} \\ A^{(1)} \\ \vdots \\ A^{(i-1)} \\ A^{(i+1)} \\ \vdots \\ A^{(n)} \end{pmatrix}.$$

By Theorem 6.2(3),

$$\det(A') = (-1)^{i-1}\det(A).$$

Moreover, for each $1 \leq j \leq n$,

$$\begin{aligned} A'_{1j} &= (-1)^{j+1}\det[A'(2\ldots n \,|\, 1\ldots \hat{\jmath}\ldots n)] \\ &= (-1)^{j+1}\det[A(1\ldots \hat{\imath}\ldots n \,|\, 1\ldots \hat{\jmath}\ldots n)] \\ &= (-1)^{i+1}A_{ij}. \end{aligned}$$

Therefore, if (6.5) holds for A' it also holds for A. This observation allows us to restrict our attention to the case that $i = 1$, i.e. we only need to prove the expansion of $\det(A)$ by the first row.

Consider the terms in (6.1) which involve a_{1j}. They are

$$\varepsilon(p)a_{1j}a_{2p(2)}\dots a_{np(n)} \tag{6.7}$$

where $p \in S_n$ is a permutation with $p(1) = j$. To each such p we can associate a permutation $q \in S_{n-1}$ defined by, for $k = 1, \dots, n-1$,

$$q(k) = \begin{cases} p(k+1) & \text{if } p(k+1) < j \\ p(k+1) - 1 & \text{if } p(k+1) > j. \end{cases} \tag{6.8}$$

One has

$$\varepsilon(p) = (-1)^{j-1}\varepsilon(q).$$

Indeed, the permutation $r \in S_n$ defined by

$$\begin{aligned} r(1) &= 1, \\ r(k) &= q(k-1) + 1, \qquad k = 2, \dots, n, \end{aligned}$$

can be obtained by composing p with $j-1$ transpositions of adjacent elements, and so it satisfies $\varepsilon(r) = (-1)^{j-1}\varepsilon(p)$. On the hand it is obvious that $\varepsilon(q) = \varepsilon(r)$.

Let $B = (b_{hk}) = A(2\dots n \,|\, 1\dots\hat{\jmath}\dots n)$. By varying $p \in S_n$ with $p(1) = j$, the permutation q defined in (6.8) varies through all of S_{n-1}. Therefore the sum of the terms in (6.7) is equal to

$$a_{1j}(-1)^{j-1} \sum_{q \in S_{n-1}} \varepsilon(q)b_{1q(1)}\dots b_{n-1q(n-1)} = a_{1j}(-1)^{j-1}\det B = a_{1j}A_{1j}.$$

Concluding, we have that the sum of the terms in (6.1) is equal to $\sum_j a_{1j}A_{1j}$, which is precisely (6.5) with $i = 1$. □

In applying Proposition 6.9 it is convenient to choose a row or column with as many zeros as possible, in order to shorten the calculations as much as possible.

The following corollary describes a pratical method for computing the inverse of a matrix $A \in \mathrm{GL}_n(\mathbf{K})$. It is an alternative to the method described in Example 3.2(8).

Corollary 6.10
For every $A = (a_{ij}) \in M_n(\mathbf{K})$,

$$A \operatorname{cof}(A)^t = \det(A)\mathbf{I}_n. \tag{6.9}$$

In particular, if A is invertible, then

$$A^{-1} = \frac{1}{\det(A)} \operatorname{cof}(A)^t. \tag{6.10}$$

Proof
Equation (6.9) is equivalent to the n^2 identities:

$$\sum_{k=1}^{n} a_{ik} A_{jk} = \det(A)\delta_{ij}, \tag{6.11}$$

where δ_{ij} is the Kronecker symbol. Indeed the left hand side is the entry in position i, j of the matrix $A \operatorname{cof}(A)^t$. If $i = j$ then (6.11) agrees with (6.5), which has already been proved. If $i \neq j$ then (6.11) is

$$\sum_{k=1}^{n} a_{ik} A_{jk} = 0. \tag{6.12}$$

By Proposition 6.9, the left-hand side of (6.12) is the expansion by the i-th row of the matrix B obtained from A by replacing the i-th row with the j-th. Since B has two equal rows, its determinant is zero, and so (6.12) holds.

Finally multiplying both sides of (6.9) by $\det(A)^{-1}A^{-1}$ gives the identity (6.10). □

The method of inversion, which we introduced in Chapter 3 to solve a system of n linear equations in n unknowns, can now be formulated in a more precise way, known as *Cramer's rule.*

Corollary 6.11 (Cramer's rule)
Let $A = a_{ij} \in \mathrm{GL}_n(\mathbf{K})$, and $\mathbf{b} = (\, b_1 \quad \dots \quad b_n \,)^t$ a column n-vector, and

$$A\mathbf{X} = \mathbf{b} \tag{6.13}$$

the corresponding system of n equations in n unknowns. The unique solution of (6.13), $\mathbf{x} = (\, x_1 \quad \dots \quad x_n \,)^t$, is given by the formula

$$x_i = \frac{\sum_{k=1}^{n} b_k A_{ki}}{\det(A)}, \qquad i = 1, \dots, n \tag{6.14}$$

Proof
It suffices to recall the method of inversion (3.16) which gives $\mathbf{x} = A^{-1}\mathbf{b}$, and substitute (6.10) for A^{-1}. □

Note that the right hand side of (6.14) has as numerator the determinant of the matrix obtained from A by replacing the i-th column with the column vector $\mathbf{b}$.

We now describe a method for calculating determinants which generalizes Proposition 6.9, for which we need a generalization of Definition 6.8.

Definition 6.12
Let $A \in M_n(\mathbf{K})$, and let

$$M = A(i_1 \dots i_k \,|\, j_1 \dots j_k)$$

be a square submatrix of A of order k. The *cofactor* of M is the product of the sign $(-1)^{i_1+\dots+i_k+j_1+\dots+j_k}$ and the determinant

$$\det(A(\{1,\dots,n\} \setminus \{i_1,\dots,i_k\} \,|\, \{1,\dots,n\} \setminus \{j_1,\dots,j_k\})).$$

That is, it is the determinant of the square submatrix of A of order $n-k$ obtained by removing the rows and columns of M, taken with the sign $\pm$ according to the parity of $i_1 + \dots + i_k + j_1 + \dots + j_k$.

In the particular case that $k = 1$ this reduces to Definition 6.8 of the cofactor of an entry of A.

We now prove a theorem which generalizes the expansion of the determinant of a matrix by a row or column.

Theorem 6.13 (Laplace)
Let $A \in M_n(\mathbf{K})$, and suppose we are given k rows (or columns) of A, where $k \leq n$. Then $\det(A)$ *is equal to the sum of the products of minors of order k taken from the given rows (or columns) with their corresponding cofactors.*

Proof
We will prove this in the rows case, leaving the columns case to the reader.

Denote the sum described in the statement of the theorem by $D(A)$, which we wish to show is equal to $\det(A)$. Any minor of order k is a sum of $k!$ terms, while its cofactor is a sum of $(n-k)!$ terms, so the product of a minor of order k with its cofactor is a sum of $k!(n-k)!$ terms. Furthermore, the number of distinct minors of order k taken from the given k rows is

$$\binom{n}{k} = \frac{n!}{k!(n-k)!}.$$

Thus $D(A)$ is a sum of $\binom{n}{k}k!(n-k)! = n!$ terms. Since $\det(A)$ is also a sum of $n!$ terms, it will suffice to show that every term in $D(A)$ appears in $\det(A)$.

Suppose first that the given rows are $A^{(1)}, A^{(2)}, \ldots, A^{(k)}$. The terms of $D(A)$ are then products of a term in $\det(A(1\ldots k \mid j_1 \ldots j_k))$ by one in $\det(A(k+1\ldots n \mid \{1,\ldots,n\} \setminus \{j_1,\ldots,j_k\}))$, multiplied by $(-1)^{1+\cdots+k+j_1+\cdots+j_k}$, for some choice of $j_1,\ldots,j_k$. Consider the particular case that $\{j_1,\ldots,j_k\} = \{1,\ldots,k\}$. A corresponding term of $D(A)$ is of the form,

$$(-1)^{1+\cdots+k+1+\cdots+k}[\varepsilon(s)a_{1s(1)}\ldots a_{k\,s(k)}][\varepsilon(t)a_{k+1\,k+t(1)}\ldots a_{n\,k+t(n-k)}]$$

$$= \varepsilon(s)\varepsilon(t)[a_{1s(1)}\ldots a_{k\,s(k)}a_{k+1\,k+t(1)}\ldots a_{n\,k+t(n-k)}] \tag{6.15}$$

where $s \in S_k$, $t \in S_{n-k}$.

To see (6.15) as a term of $\det(A)$ notice that the permution

$$u = \begin{pmatrix} 1 & \ldots & k & k+1 & \ldots & n \\ s(1) & \ldots & s(k) & k+t(1) & \ldots & k+t(n-k) \end{pmatrix}$$

has sign $\varepsilon(s)\varepsilon(t)$, because every occurrence of $s(h)$ is before every occurrence of $t(l)$ so that the number of transpositions used to produce u is just the product of the number of transpositions used to produce s with the number for t. It follows that (6.15) is a term in $\det(A)$.

To show that the product of a term of $\det(A(1\ldots k \mid j_1 \ldots j_k))$ by one in $\det(A(k+1\ldots n \mid \{1,\ldots,n\} \setminus \{j_1,\ldots,j_k\}))$, multiplied by $(-1)^{1+\cdots+k+j_1+\cdots+j_k}$, for *any* choice of $1 \le j_1 < j_2 < \cdots < j_k \le n$ is a term in $\det(A)$, we permute the columns of A in a such a way as to obtain a matrix B whose first, second, ..., k-th column is the j_1-th, j_2-th, ..., j_k-th column of A. In performing this permutation, one does $(j_1-1)+(j_2-2)+\cdots+(j_k-k) = j_1+\cdots+j_k-(1+\cdots+k)$ transpositions of the columns, and so

$$\det(B) = (-1)^{j_1+\cdots+j_k-(1+\cdots+k)}\det(A) = (-1)^{1+\cdots+k+j_1+\cdots+j_k}\det(A).$$

The term of $D(A)$ we are considering is equal to $(-1)^{1+\cdots+k+j_1+\cdots+j_k}$ times a term of $D(B)$, of the type we have already considered, and so equal to $(-1)^{1+\cdots+k+j_1+\cdots+j_k}$ times a term in $\det(B)$, and so equal to a term in $\det(A)$.

If the given rows are $A^{(i_1)}, A^{(i_2)}, \ldots, A^{(i_k)}$, consider the matrix C obtained from A by permuting the rows in such a way that the i_1-th, i_2-th, ..., i_k-th rows of A become the first, second, ..., k-th rows of C. The number of transpositions performed is $(i_1 - 1) + (i_2 - 2) + \cdots + (i_k - k) = i_1 + \cdots + i_k - (1 + \cdots + k)$, and so $\det(C) = (-1)^{i_1+\cdots+i_k-(1+\cdots+k)} \det(A)$. On the other hand, each term in $D(C)$ is equal to a term in $D(A)$ multiplied by $(-1)^{i_1+\cdots+i_k-(1+\cdots+k)}$, and so

$$D(C) = (-1)^{i_1+\cdots+i_k-(1+\cdots+k)} D(A).$$

Comparing these, we obtain $D(A) = \det(A)$ as required. □

The procedure descibed in Theorem 6.13 for calculating the determinant of a square matrix is called the *method of Laplace.* It is particularly useful when the matrix has many entries equal to 0. Consider for example the matrix

$$A = \begin{pmatrix} 1 & -2 & 1 & 0 \\ 2 & 1 & -2 & 0 \\ 0 & -3 & 1 & 4 \\ 0 & -1 & 2 & -1 \end{pmatrix}.$$

Expanding $\det(A)$ by the method of Laplace with respect to the first two rows one finds,

$$\begin{aligned} \det(A) &= (-1)^6 \begin{vmatrix} 1 & -2 \\ 2 & 1 \end{vmatrix} \begin{vmatrix} 1 & 4 \\ 2 & -1 \end{vmatrix} + (-1)^7 \begin{vmatrix} 1 & 1 \\ 2 & -2 \end{vmatrix} \begin{vmatrix} -3 & 4 \\ -1 & -1 \end{vmatrix} \\ & 5 \times (-9) - (-4) \times 7 = -17. \end{aligned}$$

The calculation of this determinant by expanding about a single row or column is more laborious.

Observations and examples 6.14

1. Let A be an upper triangular matrix. Expanding the determinant by the first column, one finds immediately, by induction on n, that $\det(A)$ is equal to the product of the entries on the diagonal. Of

course, the same is true for lower triangular matrices, though to prove it one should expand the determinant about the first row.

2. Suppose we are given a matrix $M \in M_{m,n}(\mathbf{K})$ and we wish to calculate its rank using Corollary 6.6. Assuming that M is not the zero matrix, so that its rank is at least 1, it is necessary to calculate minors of larger and larger order, starting from order 2. When there is a non-zero minor of order r, but all minors of order $r+1$ are zero (or if $r = \min\{m, n\}$), then one concludes that $\text{rk}(M) = r$. Indeed, if all minors of order $r+1$ are zero, then so are all minors of order greater than $r+1$: this follows immediately by induction on s expanding each minor of order $s > r+1$ by any one of its rows or columns.

   Calculating the rank can be simplified considerably if one takes into account the so-called *principle of augmented minors*, that is, of the following observation.

   Let $B = M(i_1 \dots i_r \mid j_1 \dots j_r)$ be a square submatrix of M of order r, for some r, such that $\det(B) \neq 0$. Suppose that every square submatrix of M of order $r+1$ obtained by adding to B one row and one column of M (Fig. 6.1) has determinant zero, that is that the so-called *augmented minors of B* are all zero. Then M has rank r.

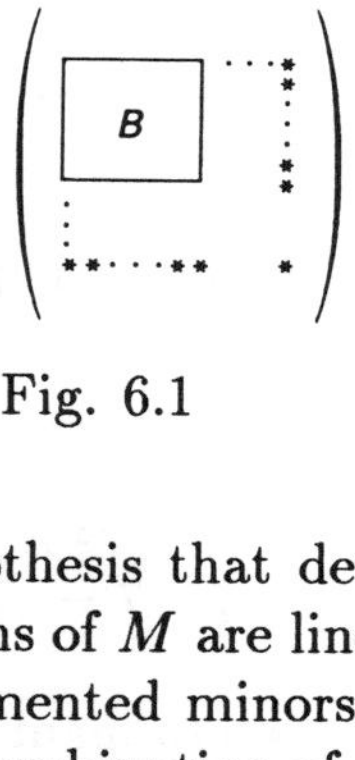

Fig. 6.1

   Indeed, from the hypothesis that $\det(B) \neq 0$, it follows that the j_1-th, ..., j_r-th columns of M are linearly independent, and so the condition on the augmented minors implies that every other column of M is a linear combination of columns $j_1, \dots, j_r$. Thus M has rank r.

   For example, the matrix

$$M = \begin{pmatrix} 1 & 1 & 2 & 3 \\ 1 & 1 & 0 & 0 \\ 0 & 0 & 2 & 3 \end{pmatrix}$$

has rank 2. Indeed the submatrix

$$B = M(12\,|\,13) = \begin{pmatrix} 1 & 2 \\ 1 & 0 \end{pmatrix}$$

has $\det(B) = -2$, and so $\mathrm{rk}(M) \geq 2$; moreover, the augmented minors of B are:

$$\det(M(123\,|\,123)) = \begin{vmatrix} 1 & 1 & 2 \\ 1 & 1 & 0 \\ 0 & 0 & 2 \end{vmatrix} = 0$$

and

$$\det(M(123\,|\,134)) = \begin{vmatrix} 1 & 2 & 3 \\ 1 & 0 & 0 \\ 0 & 2 & 3 \end{vmatrix} = 0,$$

and so $\mathrm{rk}(M) = 2$.

3. The determinant, by its definition given in (6.1), is defined even when the entries of the square matrix A lie in a domain $\mathbf{D}$. Obviously, the determinant of such a matrix is also an element of $\mathbf{D}$. For example, the determinant of a matrix in $M_n(\mathbf{K}[X])$, i.e. a matrix whose entries are polynomials in the unknown X with coefficients in $\mathbf{K}$, is itself a polynomial in $\mathbf{K}[X]$; if $A \in M_m(\mathbf{Z})$ then $\det(A)$ is an integer etc. We will make use of this observation and we will consider, for example, matrices whose entries are polynomials in one or more variables,their determinants when they are square matrices, their minors, and so on.
4. Consider the following system of equations in three unknowns:

$$\begin{aligned} 2X_1 + 3X_2 - \ X_3 &= 1 \\ X_1 + 4X_2 + 2X_3 &= 2 \\ 3X_1 - \ X_2 - \ X_3 &= 3. \end{aligned}$$

The determinant of the matrix of coefficients is

$$\begin{vmatrix} 2 & 3 & -1 \\ 1 & 4 & 2 \\ 3 & -1 & -1 \end{vmatrix} = 30 \neq 0$$

and so the system has a unique solution (x_1, x_2, x_3) given by Cramer's rule:

$$x_1' = \frac{\begin{vmatrix} 1 & 3 & -1 \\ 2 & 4 & 2 \\ 3 & -1 & -1 \end{vmatrix}}{\det(A)} = 36/30 = 6/5$$

$$x_2 = \frac{\begin{vmatrix} 2 & 1 & -1 \\ 1 & 2 & 2 \\ 3 & 3 & -1 \end{vmatrix}}{\det(A)} = -6/30 = -1/5$$

$$x_3 = \frac{\begin{vmatrix} 2 & 3 & 1 \\ 1 & 4 & 2 \\ 3 & -1 & 3 \end{vmatrix}}{\det(A)} = 24/30 = 4/5.$$

5. Suppose we are given a system of m equations in n unknowns in which the coeffiecients of the unknowns and the terms on the right hand side, are functions of one or more variable parameters in **K**. For different values of the parameter, one obtains different systems of equations with coefficients in **K**, and one wants to check the compatibility of these equations and find the solutions if any exist. The most efficient and natural way of proceeding is to use Theorem 5.7 to analyse the possible values of the rank of the matrix and of the augmented matrix as a function of the parameters. Once the values of the parameters for which the system is compatible are known, as well as the dimension of the space of solutions, one proceeds by solving the equation in each case.

   Consider, for example, the system of equations with real coefficients in the unknowns X, Y, Z:

$$\begin{aligned} X - \quad Y + mZ &= \ 0 \\ mY - \quad Z &= \ 0 \\ -X + \quad Y + \quad Z &= m. \end{aligned}$$

The determinant of the matrix of coefficients is $m^2 + m$, which is zero if $m = 0, -1$. For each $m \neq 0, -1$ the matrix of coefficients has rank 3 and so by Theorem 5.7, the system is compatible, and so has a unique solution

$$(1 - m, \frac{1}{m+1}, \frac{m}{m+1}).$$

When $m = 0$, we have the homogeneous system

$$\begin{aligned} X - Y \qquad &= 0 \\ - Z &= 0 \\ -X + Y + Z &= 0 \end{aligned}$$

which has a 1-parameter family of solutions $\{(t,t,0)\,|\,t \in \mathbf{R}\}$. When $m = -1$ we get the system

$$\begin{aligned} X - Y - Z &= 0 \\ -Y - Z &= 0 \\ -X + Y + Z &= 1 \end{aligned}$$

which is incompatible.

6. Let $x_1, x_2, \ldots, x_n \in \mathbf{K}$, with $n \geq 2$. The *Vendermonde determinant with respect to* $x_1, \ldots, x_n$, denoted $V(x_1, x_2, \ldots, x_n)$ is the determinant of the *Vandermonde matrix*

$$\begin{pmatrix} 1 & 1 & \ldots & 1 \\ x_1 & x_2 & \ldots & x_n \\ x_1^2 & x_2^2 & \ldots & x_n^2 \\ \vdots & \vdots & & \vdots \\ x_1^{n-1} & x_2^{n-1} & \ldots & x_n^{n-1} \end{pmatrix}.$$

One has,

$$\begin{aligned} V(x_1, x_2, \ldots, x_n) = (x_2 - x_1) \ldots (x_n - x_1)(x_3 - x_2) \ldots \\ (x_n - x_2)(x_4 - x_3) \cdots (x_n - x_{n-1}). \qquad (6.16) \end{aligned}$$

The expression on the right hand side is the product of all terms of the form $(x_i - x_j)$ for $1 \leq j < i \leq n$.

The identity (6.16) is proved by induction on n. For $n = 2$ it is obvious, so suppose that $n \geq 3$ and that the identity has been proved for $n-1$. Now perform row operations on the Vandermonde matrix as follows: subtract the $(n-1)$-th row multiplied by x_1 from the n-th row, next subtract the $(n-2)$-th row multiplied by x_1 from the $(n-1)$th row, and so on, to obtain

$$V(x_1, x_2, \ldots, x_n) = \begin{vmatrix} 1 & 1 & \ldots & 1 \\ 0 & x_2 - x_1 & \ldots & x_n - x_1 \\ 0 & x_2(x_2 - x_1) & \ldots & x_n(x_n - x_1) \\ \vdots & \vdots & & \vdots \\ 0 & x_2^{n-2}(x_2 - x_1) & \ldots & x_n^{n-2}(x_n - x_1) \end{vmatrix}$$

$$= (x_2 - x_1)(x_3 - x_1) \ldots (x_n - x_1) \begin{vmatrix} 1 & 1 & \ldots & 1 \\ x_2 & x_3 & \ldots & x_n \\ x_2^2 & x_3^2 & \ldots & x_n^2 \\ \vdots & \vdots & & \vdots \\ x_2^{n-1} & x_3^{n-1} & \ldots & x_n^{n-1} \end{vmatrix}$$

$$= (x_2 - x_1)(x_3 - x_1) \dots (x_n - x_1)V(x_2, \dots, x_n).$$

7. It is not difficult to show that properties (2), (3) and (5) of Theorem 6.2 are enough to characterize uniquely the determinant. More precisely, one can show that if $D : M_n(\mathbf{K}) \to \mathbf{K}$ is a map satisfying, for each $A = (\, A_{(1)} \quad \dots \quad A_{(n)} \,)$,
   a) if $A_{(j)} = cV + c'V'$, then

$$\begin{aligned} D(A_{(1)} \quad \dots \quad cV + c'V' \quad \dots \quad A_{(n)}) \\ = \quad cD(A_{(1)} \quad \dots \quad V \quad \dots \quad A_{(n)}) \\ +c'D(A_{(1)} \quad \dots \quad V' \quad \dots \quad A_{(n)}), \end{aligned}$$

   b) for each $1 \leq i < j \leq n$,

$$\begin{aligned} D(\, A_{(1)} \quad \dots \quad A_{(j)} \quad \dots \quad A_{(i)} \quad \dots \quad A_{(n)} \,) = \\ -D(\, A_{(1)} \quad \dots \quad A_{(i)} \quad \dots \quad A_{(j)} \quad \dots \quad A_{(n)} \,), \end{aligned}$$

   c) $D(\mathbf{I}_n) = 1$,

   then $D(A) = \det(A)$ for every $A \in M_n(\mathbf{K})$.

EXERCISES

6.1 Use Corollary 6.10 to calculate the inverse of each matrix in Exercise 3.4.

6.2 Discuss the following systems of equations in the real unknowns X, Y, where m is a real parameter.

a)
$$\begin{aligned} 2X - Y &= m + 1 \\ mX + Y &= 1 \end{aligned}$$

b)
$$\begin{aligned} 2X + \quad mY &= 1 \\ 2X + (1+m)Y &= 1 \\ (3-m)X + \quad 3Y &= 1 + m \end{aligned}$$

c)
$$\begin{aligned} 2X + mY &= -4 \\ mX - 3Y &= 5 \\ 3X + Y &= -5m. \end{aligned}$$

6.3 Discuss the following systems of equations in the real unknowns X, Y, Z, where m is a real parameter.

a)
$$\begin{aligned} X + mY + Z &= 2m \\ mX + Y + Z &= 2 \end{aligned}$$

b)
$$\begin{aligned} Y + mZ &= m + 1 \\ X + Y + Z &= 2 \\ mX + Y &= m + 1 \end{aligned}$$

c)
$$\begin{aligned} 2X + mY + mZ &= 1 \\ mX + 2Y + mZ &= 1 \\ mX + mY + 2Z &= 1 \end{aligned}$$

d)
$$\begin{aligned} X + Y - 2Z &= 0 \\ 2X - Y + mZ &= 0 \\ X - Y - Z &= 0 \end{aligned}$$

e)
$$\begin{aligned} X + Y + 2Z &= 1 \\ X + 2Y + 4Z &= 1 \\ 2X + 3Y + 6Z &= m \end{aligned}$$

f)
$$\begin{aligned} X - Y &= 2 \\ mY + Z &= m \\ Y + mZ &= m \end{aligned}$$

g)
$$\begin{aligned} mY + (m-2)Z &= 0 \\ mX + Y + 2Z &= 0 \\ mX + 3Z &= 0 \end{aligned}$$

h)
$$\begin{aligned} X - Y + Z &= 0 \\ -X - mY + 2mZ &= -1/3 \\ mX + mY &= -1/3. \end{aligned}$$

6.4 Solve the following systems of equations using Cramer's rule.

a) ($\mathbf{K} = \mathbf{R}$)
$$\begin{aligned} 2X - Y &= 2 - \sqrt{2} \\ -X + \sqrt{2}Z &= 1 \\ \sqrt{2}X + Y &= 2\sqrt{2} \end{aligned}$$

b) $(\mathbf{K}=\mathbf{C})$

$$\begin{aligned} 2X + iY + Z &= 1-2i \\ 2Y - iZ &= -2+2i \\ iX + iY + iZ &= 1+i \end{aligned}$$

c) $(\mathbf{K}=\mathbf{Q})$

$$\begin{aligned} X_1 \qquad + 2X_3 \qquad &= 4 \\ -X_1 + X_2 \qquad\qquad &= -1 \\ X_2 + X_3 \qquad &= 2 \\ X_1 \qquad\qquad + X_4 &= 1 \end{aligned}$$

6.5 Let $A, B, C \in M_n(\mathbf{K})$, and put

$$M = \begin{pmatrix} A & B \\ \mathbf{0} & C \end{pmatrix}, \quad N = \begin{pmatrix} A & \mathbf{0} \\ B & C \end{pmatrix}.$$

Prove that $\det(M) = \det(A)\det(C) = \det(N)$.

6.6 Let $A = (a_{ij}) \in M_{n-1,n}(\mathbf{K})$ and suppose that $\mathrm{rk}(A) = n-1$. Prove that the 1-parameter family of solutions of the homogeneous system of equations in n unknowns, $A\mathbf{X} = \mathbf{0}$ are proportional to the n-tuple of minors of maximal order of A, taken with alternating signs.

6.7 Let $a_1, a_2, \ldots, a_n \in \mathbf{K}^*$. Consider the matrix $M = (m_{ij}) \in M_n(\mathbf{K})$ defined by

$$\begin{aligned} m_{ii} &= 1 + a_i, \quad i = 1, \ldots, n \\ m_{ij} &= 1, \quad \text{if } i \neq j \end{aligned}$$

Prove that $\det(M) = a_1 a_2 \ldots a_n (1 + \frac{1}{a_1} + \frac{1}{a_2} + \cdots + \frac{1}{a_n})$.

6.8 Let $\mathbf{D}$ be a domain and let $x, a_{12}, a_{13}, \ldots, a_{1n}, a_{23}, \ldots a_{2n}, \ldots, a_{n-2\,n-1}, a_{n-2\,n}, a_{n-1\,n} \in \mathbf{D}$. Prove the following identity:

$$\begin{vmatrix} x & a_{12} & \ldots & \ldots & a_{1n} \\ x & x & a_{23} & \ldots & a_{2n} \\ \vdots & & & & \vdots \\ x & x & \ldots & x & a_{n-1\,n} \\ x & x & \ldots & x & x \end{vmatrix}$$

$$= x(x - a_{12})(x - a_{23})(x - a_{34}) \cdots (x - a_{n-1\,n}).$$

7

Affine space (I)

In this chapter we introduce the notion of 'affine space', which generalizes the ordinary plane and space we are familiar with. To any affine space there is associated a vector space which is given as part of the definition. In affine spaces (or 'affine geometry') one studies exclusively those geometric properties which can be deduced by using vectors.

The figures we use in the text refer mostly to the ordinary plane or space, and are there principally to support the reader's intuition.

Definition 7.1

Let $\mathbf{V}$ be a vector space over $\mathbf{K}$. An *affine space over* $\mathbf{V}$ (or, an *affine space with associated vector space* $\mathbf{V}$) is a non-empty set $\mathbf{A}$, whose elements are called *points of* $\mathbf{A}$, together with a map

$$\mathbf{A} \times \mathbf{A} \to \mathbf{V} \tag{7.1}$$

that associates to every ordered pair $(P, Q) \in \mathbf{A} \times \mathbf{A}$ a vector in $\mathbf{V}$ which is denoted $\overrightarrow{PQ}$, and called *the vector with base point P and end point Q*, in such a way that the following axioms are satisfied:

[AS1] For every point $P \in \mathbf{A}$ and every $\mathbf{v} \in \mathbf{V}$ there is a unique $Q \in \mathbf{A}$ such that
$$\overrightarrow{PQ} = \mathbf{v}.$$

[AS2] For every triple of points $P, Q, R \in \mathbf{A}$ there is the following identity in $\mathbf{V}$:
$$\overrightarrow{PQ} + \overrightarrow{QR} = \overrightarrow{PR}.$$

For every $(P, Q) \in \mathbf{A} \times \mathbf{A}$ we will call P the *base point* of the vector $\overrightarrow{PQ}$. If $\mathbf{K} = \mathbf{R}$ or $\mathbf{K} = \mathbf{C}$ then $\mathbf{A}$ is called a *real affine space* or *complex affine space*, respectively. The map (7.1) is said to define on the set $\mathbf{A}$ *the structure of an affine space*, or just an *affine structure.*

Taking $P = Q$ in Axiom AS2 gives that $\overrightarrow{PP} = \mathbf{0}$ for every $P \in \mathbf{A}$; taking instead $P = R$ then gives $\overrightarrow{PQ} = -\overrightarrow{QP}$ for every $P, Q \in \mathbf{A}$.

In a similar way to vector spaces, there can be more than one affine structure on a given set $\mathbf{A}$, that is, there may be different ways to associate a vector space $\mathbf{V}$ and a map $\mathbf{A} \times \mathbf{A} \to \mathbf{V}$ that satisfy Axioms AS1 and AS2.

From now on we will assume that *every affine space we consider is such that the associated vector space* $\mathbf{V}$ *is finite dimensional.*

The dimension of $\mathbf{V}$ is also called the *dimension of the affine space* A, and is denoted $\dim(\mathbf{A})$.

Affine spaces of dimension 1 or 2 are commonly called *affine lines* or *affine planes*, respectively.

Examples 7.2

1. The ordinary line, plane and space are examples of a real affine line, a real affine plane and a real 3-dimensional affine space, respectively. The associated vector spaces are the spaces of geometric vectors of the respective spaces, and the operation which associates a vector to an ordered pair of points is the one that was used in Chapter 1 to define geometric vectors. Thus affine spaces are a generalization of the ordinary line, plane and space.

2. Let $\mathbf{V}$ be a finite dimensional vector space over $\mathbf{K}$. Putting

$$\overrightarrow{\mathbf{ab}} = \mathbf{b} - \mathbf{a}$$

defines on $\mathbf{V}$ the structure of an affine space over itself.

Axiom AS1 is satisfied because for every point $\mathbf{p} \in \mathbf{V}$ and for every vector $\mathbf{v} \in \mathbf{V}$ the point $\mathbf{q} = \mathbf{p} + \mathbf{v}$ is the unique point that satisfies the equation $\mathbf{q} - \mathbf{p} = \mathbf{v}$. Axiom AS2 is satisfied because the identity

$$\mathbf{r} - \mathbf{p} = (\mathbf{q} - \mathbf{p}) + (\mathbf{r} - \mathbf{q})$$

holds for every $\mathbf{p}, \mathbf{q}, \mathbf{r} \in \mathbf{V}$.

Thus every vector space $\mathbf{V}$ can be considered as an affine space over itself. With this affine structure, $\mathbf{V}$ is denoted by $\mathbf{V}_a$.

3. A particular case of the preceding example arises when $\mathbf{V} = \mathbf{K}^n$. The resulting affine space $\mathbf{K}^n_a$ is called the *affine numerical n-space over* $\mathbf{K}$. It is usually denoted by $\mathbf{A}^n(\mathbf{K})$, or just by $\mathbf{A}^n$ if $\mathbf{K}$ is clear from the context.

Axiom AS1 implies that given any point $O \in \mathbf{A}$ the resulting map associating $P \in \mathbf{A}$ to $\overrightarrow{OP} \in \mathbf{V}$ is one-to-one and onto. This correspondence is the generalization of the one in ordinary space, which, if a point O is specified, associates to each point P the geometric vector represented by the oriented segment with initial point O and end point P.

Definition 7.3
Let $\mathbf{V}$ be a $\mathbf{K}$-vector space, and $\mathbf{A}$ an affine space over $\mathbf{V}$. An *affine system of coordinates* in the space $\mathbf{A}$ is given by a point $O \in \mathbf{A}$ and a basis $\{\mathbf{e}_1, \ldots, \mathbf{e}_n\}$ of $\mathbf{V}$. This coordinate system is denoted $O\mathbf{e}_1 \ldots \mathbf{e}_n$.

For every point $P \in \mathbf{A}$ one has $\overrightarrow{OP} = a_1\mathbf{e}_1 + \cdots + a_n\mathbf{e}_n$ for some $a_1, \ldots, a_n \in \mathbf{K}$. These scalars $a_1, \ldots, a_n$ are called the *affine coordinates* (or just *coordinates*), and $(a_1, \ldots, a_n)$ the coordinate n-tuple, of P with respect to the coordinate system $O\mathbf{e}_1 \ldots \mathbf{e}_n$.

The point O is called the *origin* of this coordinate system. It has coordinate n-tuple $(0, 0, \ldots, 0)$.

Given an affine coordinate system $O\mathbf{e}_1 \ldots \mathbf{e}_n$ on $\mathbf{A}$, we will write $P(x_1, \ldots, x_n)$ for the point $P \in \mathbf{A}$ with coordinates $x_1, \ldots, x_n$ (Fig. 7.1 refers to the ordinary plane).

If $A(a_1, \ldots, a_n)$, $B(b_1, \ldots, b_n) \in \mathbf{A}$, the vector $\overrightarrow{AB}$ has coordinate n-tuple $(b_1 - a_1, \ldots, b_n - a_n)$ with repect to the basis $\{\mathbf{e}_1, \ldots, \mathbf{e}_n\}$. This follows from the identity $\overrightarrow{AB} = \overrightarrow{OB} - \overrightarrow{OA}$.

If $\mathbf{A} = \mathbf{A}^n$, then the affine coordinate system $O\mathbf{E}_1 \ldots \mathbf{E}_n$ in which $O = (0, \ldots, 0)$ and $\{\mathbf{E}_1, \ldots, \mathbf{E}_n\}$ is the canonical basis of $\mathbf{K}^n$ is called the standard affine coordinate system. In this coordinate system, every point $(x_1, \ldots, x_n) \in \mathbf{A}^n$ has itself as coordinate n-tuple.

The most important subsets of affine spaces are the 'affine subspaces':

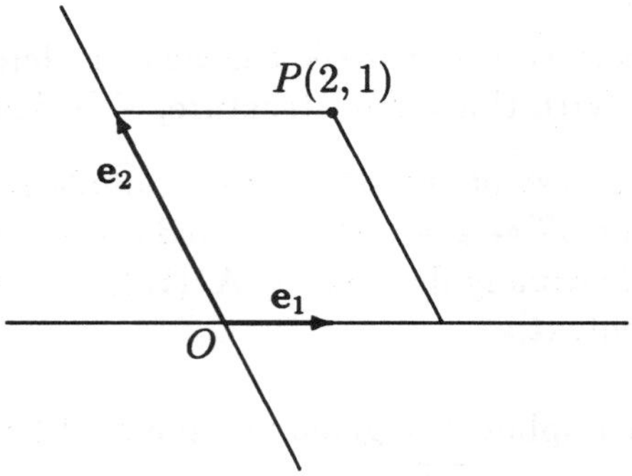

Fig. 7.1

Definition 7.4
Let $\mathbf{V}$ be a $\mathbf{K}$-vector space and let $\mathbf{A}$ be an affine space over $\mathbf{V}$. Given a point $Q \in \mathbf{A}$ and a vector subspace $\mathbf{W}$ of $\mathbf{V}$, the *affine subspace of* $\mathbf{A}$ *passing through* Q *and parallel to* $\mathbf{W}$ is the subset $S \subset \mathbf{A}$ consisting of points $P \in \mathbf{A}$ with $\overrightarrow{QP} \in \mathbf{W}$.

Note that $Q \in S$ because the vector subspace $\mathbf{W}$ contains the zero vector and $\mathbf{0} = \overrightarrow{QQ}$, so S is non-empty.

The subspace $\mathbf{W} \subset \mathbf{V}$ is called the *vector subspace associated to* S. The number $\dim(\mathbf{W})$ is called the *dimension* of S and written $\dim(S)$.

If $\dim(S) = 0$ then $S = \{Q\}$ is a single point; conversely, every subset of $\mathbf{A}$ consisting of one point is an affine subspace of dimension 0.

If $\dim(S) = 1$ then S is said to be a *line* in $\mathbf{A}$, and $\mathbf{W}$ the *direction* of S; any non-zero vector $\mathbf{a} \in \mathbf{W}$ is called a *direction vector* for the line. It follows from the definition that the line S consists of all points $P \in \mathbf{A}$ such that $\overrightarrow{QP} = t\mathbf{a}$ for some $t \in \mathbf{K}$.

If $\dim(S) = 2$ then S is said to be a *plane* in $\mathbf{A}$. Fig. 7.2 represents a plane in ordinary space.

The dimension of an affine subspace of $\mathbf{A}$ cannot be more than $\dim(\mathbf{A})$. If $\dim(S) = \dim(\mathbf{A})$ then $S = \mathbf{A}$ because in this case $\mathbf{W} = \mathbf{V}$ and so $\overrightarrow{QP} \in \mathbf{W}$ for every $P \in \mathbf{A}$.

If $\dim(S) = \dim(\mathbf{A}) - 1$, then S is said to be a *hyperplane*. For example, a line in the affine plane is a hyperplane, as is a plane in a 3-dimensional affine space.

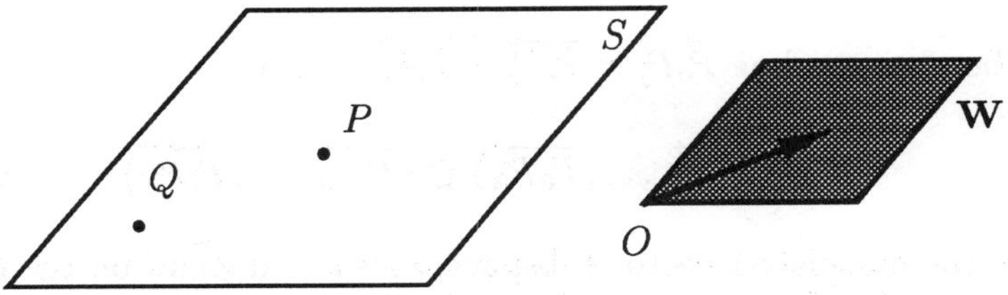

Fig. 7.2

Examples 7.5

1. Let $\mathbf{A}$ be ordinary space. The affine subspaces of $\mathbf{A}$ are the points, the lines, the planes and $\mathbf{A}$ itself.

2. Let $\mathbf{V}$ be a non-zero finite dimensional vector space over $\mathbf{K}$. Consider a vector subspace $\mathbf{W} \subset \mathbf{V}$ and a point $\mathbf{q} \in \mathbf{V}_a$. The affine subspace of $\mathbf{V}_a$ passing through $\mathbf{q}$ and parallel to $\mathbf{W}$ is the set

$$\mathbf{q} + \mathbf{W} = \{\mathbf{q} + \mathbf{w} \mid \mathbf{w} \in \mathbf{W}\}.$$

Indeed, by definition $\mathbf{q} + \mathbf{W}$ consists of all $\mathbf{v} \in \mathbf{V}_a$ satisfying $\mathbf{v} - \mathbf{q} \in \mathbf{W}$.

If in particular $\mathbf{q} \in \mathbf{W}$ then $\mathbf{q} + \mathbf{W} = \mathbf{W}$. We thus see that all vector subspaces of a vector space of $\mathbf{V}$ are affine subspaces of $\mathbf{V}_a$, and that every affine subspace is of the form $\mathbf{q} + \mathbf{W}$ for some $q \in \mathbf{V}_a$ and for some subspace $\mathbf{W} \subset \mathbf{V}$, i.e. it is a *translate* of a vector subspace of $\mathbf{V}$.

3. Given $N + 1 \geq 2$ points $P_0, \ldots, P_N$ in an affine space $\mathbf{A}$, the affine subspace passing through P_0 and having associated vector subspace $\langle \overrightarrow{P_0P_1}, \overrightarrow{P_0P_2}, \ldots, \overrightarrow{P_0P_N} \rangle$ is written $\overline{P_0P_1 \ldots P_N}$, and called the *subspace generated by* (or *the span of*) $P_0, P_1, \ldots, P_N$.

Although the role of P_0 in this definition is different from the others, the affine subspace $\overline{P_0P_1 \ldots P_N}$ does not depend on the order in which the points are taken. To see this, note that the vector subspace $\langle \overrightarrow{P_0P_1}, \overrightarrow{P_0P_2}, \ldots, \overrightarrow{P_0P_N} \rangle$ contains all the vectors $\overrightarrow{P_iP_j} = \overrightarrow{P_0P_j} - \overrightarrow{P_0P_i}$, and so

$$\langle \overrightarrow{P_0P_1}, \overrightarrow{P_0P_2}, \ldots, \overrightarrow{P_0P_N} \rangle \supset \langle \overrightarrow{P_iP_0}, \ldots, \overrightarrow{P_iP_N} \rangle \tag{7.2}$$

for every $i = 0, 1, \ldots, N$. Conversely, for each i, every vector $\overrightarrow{P_0P_j}$

can be expressed as $\overrightarrow{P_0P_j} = \overrightarrow{P_iP_j} - \overrightarrow{P_iP_0}$, and so

$$\langle \overrightarrow{P_0P_1}, \overrightarrow{P_0P_2}, \ldots, \overrightarrow{P_0P_N} \rangle \subset \langle \overrightarrow{P_iP_0}, \ldots, \overrightarrow{P_iP_N} \rangle \qquad (7.3).$$

Thus the associated vector subspace does not depend on the order of the points. Furthermore, if $P \in \mathbf{A}$ satisfies

$$\overrightarrow{P_0P} \in \langle \overrightarrow{P_0P_1}, \ldots, \overrightarrow{P_0P_N} \rangle$$

then

$$\overrightarrow{P_iP} = \overrightarrow{P_0P} - \overrightarrow{P_0P_i} \in \langle \overrightarrow{P_iP_0}, \ldots, \overrightarrow{P_iP_N} \rangle$$

by (7.3), and if $P \in \mathbf{A}$ satisfies $\overrightarrow{P_iP} \in \langle \overrightarrow{P_iP_0}, \ldots, \overrightarrow{P_iP_N} \rangle$ then, by (7.2),

$$\overrightarrow{P_0P} = \overrightarrow{P_iP} - \overrightarrow{P_iP_0} \in \langle \overrightarrow{P_0P_1}, \ldots, \overrightarrow{P_0P_N} \rangle.$$

By the definition of $\overline{P_0P_1 \ldots P_N}$ it follows that

$$\dim(\overline{P_0P_1 \ldots P_N}) \leq N.$$

If there is equality, then the points $P_0, P_1, \ldots, P_N$ are said to be *independent*, otherwise $P_0, P_1, \ldots, P_N$ are *dependent.* By definition, therefore, the points $P_0, P_1, \ldots, P_N$ are independent if and only if the vectors $\overrightarrow{P_0P_1}, \overrightarrow{P_0P_2}, \ldots, \overrightarrow{P_0P_N}$ are linearly independent. Clearly, if $P_0, P_1, \ldots, P_N$ are independent then $N \leq \dim(\mathbf{A})$.

Two points $P_0, P_1 \in \mathbf{A}$ are independent if and only if they are distinct, in which case $\overline{P_0P_1}$ is a line. Three points P_0, P_1, P_2 are independent if and only if they do not belong to a line, in which case $\overline{P_0P_1P_2}$ is a plane (Fig. 7.3).

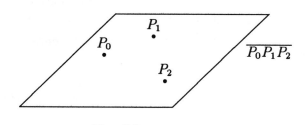

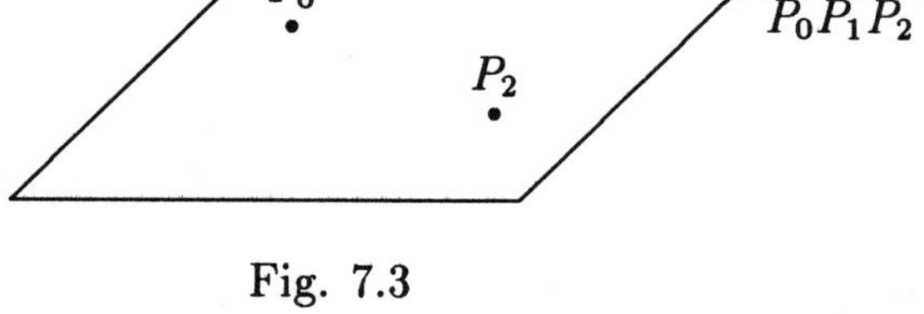

Fig. 7.3

The points $P_0, P_1, \ldots, P_N \in \mathbf{A}$ are said to be *collinear* if there is a line containing them, or equivalently if

$$\dim(\overline{P_0P_1\ldots P_N}) \leq 1.$$

The points $P_0, P_1, \ldots, P_N \in \mathbf{A}$ are said to be *coplanar* if there is a plane containing them, or equivalently if

$$\dim(\overline{P_0P_1\ldots P_N}) \leq 2.$$

4. Let $\mathbf{A}$ be an affine space over $\mathbf{V}$, and let $C \in \mathbf{A}$. For each $P \in \mathbf{A}$, the *reflexion of P in C* is the point $\sigma_C(P)$ satisfying the vector identity

$$\overrightarrow{C\sigma_C(P)} = -\overrightarrow{CP}.$$

Notice that $\sigma_C(\sigma_C(P)) = P$ for every P, because

$$\overrightarrow{C\sigma_C(\sigma_C(P))} = -\overrightarrow{C\sigma_C(P)} = \overrightarrow{CP}.$$

If an affine system of coordinates $O\mathbf{e}_1\ldots\mathbf{e}_n$ is given, and the points C and P have respective coordinates $c_1, \ldots, c_n$ and $x_1, \ldots, x_n$ then the point $\sigma_C(P)$ has coordinates $2c_1 - x_1, \ldots, 2c_n - x_n$. For,

$$(2c_1 - x_1, \ldots, 2c_n - x_n) - (c_1, \ldots, c_n) = (c_1 - x_1, \ldots, c_n - x_n)$$

which is the n-tuple of coordinates of $\overrightarrow{PC} = -\overrightarrow{CP}$. In the special case that $C = O$ the coordinates of $\sigma_C(P)$ are $-x_1, \ldots, -x_n$.

The map $\sigma_C : \mathbf{A} \to \mathbf{A}$ maps affine subspaces to affine subspaces. More precisely, if $S \subset \mathbf{A}$ is the affine subspace passing through Q and having associated vector subspace $\mathbf{W} \subset \mathbf{V}$ then $\sigma_C(S) = \{\sigma_C(P) \,|\, P \in S\}$ is the affine subspace S' having the same associated vector subspace $\mathbf{W}$ and passing through $\sigma_C(Q)$.

Indeed, for every $P \in S$, one has

$$\overrightarrow{\sigma_C(Q)\sigma_C(P)} = \overrightarrow{\sigma_C(Q)C} + \overrightarrow{C\sigma_C(P)} = \overrightarrow{CQ} - \overrightarrow{CP} = -\overrightarrow{QP} \in \mathbf{W},$$

and so $\sigma_C(P) \in S'$. Conversely, if $P' \in S'$ then putting $P = \sigma_C(P')$ gives

$$\overrightarrow{QP} = \overrightarrow{QC} + \overrightarrow{CP} = -\overrightarrow{\sigma_C(Q)C} - \overrightarrow{CP'} = \overrightarrow{\sigma_C(Q)P'} \in \mathbf{W}$$

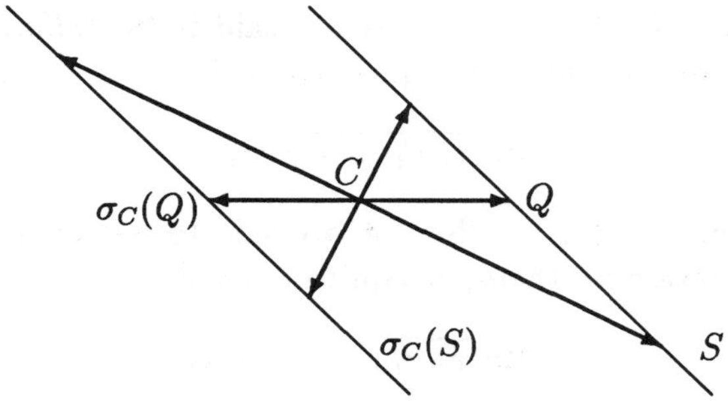

Fig. 7.4

and so $P \in S$. Thus $P' = \sigma_C(P) \in \sigma_C(S)$. (Fig. 7.4 refers to the ordinary plane.)

Proposition 7.6

1) *An affine subspace is determined by its associated vector subspace and any one of its points.*
2) *Let S be an affine subspace of $\mathbf{A}$ with associated vector subspace $\mathbf{W}$. Associating to any pair of points $P, Q \in S$ the vector $\overrightarrow{PQ}$ defines on S the structure of an affine space over $\mathbf{W}$.*

Proof

1) Let S be the affine subspace of $\mathbf{A}$ passing through Q and having associated vector subspace $\mathbf{W}$. Let $M \in S$ and let T be the affine subspace passing through M and having associated vector subspace $\mathbf{W}$. If $P \in S$ then

$$\overrightarrow{MP} = \overrightarrow{MQ} + \overrightarrow{QP} = -\overrightarrow{QM} + \overrightarrow{QP}$$

which is a vector in $\mathbf{W}$ since both of the summands are. Thus $P \in T$. If, conversely, $P \in T$ then

$$\overrightarrow{QP} = \overrightarrow{QM} + \overrightarrow{MP} = -\overrightarrow{MQ} + \overrightarrow{MP} \in \mathbf{W},$$

and so $P \in S$. Therefore, $T = S$.

2) If $P, Q \in S$ then $\overrightarrow{PQ} \in \mathbf{W}$: by part (1), S coincides with the affine subspace passing through P parallel to $\mathbf{W}$. There is thus a map,

$$\begin{array}{rcl} S \times S & \rightarrow & \mathbf{W} \\ (P, Q) & \mapsto & \overrightarrow{PQ} \end{array}$$

which satisfies axioms AS1 and AS2, since they are satisfied in **A**.

Observations and examples 7.7

1. Let **A** be a *real* affine space with associated vector space **V**. Fix a point $Q \in \mathbf{A}$ and a non-zero vector $\mathbf{a} \in \mathbf{V}$. The set of points $P \in \mathbf{A}$ such that

$$\overrightarrow{QP} = t\mathbf{a}$$

for some $t \geq 0$, is called the *half line with origin Q and direction* **a**.

If A and B are two distinct points of **A** the *segment with ends A and B* is the set of points $P \in \mathbf{A}$ such that

$$\overrightarrow{AP} = t\overrightarrow{AB}$$

for some $t \in \mathbf{R}$ such that $0 \leq t \leq 1$. It is denoted AB.

The points $P_1, \ldots, P_{s-1} \in AB$ which divide the segment AB in s equal parts are defined by the conditions

$$\overrightarrow{AP_i} = \tfrac{i}{s}\overrightarrow{AB}, \quad i = 1, 2, \ldots, s-1.$$

Let $O\mathbf{e}_1 \ldots \mathbf{e}_n$ be an affine coordinate system on **A** and let $A = A(a_1, \ldots, a_n)$, $B = B(b_1, \ldots, b_n)$ and $P_i = P_i(x_1, \ldots, x_n)$. The previous condition is equivalent to the following:

$$(x_1 - a_1, \ldots, x_n - a_n) = \tfrac{i}{s}(b_1 - a_1, \ldots, b_n - a_n)$$

and so

$$(x_1, \ldots, x_n) = \tfrac{1}{s}(ib_1 + (s-i)a_1, \ldots, ib_n + (s-i)a_n).$$

In particular the *midpoint of the segment AB* has coordinates

$$(x_1, \ldots, x_n) = \left(\frac{a_1 + b_1}{2}, \frac{a_2 + b_2}{2}, \ldots, \frac{a_n + b_n}{2}\right).$$

2. Let $\mathbf{A}$ be a real affine space, and let $A, B, C \in \mathbf{A}$ be three points which are not collinear. The *triangle with vertices* A, B, C is the set of points $P \in \mathbf{A}$ such that

$$\overrightarrow{AP} = t\overrightarrow{AB} + u\overrightarrow{AC}$$

for some $t, u \in \mathbf{R}$ with $t, u \geq 0$ and $t + u \leq 1$.

The *parallelogram determined by* A, B, C is the set of points $P \in \mathbf{A}$ such that

$$\overrightarrow{AP} = t\overrightarrow{AB} + u\overrightarrow{AC}$$

for some $t, u \in \mathbf{R}$ with $0 \leq t, u \leq 1$. (Note that this set depends on the order of A, B, C whereas the triangle defined above does not.) See Fig. 7.5.

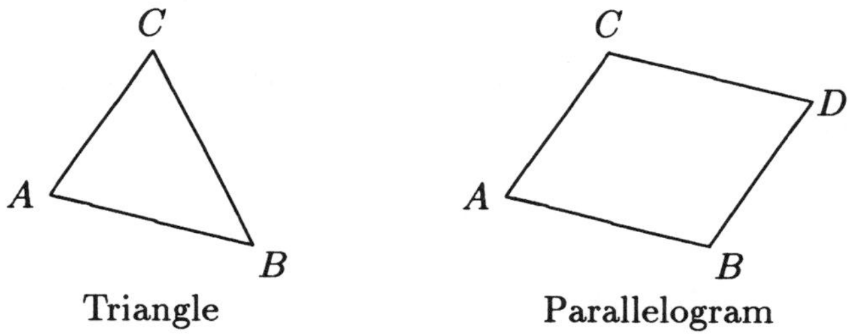

Fig. 7.5

3. Let $\mathbf{A}$ be a real affine space, and let $A, B, C, D \in \mathbf{A}$ be four points which are not coplanar. The *tetrahedron with vertices* A, B, C, D is the set of points $P \in \mathbf{A}$ such that

$$\overrightarrow{AP} = t\overrightarrow{AB} + u\overrightarrow{AC} + v\overrightarrow{AD}$$

for some $t, u, v \in \mathbf{R}$ with $t, u, v \geq 0$ and $t + u + v \leq 1$.

The *parallelopiped determined by* A, B, C, D is the set of points $P \in \mathbf{A}$ such that

$$\overrightarrow{AP} = t\overrightarrow{AB} + u\overrightarrow{AC} + v\overrightarrow{AD}$$

for some $t, u, v \in \mathbf{R}$ with $0 \leq t, u, v \leq 1$. (Again, this set depends on the order of A, B, C, D whereas the tetrahedron does not.) See Fig. 7.6.

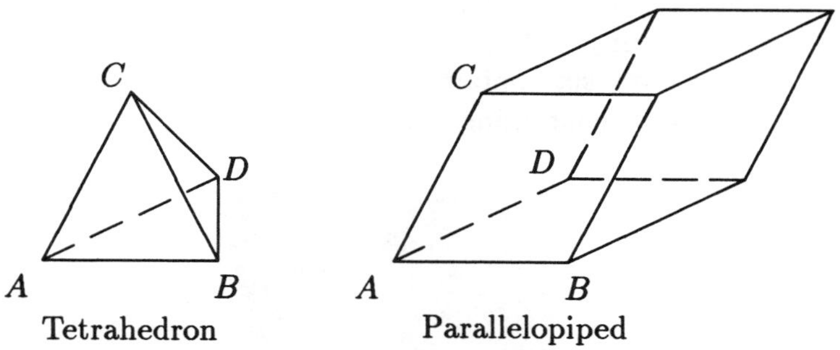

Fig. 7.6

4. Segements, triangles and tetrahedra are particular cases of subsets of real affine spaces called 'simplices'.

   Let $\mathbf{A}$ be a real affine space, and let $P_0, \ldots, P_k \in \mathbf{A}$ be independent points. The *k-simplex with vertices* $P_0, \ldots, P_k$ is the set of points $P \in \mathbf{A}$ satisfying

$$\overrightarrow{P_0P} = t_1\overrightarrow{P_0P_1} + t_2\overrightarrow{P_0P_2} + \cdots + t_k\overrightarrow{P_0P_k}$$

   for some $t_1, t_2 \ldots, t_k \in \mathbf{R}$ with $t_1, t_2 \ldots, t_k \geq 0$ and $\sum_{i=1}^{k} t_i \leq 1$.

   For $k = 1, 2, 3$ a k-simplex is a segment, a triangle and a tetrahedron respectively.

5. Let $\mathbf{A}$ be a real affine space. A subset S is said to be *convex* if for every $A, B \in S$ the segment AB lies in S (Fig. 7.7).

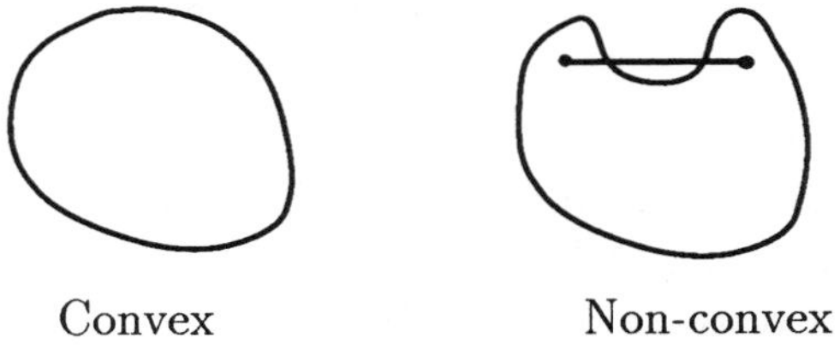

Fig. 7.7

It is easy to check that every affine subspace and every simplex are convex subsets.

From the definition it follows that the intersection of any family of convex sets is itself convex.

If S is a subset of $\mathbf{A}$, the ***convex hull*** of S is defined to be the smallest convex set containing S. It is thus the intersection of all convex subsets containing S (Fig. 7.8).

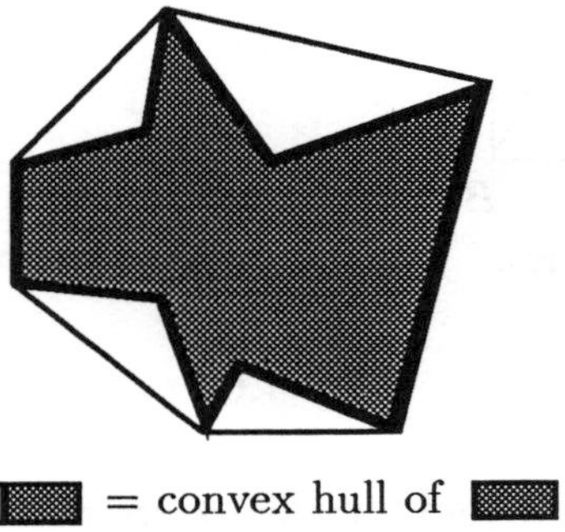

The whole figure is the convex hull of the shaded region

Fig. 7.8

6. Let $\mathbf{A}$ be an affine space with associated vector space $\mathbf{V}$. Axiom AS1 defines a map

$$t : \mathbf{A} \times \mathbf{V} \to \mathbf{A}$$

that associates to each pair of points $(A, \mathbf{a})$ the point $B = t(A, \mathbf{a})$ such that $\overrightarrow{AB} = \mathbf{a}$.

The map t enjoys the following properties:

a) $t(t(A, \mathbf{a}), \mathbf{b}) = t(A, \mathbf{a} + \mathbf{b})$, for every $A \in \mathbf{A}$ and $\mathbf{a}, \mathbf{b} \in \mathbf{V}$;
b) for every $A, B \in \mathbf{A}$ there is a unique $\mathbf{a} \in \mathbf{V}$ such that $B = t(A, \mathbf{a})$.

To see (a), put $B = t(A, \mathbf{a})$ and $C = t(t(A, \mathbf{a})\mathbf{b})$. One has, $\overrightarrow{AB} = \mathbf{a}$ and $\overrightarrow{BC} = \mathbf{b}$, and (a) now becomes $\overrightarrow{AC} = \overrightarrow{AB} + \overrightarrow{BC}$, which holds by Axiom AS2.

(b) is just a restatement of AS1.

Conversely, given any vetor space $\mathbf{V}$, a set $\mathbf{A}$ and a map $t : \mathbf{A} \times \mathbf{V} \to \mathbf{A}$ which satisfies conditions (a) and (b), then there is defined on $\mathbf{A}$ the structure of an affine space over $\mathbf{V}$. Indeed, by (b), t determines a map $\mathbf{A} \times \mathbf{A} \to \mathbf{V}$ which satisfies AS1, and by (a) it also satisfies AS2.

7. Let $\mathbf{A}$ be an affine space and $A, B, C, D \in \mathbf{A}$. If $\overrightarrow{AB} = \overrightarrow{CD}$ then $\overrightarrow{AC} = \overrightarrow{BD}$.

This is because,

$$\overrightarrow{AC} = \overrightarrow{AB} + \overrightarrow{BC} = \overrightarrow{CD} + \overrightarrow{BC} = \overrightarrow{BD}.$$

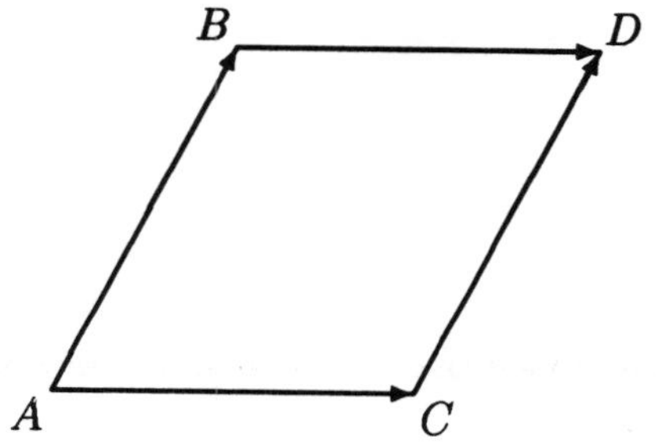

Fig. 7.9

8

Affine space (II)

Let $\mathbf{A}$ be an affine space over $\mathbf{V}$ in which there is a given coordinate system $O\mathbf{e}_1 \dots \mathbf{e}_n$.

Let S be the subspace passing throught the point $Q(q_1, \dots, q_n) \in \mathbf{A}$ and parallel to the vector subspace $\mathbf{W}$ of $\mathbf{V}$. Choose a basis $\{\mathbf{w}_1, \dots, \mathbf{w}_s\}$ of $\mathbf{W}$ with $\mathbf{w}_i(w_{1i}, \dots, w_{ni})$, for $i = 1, \dots, s = \dim(S)$.

For each point $P(x_1, \dots, x_n) \in S$ one has,

$$\overrightarrow{QP} = t_1\mathbf{w}_1 + \cdots + t_s\mathbf{w}_s \tag{8.1}$$

for some $t_1, \dots, t_s \in \mathbf{K}$. Equating the coordinates of the left and right hand sides of (8.1) gives

$$\begin{aligned} x_1 &= q_1 + t_1 w_{11} + \cdots + t_s w_{1s} \\ x_2 &= q_2 + t_1 w_{21} + \cdots + t_s w_{2s} \\ \vdots & \qquad\qquad \vdots \\ x_n &= q_n + t_1 w_{n1} + \cdots + t_s w_{ns} \end{aligned} \tag{8.2}$$

By varying the parameters $t_1, \dots, t_s \in \mathbf{K}$ in (8.2) one obtains every point P in S. Equations (8.2) are called *parametric equations for* S.

Notice that the parametric equations (8.2) are not uniquely determined, but depend upon the choice of Q and of $\mathbf{w}_1, \dots, \mathbf{w}_s$.

In the case that the affine subspace is a line ℓ passing through $Q(q_1, \dots, q_n)$ having direction vector $\mathbf{a}(a_1, \dots, a_n)$ (see Fig. 8.1), equations (8.2) take the form

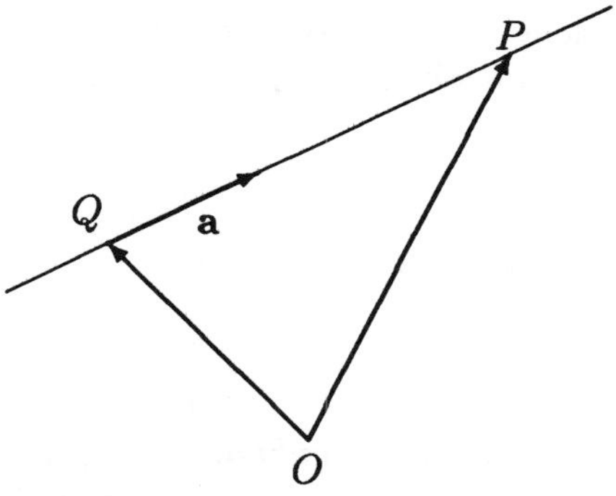

Fig. 8.1

$$\begin{aligned} x_1 &= q_1 + a_1 t \\ x_2 &= q_2 + a_2 t \\ \vdots &\quad\ \ \vdots \\ x_n &= q_n + a_n t. \end{aligned} \tag{8.3}$$

These are parametric equations for the line ℓ.

When the line ℓ is given by two distinct points $Q(q_1, \ldots, q_n)$ and $Q'(q'_1, \ldots, q'_n)$, then $\overrightarrow{QQ'}$ is a vector in the direction of ℓ, and so in (8.3) one can take $\mathbf{a} = \overrightarrow{QQ'}$, that is $(a_1, \ldots, a_n) = (q'_1 - q_1, \ldots, q'_n - q_n)$.

If $n = 2$, that is if $\mathbf{A}$ is an affine plane, then (8.3) reduces to two equations. The coordinates of $\mathbf{a}$ are usually denoted by l, m, the coordinates of the variable point by x, y and those of the point Q by a, b. Thus the parametric equations of a line in an affine plane is usually written as

$$\begin{aligned} x &= a + lt \\ y &= b + mt. \end{aligned} \tag{8.4}$$

In the case that $\dim(\mathbf{A}) = 3$ the parametric equations of a line are usually written as

$$\begin{aligned} x &= a + lt \\ y &= b + mt \\ z &= c + nt, \end{aligned} \tag{8.5}$$

where now x, y, z are the coordinates of the variable point P, l, m, n those of the direction vector and a, b, c the coordinates of the point $Q \in \ell$.

Let us return to the general case. Another way of representing an affine subspace by equations is given by the following theorem.

Theorem 8.1
1) Let

$$\begin{aligned} a_{11}X_1 + \cdots + a_{1n}X_n &= b_1 \\ \vdots \qquad\qquad & \quad \vdots \\ a_{t1}X_1 + \cdots + a_{tn}X_n &= b_t \end{aligned} \tag{8.6}$$

be a system of linear equations in the unknowns $X_1, \ldots, X_n$. *The set* S *of points of* $\mathbf{A}$ *whose coordinates are solutions of (8.6), if there are any, is an affine subspace of dimension* $n - r$, *where* r *is the rank of the matrix of coefficients of the system. The vector subspace associated to* S *is the vector subspace* $\mathbf{W}$ *of* $\mathbf{V}$ *whose equations are given by the associated homogeneous system,*

$$\begin{aligned} a_{11}X_1 + \cdots + a_{1n}X_n &= 0 \\ \vdots \qquad\qquad & \quad \vdots \\ a_{t1}X_1 + \cdots + a_{tn}X_n &= 0. \end{aligned} \tag{8.7}$$

2) For every affine subspace S *of* $\mathbf{A}$ *of dimension* s *there is a system of* $n - s$ *linear equations in* n *unknowns whose solutions correspond precisely to the coordinates of the points in* S.

Proof
1) By hypothesis there is a point $Q(q_1, \ldots, q_n) \in S$. For every point $P(p_1, \ldots, p_n) \in S$ one has

$$\begin{aligned} a_{j1}(x_1 - q_1) + \cdots + a_{jn}(x_n - q_n) & \\ &= a_{j1}x_1 + \cdots + a_{jn}x_n - (a_{j1}q_1 + \cdots + a_{jn}q_n) \\ &= b_j - b_j = 0 \end{aligned}$$

for each $j = 1, \ldots, t$, that is $(x_1 - q_1, \ldots, x_n - q_n)$ is a solution of (8.7), i.e. $\overrightarrow{QP} \in \mathbf{W}$. Thus S is contained in the affine subspace Σ passing through Q and parallel to $\mathbf{W}$.

Conversely, if $P(x_1, \ldots, x_n) \in \Sigma$, then $\overrightarrow{QP} \in \mathbf{W}$ and so the coordinates $(x_1 - q_1, \ldots, x_n - q_n)$ of $\overrightarrow{QP}$ are solutions of (8.7). Thus,

$$\begin{aligned} 0 &= a_{j1}(x_1 - q_1) + \cdots + a_{jn}(x_n - q_n) \\ &= a_{j1}x_1 + \cdots + a_{jn}x_n - (a_{j1}q_1 + \cdots + a_{jn}q_n) \end{aligned}$$

or,

$$a_{j1}x_1 + \cdots + a_{jn}x_n = a_{j1}q_1 + \cdots + a_{jn}q_n = b_j$$

for each $j = 1, \ldots, t$, that is, $P \in S$. Thus $S = \Sigma$, and S is an affine subspace.

2) Suppose that $S \subset \mathbf{A}$ is the affine subspace passing through the point $Q(q_1, \ldots, q_n)$ and having associated vector subspace $\mathbf{W}$. Let

$$\begin{aligned} a_{11}X_1 + \cdots + \quad a_{1n}X_n &= 0 \\ \vdots \qquad\qquad\quad & \quad \vdots \\ a_{n-s\,1}X_1 + \cdots + a_{n-s\,n}X_n &= 0. \end{aligned}$$

be equations for $\mathbf{W}$. The points $P(x_1, \ldots, x_n)$ of S are characterized by the condition that $\overrightarrow{QP} \in \mathbf{W}$, that is that

$$a_{j1}(x_1 - q_1) + \cdots + a_{jn}(x_n - q_n) = 0, \qquad j = 1, \ldots, n - s,$$

or, alternatively,

$$a_{j1}x_1 + \cdots + a_{jn}x_n = b_j, \qquad j = 1, \ldots, n - s$$

where we have put $b_j = a_{j1}q_1 + \cdots + a_{jn}q_n$.

Thus, the points $P(x_1, \ldots, x_n) \in S$ are precisely those points in $\mathbf{A}$ whose coordinates satisfy the equations

$$\begin{aligned} a_{11}X_1 + \cdots + \quad a_{1n}X_n &= 0 \\ \vdots \qquad\qquad\quad & \quad \vdots \\ a_{n-s\,1}X_1 + \cdots + a_{n-s\,n}X_n &= 0. \end{aligned}$$

□

Equations (8.6) are called the (*Cartesian*) *equations for the subspace S with respect to the coordinate system $O\mathbf{e}_1 \cdots \mathbf{e}_n$*. Clearly, this system of equations is not uniquely determined: two systems of equations determine the same affine subspace of $\mathbf{A}$ if and only if they are equivalent.

Observations 8.2

1. The subspace S with equations (8.6) contains the origin if and only if $b_1 = \cdots = b_t = 0$, that is, if and only if the equations

are homogeneous. Thus every homogeneous system of equations (8.7) defines not only a vector subspace **W** of **V** but also an affine subspace S of **A** passing through the origin and having associated vector subspace **W**. For each point $P(x_1, \ldots, x_n) \in S$ the vector $\mathbf{w}(x_1, \ldots, x_n)$ having the same coordinates as P belongs to the vector subspace **W** associated to S. There is thus a one-one correspondence between S and its associated vector subspace **W**.

2. From Theorem 8.1(2) it follows that every hyperplane H of **A** is represented by a single equation,

$$a_1X_1 + \cdots + a_nX_n = b,$$

in which $a_1, \ldots, a_n \in \mathbf{K}$ are not all zero. In particular, if $\dim(\mathbf{A}) = 2$ then every line in **A** has an equation of the form

$$a_1X_1 + a_2X_2 = b.$$

Similarly, if $\dim(\mathbf{A}) = 3$ then every plane in **A** has an equation of the form

$$a_1X_1 + a_2X_2 + a_3X_3 = b.$$

The hyperplane H in **A** contains the origin if and only if $b = 0$. In particular, for each $j = 1, \ldots, n$, the hyperplane with equation

$$X_j = 0$$

passes through the origin; it is called *the j-th coordinate hyperplane* and denoted H_j.

Definition 8.3

Let S and T be two affine subspaces of **A** of positive dimension, with respective associated vector subspaces **W** and **U**. S and T are said to be *parallel* if $\mathbf{W} \subset \mathbf{U}$ or $\mathbf{U} \subset \mathbf{W}$. We will write $S \parallel T$ to mean that S and T are parallel.

If $\dim(S) = \dim(T)$, then S and T are parallel if and only if $\mathbf{U} = \mathbf{W}$ (in particular, if $S = T$ then $S \parallel T$). Fig. 8.2 shows two parallel planes in ordinary space.

In the case that S and T are lines, saying they are parallel means that they have the same direction: their direction vectors are proportional.

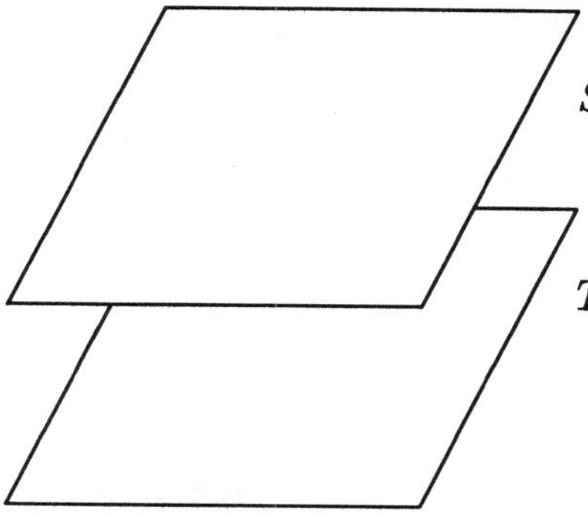

Fig. 8.2

If the affine subspace S is contained in the affine subspace T, then S is parallel to T. Indeed, as P and Q vary in S, the vector $\overrightarrow{QP}$ describes the vector subspace $\mathbf{W}$ associated to S, but it is also contained in $\mathbf{U}$ since $P, Q \in T$. Therefore $\mathbf{W} \subset \mathbf{U}$.

Proposition 8.4
Let S and T be parallel affine subspaces of $\mathbf{A}$ with $\dim(S) \leq \dim(T)$.

1) *If S and T have a point in common then $S \subset T$.*
2) *If $\dim(S) = \dim(T)$, and S and T have a point in common then $S = T$.*

Proof
1) Let $Q \in S \cap T$. For every $P \in S$ we have $\overrightarrow{QP} \in \mathbf{W} \subset \mathbf{U}$, and so $P \in T$; thus $S \subset T$.
2) If $\dim(S) = \dim(T)$ then $\mathbf{W} = \mathbf{U}$, so (1) implies both $S \subset T$ and $T \subset S$; thus $S = T$. □

Corollary 8.5
If S is an affine subspace of $\mathbf{A}$ and $P \in \mathbf{A}$, there is a unique affine subspace T of $\mathbf{A}$ which contains P, is parallel to S and has the same dimension as S.

Proof
This follows immediately from Proposition 8.4(2). □

In the case that $\mathbf{A}$ is the ordinary affine plane and S is a line, Corollary 8.5 is equivalent to the 'parallel postulate' of Euclidean geometry. The axioms of affine spaces therefore imply the validity of this postulate in affine planes.

Example 8.6

Let $\mathbf{V}$ be a vector space over $\mathbf{K}$ and $\mathbf{W}$ a vector subspace of $\mathbf{V}$. By Proposition 8.4(2) it follows that, if $\mathbf{v}, \mathbf{v}' \in \mathbf{V}$, the two affine subspaces $\mathbf{v} + \mathbf{W}$ and $\mathbf{v}' + \mathbf{W}$ of $\mathbf{V}_a$ either coincide or are disjoint. From this it follows that the family of all affine subspaces of $\mathbf{V}_a$ having associated vector subspace $\mathbf{W}$ forms a partition of $\mathbf{V}$. The quotient set of $\mathbf{V}$ by this partition is the set whose elements are the affine subspaces of $\mathbf{V}_a$ having associated vector subspace $\mathbf{W}$; it is denoted $\mathbf{V}/\mathbf{W}$.

Note that $\mathbf{v} + \mathbf{W} = \mathbf{v}' + \mathbf{W}$ if and only if $\mathbf{v}' \in \mathbf{v} + \mathbf{W}$ since the former is equivalent to

$$(\mathbf{v} + \mathbf{W}) \cap (\mathbf{v}' + \mathbf{W}) \neq \emptyset,$$

that is to the existence of $\mathbf{w}, \mathbf{w}' \in \mathbf{W}$ such that $\mathbf{v} + \mathbf{w} = \mathbf{v}' + \mathbf{w}'$. This in turn is equivalent to the condition

$$\mathbf{v}' = \mathbf{v} + (\mathbf{w} - \mathbf{w}') \in \mathbf{v} + \mathbf{W}.$$

Define an operation of addition on $\mathbf{V}/\mathbf{W}$ by

$$(\mathbf{v}_1 + \mathbf{W}) + (\mathbf{v}_2 + \mathbf{W}) = (\mathbf{v}_1 + \mathbf{v}_2) + \mathbf{W}.$$

This operation is well defined because the sum does not depend on which vectors are chosen to represent the two subspaces: if

$$\mathbf{v}_1 + \mathbf{W} = \mathbf{v}_1' + \mathbf{W} \qquad \mathbf{v}_2 + \mathbf{W} = \mathbf{v}_2' + \mathbf{W},$$

then $\mathbf{v}_1' = \mathbf{v}_1 + \mathbf{w}_1$ and $\mathbf{v}_2' = \mathbf{v}_2 + \mathbf{w}_2$ for some $\mathbf{w}_1, \mathbf{w}_2 \in \mathbf{W}$, and so

$$(\mathbf{v}_1' + \mathbf{v}_2') + \mathbf{W} = (\mathbf{v}_1 + \mathbf{v}_2 + \mathbf{w}_1 + \mathbf{w}_2) + \mathbf{W} = (\mathbf{v}_1 + \mathbf{v}_2) + \mathbf{W}.$$

Define scalar multiplication on $\mathbf{V}/\mathbf{W}$ by

$$c(\mathbf{v} + W) = c\mathbf{v} + \mathbf{W}$$

where $c \in \mathbf{K}$. This too is well defined because if $\mathbf{v} + \mathbf{W} = \mathbf{v}' + \mathbf{W}$ then $\mathbf{v}' = \mathbf{v} + \mathbf{w}$ for some $\mathbf{w} \in \mathbf{W}$, and so

$$c\mathbf{v}' + \mathbf{W} = c(\mathbf{v} + \mathbf{w}) + \mathbf{W} = (c\mathbf{v} + c\mathbf{w}) + \mathbf{W} = c\mathbf{v} + \mathbf{W}.$$

The zero for addition in $\mathbf{V}/\mathbf{W}$ is then easily seen to be the subspace $\mathbf{W} = \mathbf{0} + \mathbf{W}$. It is also easy to check that these operations define on $\mathbf{V}/\mathbf{W}$ the structure of a $\mathbf{K}$-vector space, called the *quotient vector space of* $\mathbf{V}$ *by* $\mathbf{W}$ (or *modulo* $\mathbf{W}$).

Notice that if $\mathbf{W} = \mathbf{V}$, then $\mathbf{V}/\mathbf{W}$ contains only one element, namely $\mathbf{V}$ itself, and so it is identified with the zero vector space. If instead $\mathbf{W} = \langle \mathbf{0} \rangle$ then the affine subspaces of $\mathbf{V}_a$ with associated vector subspace $\mathbf{W}$ are just the points of $\mathbf{V}_a$, so in this case the space $\mathbf{V}/\mathbf{W}$ coincides with $\mathbf{V}_a$.

Definition 8.7

If two affine subspaces S and T of $\mathbf{A}$ are not parallel they are said to be either *skew* if they do not meet, or *incident* if they do meet (i.e. have a point in common). See Fig. 8.3.

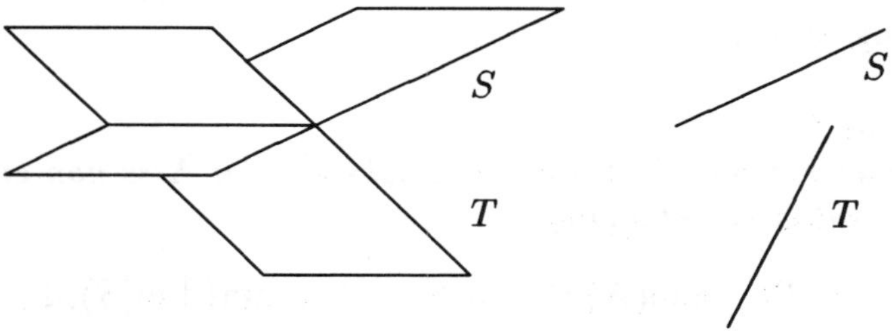

Fig. 8.3

Consider now two subspaces S and T with $\dim(S) = s$ and $\dim(T) = t$. Suppose they have equations

$$S: \quad \sum_{j=1}^{n} m_{ij} X_j = b_i, \qquad i = 1, \ldots, n - s \tag{8.8}$$

$$T: \quad \sum_{j=1}^{n} n_{kj} X_j = c_k, \qquad k = 1, \ldots, n - t. \tag{8.9}$$

The intersection $S \cap T$ is the locus of points in $\mathbf{A}$ whose coordinates are simultaneously solutions of both (8.8) and (8.9); that is, they are solutions of the system

$$\begin{aligned} \textstyle\sum_{j=1}^{n} m_{ij}X_j &= b_i, \qquad i = 1,\dots,n-s, \\ \textstyle\sum_{j=1}^{n} n_{kj}X_j &= c_k, \qquad k = 1,\dots,n-t. \end{aligned} \tag{8.10}$$

By Theorem 8.1, if system (8.10) has solutions, then it represents an affine space. Thus, if $S \cap T$ is non-empty it is an affine subspace of **A**.

The dimension of $S \cap T$ is $n - r$ where r is the rank of the matrix of coefficients of system (8.10). Since,

$$r \leq n - s + n - t = 2n - (s + t),$$

we also have that, if $S \cap T \neq \emptyset$, then

$$\dim(S \cap T) = n - r \geq n - [2n - (s + t)] = s + t - n. \tag{8.11}$$

On the other hand, since $S \cap T$ is a subspace of both S and T, its dimension cannot be greater than those of S and T. In sum, we have the following result.

Proposition 8.8
If the intersection $S \cap T$ of two affine subspaces of **A** *is non-empty it is an affine subspace satisfying,*

$$\dim(S) + \dim(T) - \dim(\mathbf{A}) \leq \dim(S \cap T) \leq \min\{\dim(S), \dim(T)\}. \tag{8.12}$$

From the proof it follows that the second inequality in (8.12) is an equality if and only if $S \subset T$ or $T \subset S$. The case that the first inequality is actually an equality is considered in the next proposition.

Proposition 8.9
Let S and T be two affine subspaces of **A** *with associated vector subspaces* **W** *and* **U** *respectively. Then* $\mathbf{V} = \mathbf{W} + \mathbf{U}$ *if and only if* $S \cap T \neq \emptyset$ *and*

$$\dim(S \cap T) = \dim(S) + \dim(T) - \dim(\mathbf{A}). \tag{8.13}$$

Proof
Suppose that S and T have equations (8.8) and (8.9) respectively. Then (8.13) holds precisely when $S \cap T$ is non-empty and equality

holds in (8.11). Now, (8.11) can be viewed as a relation between $s = \dim(\mathbf{W})$, $t = \dim(\mathbf{U})$ and $\dim(\mathbf{W} \cap \mathbf{U}) = \dim(S \cap T)$: it is a consequence of Grassman's formula. In more detail, Grassman's formula asserts that (8.11) is an equality if and only if $n = \dim(\mathbf{W} + \mathbf{U})$, that is, if and only if $\mathbf{V} = \mathbf{W} + \mathbf{U}$. It is therefore sufficient to show that in this case system (8.10) is compatible. The matrix of coefficients of (8.10) has rank $r = (n - s) + (n - t) = 2n - (s + t)$, which is also the number of rows of the matrix. Moreover, the augmented matrix has the same number of rows and so has the same rank: the system is therefore compatible, by Theorem 5.7. □

When $\dim(\mathbf{A}) = 2$ or 3, Propositions 8.8 and 8.9 recover results on the relative positions of lines and planes which we know to be true in ordinary plane and space. In Chapters 9 and 10 we will treat these cases directly, without using these results.

A particularly important case of Proposition 8.9 is that in which the associated vector subspaces $\mathbf{W}$ and $\mathbf{U}$ of the affine subspaces are supplementary, i.e. are such that $\mathbf{V} = \mathbf{W} \oplus \mathbf{U}$. Since in this case $\dim(S) + \dim(T) = \dim(\mathbf{A})$, (8.13) asserts that S and T have only one point in common. This occurs, for example, for two non-parallel

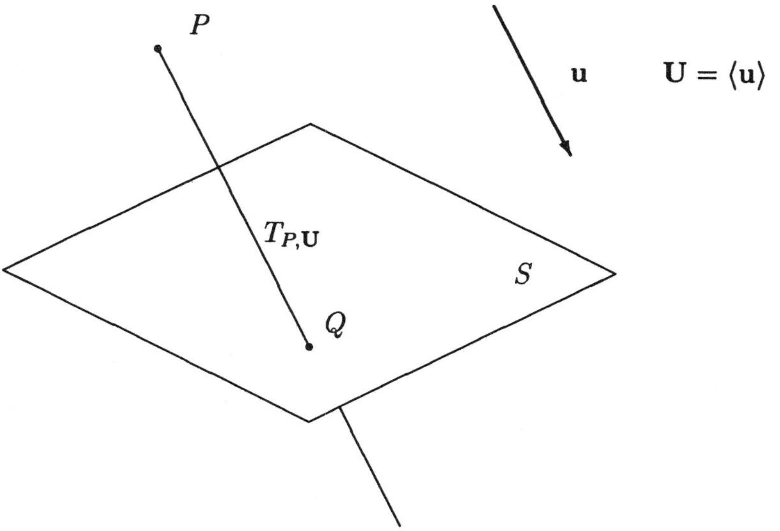

Fig. 8.4

lines in the plane, and for a line and a plane which are not parallel in 3-dimensional space.

From this observation, we can make the following geometrical construction. Let S be an affine subspace of $\mathbf{A}$ with associated vector subspace $\mathbf{W}$. Fix a subspace $\mathbf{U}$ of $\mathbf{V}$ satisfying $\mathbf{V} = \mathbf{W} \oplus \mathbf{U}$. By Corollary 8.5, for each point $P \in \mathbf{A}$ there is a unique affine subspace $T_{P,\mathbf{U}}$ passing through P and parallel to $\mathbf{U}$. From (8.13) it follows that $S \cap T_{P,\mathbf{U}} = \{Q\}$, say. The map $p_{S,\mathbf{U}} : \mathbf{A} \to S$ defined by putting $p_{S,\mathbf{U}}(P) = Q$, is called the *projection of* $\mathbf{A}$ *on to* S *parallel to* $\mathbf{U}$.

If $\dim(\mathbf{U}) = 1$, so S is a hyperplane, then $p_{S,\mathbf{U}}$ is called the projection of $\mathbf{A}$ on to S in the *direction* $\mathbf{U}$. Fig. 8.4 demonstrates this projection in ordinary space.

EXERCISES

8.1 Let $\mathcal{P}$ be the affine plane in $\mathbf{A}^3(\mathbf{C})$ with equation $2X + Y - 1 = 0$. In each of the following cases calculate the coordinates of $p_{\mathbf{u}}(x, y, z)$, where $(x, y, z) \in \mathbf{A}^3$ and $p_{\mathbf{u}} : \mathbf{A}^3 \to \mathcal{P}$ is the projection in the direction $\mathbf{u} \in \mathbf{C}^3$, where

a) $\mathbf{u} = (1, 0, 0)$ b) $\mathbf{u} = (i, 0, 0)$
c) $\mathbf{u} = (2i, i, 1)$ d) $\mathbf{u} = (0, i, 2)$.

8.2 Let $\mathbf{A}$ be a *real* affine space with associated vector space $\mathbf{V}$, and suppose we are given a coordinate frame $O\mathbf{e}_1 \ldots \mathbf{e}_n$. Let $H \subset \mathbf{A}$ be the hyperplane with equation

$$a_1X_1 + \cdots + a_nX_n + c = 0.$$

The subsets of $\mathbf{A}$,

$$\Sigma_+ = \{P(x_1, \ldots, x_n) \mid a_1x_1 + \cdots + a_nx_n + c \geq 0\}$$

$$\Sigma_- = \{P(x_1, \ldots, x_n) \mid a_1x_1 + \cdots + a_nx_n + c \leq 0\}$$

are the *halfspaces* of $\mathbf{A}$ defined by H. From the definition there follows that,

$$\Sigma_+ \cap \Sigma_- = H, \quad \text{and} \quad \Sigma_+ \cup \Sigma_- = \mathbf{A}.$$

Show that the definition of halfspace does not depend on the equation of H nor on the choice of coordinate frame.

Prove that the halfspaces are convex subsets of **A**.

If dim(**A**) $= 1$, the half spaces are called *half lines*; if dim(**A**) $= 2$ they are called *half planes*. Show that this definition of half line coincides with that given in Example 7.7(1).

[*Hint*. Show that Σ_+ and Σ_- are characterized by the following geometric property: given any $P_0 \in \Sigma_+$ (or $P_0 \in \Sigma_-$), the point $P \in \Sigma_+$ (respectively, $P \in \Sigma_-$) if and only if the segment P_0P is completely contained in Σ_+ (respectively, $P_0P \subset \Sigma_-$).]

8.3 Let **A** be a real affine space and let A, B, C, D be independent points of **A**. Prove that the triangle with vertices A, B, C is the convex hull of the set $\{A, B, C\}$, and that the tetrahedron with vertices A, B, C, D is the convex hull of the set $\{A, B, C, D\}$.

9

Geometry in affine planes

In this chapter, we consider in some detail the case of affine planes, that is, of affine spaces of dimension 2 over $\mathbf{K}$. Among these there is of course the ordinary plane, which is a real affine plane.

Consider, therefore, a $\mathbf{K}$-vector space $\mathbf{V}$ of dimension 2, and an affine plane $\mathbf{A}$ over $\mathbf{V}$. Fix an affine reference frame (coordinate system) $O\mathbf{e}_1\mathbf{e}_2$ in $\mathbf{A}$. Let x, y denote the coordinates of the variable point $P \in \mathbf{A}$, and X, Y denote unknowns. The affine subspaces of $\mathbf{A}$ are $\mathbf{A}$ itself, the lines in $\mathbf{A}$ and the points of $\mathbf{A}$.

Every line ℓ in $\mathbf{A}$ has parametric equations

$$\begin{aligned} x &= a + lt \\ y &= b + mt \end{aligned} \tag{9.1}$$

in which $Q(a, b)$ is any point of ℓ and $\mathbf{v}(l, m)$ is a direction vector for ℓ. Since it has codimension 1 in $\mathbf{A}$, ℓ can be represented by a single Cartesian equation

$$AX + BY + C = 0, \tag{9.2}$$

for some $A, B, C \in \mathbf{K}$ with $(A, B) \neq (0, 0)$. The constants A, B, C are determined by ℓ only up to a non-zero common factor, so a line has infinitely many equations, each proportional to the others. The line ℓ contains the origin O if and only if $C = 0$. The lines with equations $X = 0$ and $Y = 0$ are called *coordinate axes*: the Y-*axis* and the X-*axis* respectively.

Suppose we are given a parametric equation (9.1) for a line ℓ, so we know a point $Q(a, b)$ on the line and a direction vector $\mathbf{v}(l, m)$. To obtain the Cartesian equations for ℓ one should view (9.1) as saying that a point $P(x,y)$ lies in ℓ if and only if the vector $\overrightarrow{QP}$ is parallel to $\mathbf{v}$. This last condition can also be expressed by requiring that (x,y) be a solution of the equation in X, Y:

$$\begin{vmatrix} X-a & Y-b \\ l & m \end{vmatrix} = 0 \tag{9.3}$$

that is

$$m(X-a) - l(Y-b) = 0, \tag{9.4}$$

or, equivalently,

$$mX - lY + lb - ma = 0. \tag{9.5}$$

Equation (9.3) is satisfied by, and only by, the points $P(x,y) \in \ell$. Thus, (9.3), or (9.4) or (9.5), is the Cartesian equation for ℓ.

If a line ℓ is given in terms of two of its points $Q(a,b)$ and $Q'(a',b')$, then $\overrightarrow{QQ'} = \mathbf{v}(a'-a, b'-b)$ can be taken as a direction vector, and (9.3) becomes

$$\begin{vmatrix} X-a & Y-b \\ a'-a & b'-b \end{vmatrix} = 0. \tag{9.6}$$

Proposition 9.1
Let ℓ and ℓ' be two lines in $\mathbf{A}$ with Cartesian equations

$$AX + BY + C = 0$$

and

$$A'X + B'Y + C' = 0$$

repectively. Then:

1) *ℓ and ℓ' are parallel if and only if the matrix*

$$\begin{pmatrix} A & B \\ A' & B' \end{pmatrix} \tag{9.7}$$

has rank 1, that is, if and only if $AB' - A'B = 0$.

2) *If ℓ and ℓ' are parallel, then they are disjoint or they coincide accordingly as the matrix*

$$\begin{pmatrix} A & B & C \\ A' & B' & C' \end{pmatrix} \tag{9.8}$$

has rank 2 or 1.

3) ***ℓ and ℓ' have precisely one point in common if and only if the matrix (9.7) has rank 2. In this case the point $\ell \cap \ell'$ has coordinates***

$$x_0 = \frac{CB' - C'B}{AB' - A'B}, \quad y_0 = \frac{AC' - A'C}{AB' - A'B}.$$

Proof

The directions of ℓ and ℓ' are determined by the following homogeneous equations

$$AX + BY = 0,$$

and

$$A'X + B'Y = 0$$

respectively. Thus ℓ and ℓ' are parallel if and only if $A' = \rho A$ and $B' = \rho B$ for some $\rho \in \mathbf{K}^*$, that is if and only if (9.7) has rank 1. This proves part (1).

The matrix (9.8) has rank 1 if and only if the equations of ℓ and ℓ' are proportional, which is equivalent to $\ell = \ell'$. Thus, if (9.8) has rank 2 and (9.7) has rank 1, ℓ and ℓ' are distinct and, by part (1), parallel, and so have no points in common. This proves part (2).

Finally, (9.7) has rank 2 if and only if ℓ and ℓ' are not parallel. In this case it follows from Cramer's rule that the system of equations consisting of the Cartesian equations for ℓ and ℓ' has the unique solution given in (9.9). This proves part (3). □

The cases considered in Proposition 9.1 cover all possible relative positions of two lines in $\mathbf{A}$, because they correspond to all possibilities for two equations in two unknowns. In particular, we see that the only possibility that two lines in an affine plane do not intersect is that (9.7) has rank 1 while (9.8) has rank 2; this corresponds to ℓ and ℓ' being parallel. Thus, *two lines in an affine plane cannot be skew.*

Let $Q(x_0, y_0) \in \mathbf{A}$. The set Φ of all lines in $\mathbf{A}$ passing through Q is called a *proper pencil of lines* and Q is called its *centre* — Fig. 9.1.

Let

$$AX + BY + C = 0$$
$$A'X + B'Y + C' = 0$$

be the Cartesian equations of a pair of distinct lines intersecting at the centre Q of Φ. Let $\lambda, \mu \in \mathbf{K}$, with at least one non-zero; the line

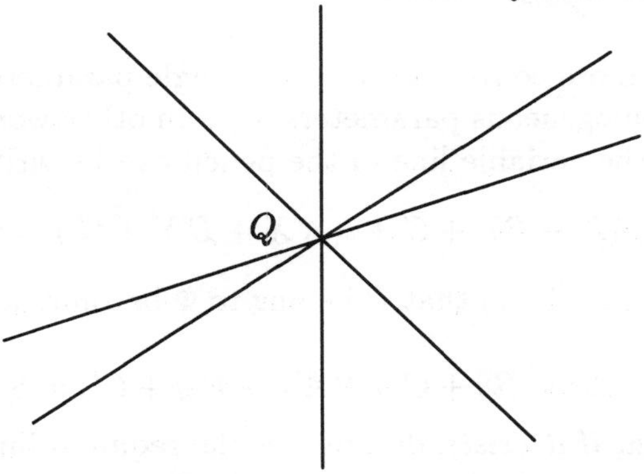

Fig. 9.1

with Cartesian equation

$$\lambda(AX + BY + C) + \mu(A'X + B'Y + C') = 0 \tag{9.10}$$

passes through Q and so belongs to Φ. Conversely, if $\ell \in \Phi$ and $P(x, y) \in \ell$ is a point distinct from Q, the condition

$$\lambda(Ax + By + C) + \mu(A'x + B'y + C') = 0 \tag{9.11}$$

is a linear homogeneous equation in λ and μ which has a unique solution, determined up to a common factor. The line (9.10) in which λ and μ satisfy (9.11) belongs to Φ and contains P, so coincides with ℓ.

Thus every line in the pencil Φ has an equation of the form (9.10); a line in Φ determines λ and μ up to a common factor.

A particular case of (9.10) is obtained by taking ℓ and ℓ' parallel to the axes: then the the equations of every line in the pencil Φ can be written in the form

$$\lambda(X - x_0) + \mu(Y - y_0) = 0$$

for some λ, μ.

Pencils of lines are useful in practice when a point Q is given as the intersection of two lines, but its coordinates are not known explicitly, and one wants to find the equation of a line passing through Q and satisfying some other condition. For example, the condition that it contain some point P distinct from Q, or that it be parallel to a given line.

It is often more convenient to use a single parameter t rather than the pair of homogeneous parameters λ, μ. In other words, rather than using (9.10), the variable line in the pencil can be written as

$$AX + BY + C + t(A'X + B'Y + C') = 0, \qquad (9.12)$$

and the condition (9.11) that P belong to Φ becomes a linear equation for t,

$$Ax + By + C + t(A'x + B'y + C') = 0, \qquad (9.13)$$

whose solution, *if it exists*, determines the required line.

Some care is required, however, as (9.13) represents all lines in the pencil Φ except the line

$$A'X + B'Y + C' = 0. \qquad (9.14)$$

This corresponds to the fact that (9.13) is incompatible if (and only if)

$$\begin{aligned} A'x + B'y + C' &= 0, \\ Ax + By + C &\neq 0, \end{aligned}$$

that is, if $P \neq Q$ and P belongs to the line (9.14).

Bearing this in mind, one can always use a single (non-homogeneous) parameter to determine the equation of a pencil, provided (9.13) is interpreted as described above if it is incompatible.

There is also a very similar idea of 'improper pencils' of lines. Let $\mathbf{v} \in \mathbf{V}$ be a given non-zero vector. The set of lines in $\mathbf{A}$ with direction $\langle \mathbf{v} \rangle$ is called an *improper pencil of lines*, and $\langle \mathbf{v} \rangle$ is its *direction* — see Fig. 9.2.

Given any line in an improper pencil Φ, with equation

$$AX + BY + C = 0,$$

then every other line in the pencil has the form

$$AX + BY + t = 0$$

as the parameter $t \in \mathbf{K}$ varies. This follows immediately from Proposition 9.1. For every point $P(x, y) \in \mathbf{A}$ there is a unique line in Φ containing P, which is the one corresponding to

$$t = -(Ax + By).$$

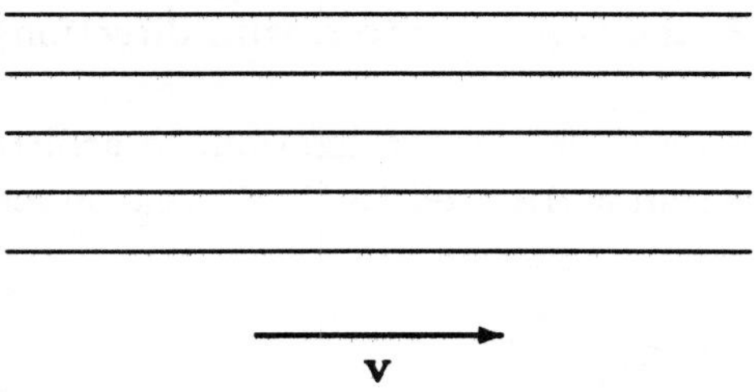

Fig. 9.2

Notice that to define a proper pencil of lines it is necessary to specify two distinct lines. On the other hand, to define an improper pencil, it is only necessary to specify a single line. The reason for this is best understood by interpreting pencils from the point of view of projective geometry.

Example 9.2

Let $C(x_0, y_0) \in \mathbf{A}$ and let $P(x, y)$ be any other point. The reflexion of P in C is the point $\sigma_C(P)$ satisfying $\overrightarrow{C\sigma_C(P)} = \overrightarrow{PC}$ — see Example 7.5(4). The coordinates x', y' of $\sigma_C(P)$ satisfy

$$(x' - x_0, y' - y_0) = (x_0 - x, y_0 - y)$$

and so are given by

$$(x', y') = (2x_0 - x, 2y_0 - y).$$

Let ℓ be the line with parametric equations

$$\begin{aligned} x &= a + lt \\ y &= b + mt. \end{aligned}$$

If $P(a + lt, b + mt) \in \ell$, then the point $\sigma_C(P)$ has coordinates

$$(2x_0 - a - lt, 2y_0 - b - mt).$$

It follows that as $P \in \ell$ varies, the point $\sigma_C(P)$ describes the line whose parametric equation is

$$\begin{aligned} x &= 2x_0 - a - lt \\ y &= 2y_0 - b - mt, \end{aligned}$$

which we denote by $\sigma_C(\ell)$. Note that $\sigma_C(\ell)$ passes through $\sigma_C(Q)$, where $Q(a,b) \in \ell$, and is parallel to ℓ, with direction vector $(-l,-m)$.

The following result is a generalization, to arbitrary affine planes, of a classical theorem attributed to the Greek philosopher, Thales of Miletus.

Theorem 9.3 (Thales)
Let H, H', H'' be three distinct parallel lines in the affine plane $\mathbf{A}$, and let ℓ_1 and ℓ_2 be two lines not parallel to H, H', H''. For $i = 1, 2$ let

$$\begin{aligned} P_i &= \ell_i \cap H \\ P_i' &= \ell_i \cap H' \\ P_i'' &= \ell_i \cap H'' \end{aligned}$$

and let $k_1, k_2 \in \mathbf{K}$ be such that

$$\overrightarrow{P_iP_i''} = k_i\overrightarrow{P_iP_i'}.$$

Then $k_1 = k_2$.

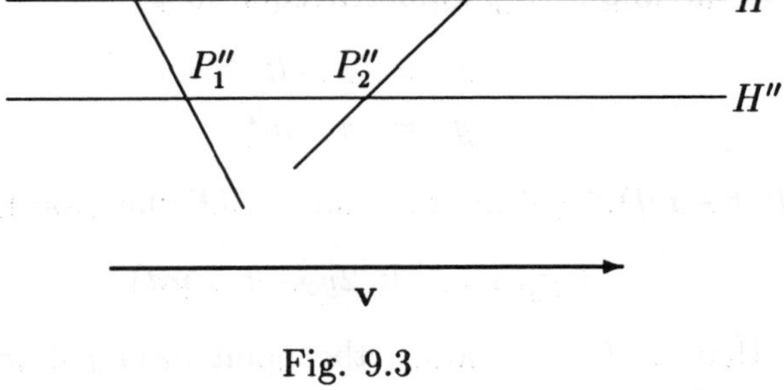

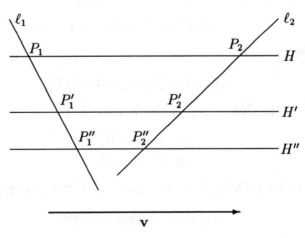

Fig. 9.3

Proof

If $\ell_1 = \ell_2$ the therorem is trivial, so suppose that $\ell_1 \neq \ell_2$, as in Fig. 9.3. We can also suppose that $P_1 \neq P_2$: if necessary, swap H and H'. Let $\mathbf{v} = \overrightarrow{P_1P_2}$. Then,

$$\begin{aligned}
\overrightarrow{P_2P_2'} - \overrightarrow{P_1P_1'} = \overrightarrow{P_1'P_2'} - \overrightarrow{P_1P_2} &= \alpha\mathbf{v} \\
\overrightarrow{P_2P_2''} - \overrightarrow{P_1P_1''} = \overrightarrow{P_1''P_2''} - \overrightarrow{P_1P_2} &= \beta\mathbf{v}
\end{aligned}$$

for some $\alpha, \beta \in \mathbf{K}$.

If $\alpha = 0$ then ℓ_1 and ℓ_2 are parallel, and $\beta = 0$, for if β were non-zero then $\mathbf{v}$ would be parallel to ℓ_1 and ℓ_2, contrary to the hypotheses. In this case, one has

$$\begin{aligned}
\overrightarrow{P_2P_2''} &= k_2\overrightarrow{P_2P_2'} = k_2\overrightarrow{P_1P_1'} \\
\overrightarrow{P_1P_1''} &= k_1\overrightarrow{P_1P_1'}
\end{aligned}$$

and since $\overrightarrow{P_1P_1''} = \overrightarrow{P_2P_2''}$ we see that $k_1 = k_2$.

If $\alpha \neq 0$, then

$$\overrightarrow{P_2P_2''} - \overrightarrow{P_1P_1''} = \beta\mathbf{v} = \alpha^{-1}\beta(\alpha\mathbf{v}) = \alpha^{-1}\beta\overrightarrow{P_2P_2'} - \alpha^{-1}\beta\overrightarrow{P_1P_1'}. \quad (9.15)$$

On the other hand,

$$\overrightarrow{P_2P_2''} - \overrightarrow{P_1P_1''} = k_2\overrightarrow{P_2P_2'} - k_1\overrightarrow{P_1P_1'} \quad (9.16)$$

Since $\alpha \neq 0$, $\overrightarrow{P_1P_1'}$ and $\overrightarrow{P_2P_2'}$ are not parallel and so they are linearly independent. Comparing (9.15) and (9.16) gives that $k_2 = \alpha^{-1}\beta = k_1$.

□

This theorem of Thales asserts essentially that the scalar $k = k_1 = k_2$ depends only on H, H', H'' and not on the transverse lines ℓ_1 and ℓ_2. A particular case arises when ℓ_1 and ℓ_2 meet at a point $O \in H$. In this case $P_1 = O = P_2$ and the theorem asserts that $\overrightarrow{OP_1''} = k\overrightarrow{OP_1'}$, and $\overrightarrow{OP_2''} = k\overrightarrow{OP_2'}$.

Using Thales' theorem, one can deduce two other important classical results in plane affine geometry: Pappus' theorem and Desargues' theorem.

Theorem 9.4 (Pappus)
Let H, H' be two distinct lines in the affine plane $\mathbf{A}$. Let $P, Q, R \in H$ and $P', Q', R' \in H'$ be distinct points, none of which lies at the intersection $H \cap H'$. If $\overline{PQ'} \parallel \overline{P'Q}$ and $\overline{QR'} \parallel \overline{Q'R}$ then $\overline{PR'} \parallel \overline{P'R}$.

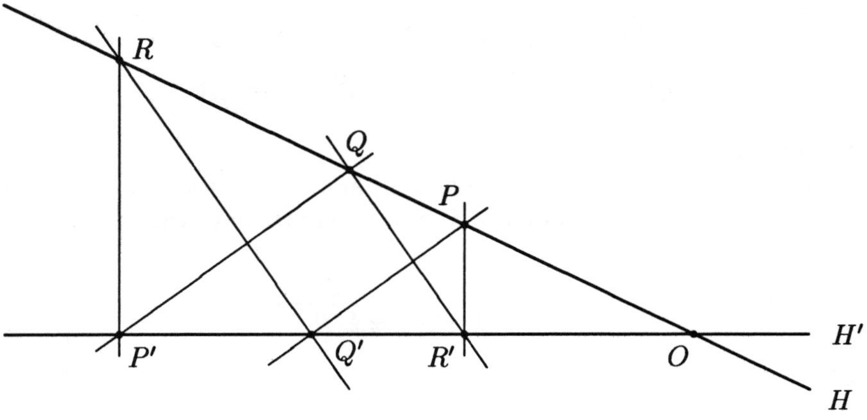

Fig. 9.4

Proof
Suppose H and H' are not parallel , and let $H \cap H' = \{O\}$, as in Fig. 9.4. By Thales' theorem, for some $h, k \in \mathbf{K}^*$,

$$\overrightarrow{OP'} = k\overrightarrow{OQ'}, \quad \overrightarrow{OQ} = k\overrightarrow{OP}, \quad \text{since} \quad \overline{PQ'} \parallel \overline{P'Q},$$

$$\overrightarrow{OQ'} = h\overrightarrow{OR'}, \quad \overrightarrow{OR} = h\overrightarrow{OQ}, \quad \text{since} \quad \overline{QR'} \parallel \overline{Q'R}.$$

But then,

$$\overrightarrow{PR'} = \overrightarrow{OR'} - \overrightarrow{OP} = h^{-1}\overrightarrow{OQ'} - k^{-1}\overrightarrow{OQ}$$

$$\overrightarrow{RP'} = \overrightarrow{OP'} - \overrightarrow{OR} = k\overrightarrow{OQ'} - h\overrightarrow{OQ},$$

and so $\overrightarrow{RP'} = hk\overrightarrow{PR'}$, that is, $\overline{RP'} \parallel \overline{PR'}$.

If $H \parallel H'$, then

$$\overrightarrow{PQ} = \overrightarrow{Q'P'}, \quad \text{since} \quad \overline{PQ} \parallel \overline{Q'P'},$$

$$\overrightarrow{QR} = \overrightarrow{R'Q'}, \quad \text{since} \quad \overline{QR} \parallel \overline{R'Q'},$$

and so

$$\overrightarrow{PR} = \overrightarrow{PQ} + \overrightarrow{QR} = \overrightarrow{Q'P'} + \overrightarrow{R'Q'} = \overrightarrow{R'P'}.$$

Thus $\overline{PR'} \parallel \overline{P'R}$. □

There are other versions of Pappus' theorem, which the interested reader can find in [5] and [2].

Theorem 9.5 (Desargues)
Let $A, B, C, A', B', C' \in \mathbf{A}$ be points such that no three are collinear, and such that $\overline{AB} \parallel \overline{A'B'}$, $\overline{BC} \parallel \overline{B'C'}$ and $\overline{AC} \parallel \overline{A'C'}$. Then the three lines $\overline{AA'}$, $\overline{BB'}$, $\overline{CC'}$ are either parallel or have a point in common.

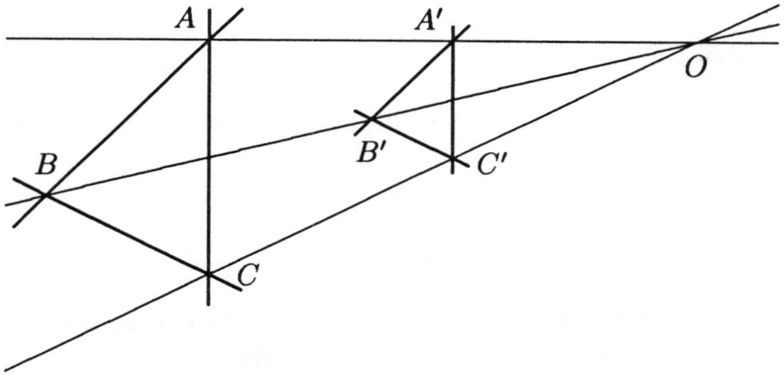

Fig. 9.5

Proof
Suppose that $\overline{AA'}$, $\overline{BB'}$, $\overline{CC'}$ are not parallel. Then two of them meet, and we may suppose that $\overline{AA'} \cap \overline{BB'} = \{O\}$, as in Fig. 9.5.

By Thales' theorem applied to $\overline{AB}$ and $\overline{A'B'}$ one has

$$\overrightarrow{OA'} = k\overrightarrow{OA}, \quad \overrightarrow{OB'} = k\overrightarrow{OB}, \quad k \in \mathbf{K}.$$

Let $\{C''\} = \overline{OC} \cap \overline{A'C'}$. By Thales' theorem, now applied to $\overline{AC}$ and $\overline{A'C'}$ one has

$$\overrightarrow{OC''} = k\overrightarrow{OC} \tag{9.17}$$

since $\overrightarrow{OA'} = k\overrightarrow{OA}$. On the other hand, putting $\{C'''\} = \overline{OC} \cap \overline{B'C'}$, Thales' theorem applied to the lines $\overline{BC}$ and $\overline{B'C'}$ implies that

$$\overrightarrow{OC'''} = k\overrightarrow{OC} \tag{9.18}$$

since $\overrightarrow{OB'} = k\overrightarrow{OB}$. A comparison of (9.17) and (9.18) implies that $C'' = C''' = C'$, and so O, C and C' are collinear. □

These two theorems also have counterparts in Projective Geometry, which can be found in many texts on that subject. Their importance lies chiefly in the relationship they have with characterization of affine space by means of graphical properties. (For more details on this topic, see [5] and [9].)

EXERCISES

9.1 Establish which of the following are triples of collinear points in $\mathbf{A}^2(\mathbf{R})$:
a) $\{(\frac{1}{2}, 2), (\frac{1}{2}, 100), (\frac{1}{2}, \frac{\pi}{4})\}$
b) $\{(1,1), (1,-1), (-1,1)\}$
c) $\{(\frac{5}{4}, \frac{9}{4}), (-\frac{3}{2}, -\frac{1}{2}), (\frac{1}{5}, \frac{6}{5})\}$.

9.2 Find the Cartesian equation of the line ℓ in $\mathbf{A}^2(\mathbf{R})$ containing the points P and Q in each of the following cases:
a) $P = (1, \frac{4}{3})$, $Q = (\frac{3}{2}, 1)$
b) $P = (0, 172)$, $Q = (\sqrt{7}, 0)$
c) $P = \mathcal{S} \cap \mathcal{S}'$, $Q = \mathcal{T} \cap \mathcal{T}'$, where $\mathcal{S}, \mathcal{S}', \mathcal{T}, \mathcal{T}'$ are the lines

$$\mathcal{S} : X + 5Y - 8 = 0, \quad \mathcal{S}' : 3X + 6 = 0$$

$$\mathcal{T} : 5X - \tfrac{1}{2}Y = 1, \quad \mathcal{T}' : X - Y = 5.$$

9.3 Determine parametric equations for the line in $\mathbf{A}^2(\mathbf{C})$ parallel to $\mathbf{v}$ and passing through $\mathcal{S} \cap \mathcal{T}$ in each of the following cases:
a) $\mathbf{v} = (2,4)$, $\quad \mathcal{S} : 3X - 2Y - 7 = 0$, $\quad \mathcal{T} : 2X + 3Y = 0$
b) $\mathbf{v} = (-5\sqrt{2}, 7)$, $\quad \mathcal{S} : X - Y = 0$, $\quad \mathcal{T} : X + Y = 1$.

9.4 Let $P = (2,3), Q = (11, -\sqrt{5}) \in \mathbf{A}^2(\mathbf{R})$. Find the midpoint of the segment PQ.

9.5 In $\mathbf{A}^2(\mathbf{R})$ let $P = (1, -1)$ and $Q = (2, 15/2)$. Find the points which divide PQ into 4 equal parts.

9.6 Prove the following generalization of Thales' theorem.

In an affine space $\mathbf{A}$ over $\mathbf{K}$ let H, H', H'' be three distinct parallel hyperplanes, and let ℓ_1 and ℓ_2 be lines which are not parallel to H, H', H''. Let $P_i = \ell_i \cap H, P_i' = \ell_i \cap H', P_i'' = \ell_i \cap H''$ (for $i = 1, 2$), and let k_1, k_2 be the scalars such that

$$\overrightarrow{P_iP_i''} = k_i \overrightarrow{P_iP_i'} \quad i = 1, 2.$$

Then $k_1 = k_2$.

10

Geometry in 3-dimensional affine space

We now consider in detail the case of 3-dimensional affine spaces. A particular case is of course what we call ordinary space.

Let $\mathbf{V}$ be a given 3-dimensional vector space over $\mathbf{K}$, and $\mathbf{A}$ be an affine space over $\mathbf{V}$. Fix a reference frame $O\mathbf{e}_1\mathbf{e}_2\mathbf{e}_3$ in $\mathbf{A}$, and denote the coordinates of the variable point $P \in \mathbf{A}$ by x, y, z, and let X, Y, Z be unknowns.

The affine subspaces of $\mathbf{A}$ are $\mathbf{A}$ itself, the planes in $\mathbf{A}$, the lines in $\mathbf{A}$ and the points of $\mathbf{A}$.

A plane $\mathcal{P}$ of $\mathbf{A}$ passing through the point $Q(q_1, q_2, q_3)$ has parametric equations of the form

$$\begin{aligned} x &= q_1 + a_1 u + b_1 v \\ y &= q_2 + a_2 u + b_2 v \\ z &= q_3 + a_3 u + b_3 v, \end{aligned} \tag{10.1}$$

where $\{\mathbf{a}(a_1, a_2, a_3),\ \mathbf{b}(b_1, b_2, b_3)\}$ is a basis for the vector subspace $\mathbf{W}$ associated to the plane — Fig. 10.1.

The plane $\mathcal{P}$ has codimension 1, so can also be described by a single Cartesian equation

$$AX + BY + CZ + D = 0 \tag{10.2}$$

for some $A, B, C, D \in \mathbf{K}$. The constants A, B, C and D are determined by $\mathcal{P}$ only up to a non-zero common factor.

The plane $\mathcal{P}$ contains the origin if and only if $D = 0$. The planes with equations $X = 0$, $Y = 0$ and $Z = 0$ are known as the *coordinate planes*: the YZ-plane, the XZ-plane and the XY-plane respectively.

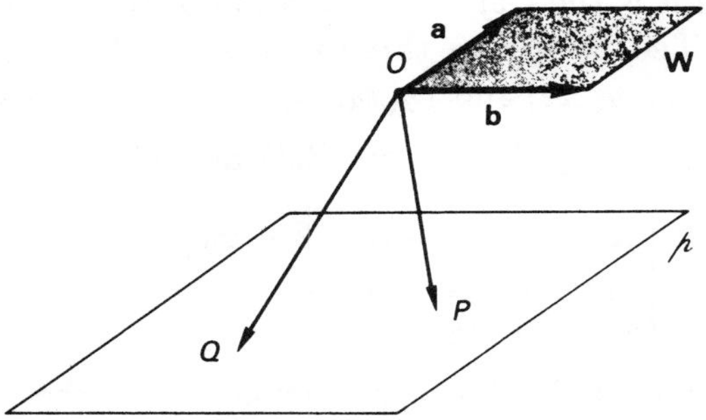

Fig. 10.1

The equation (10.2) for $\mathcal{P}$ can be derived from (10.1) by observing that the latter expresses the linear dependence of the vectors $\overrightarrow{QP}, \mathbf{a}, \mathbf{b}$. Equation (10.1) is thus equivalent to the vanishing of the determinant of the matrix whose rows are the coordinates of these vectors. Therefore the point $P(x, y, z)$ belongs to the plane passing through $Q(q_1, q_2, q_3)$ and parallel to $\langle \mathbf{a}(a_1, a_2, a_3), \mathbf{b}(b_1, b_2, b_3) \rangle$ if and only if its coordinates satisfy the equation in X, Y, Z:

$$\begin{vmatrix} X - q_1 & Y - q_2 & Z - q_3 \\ a_1 & a_2 & a_3 \\ b_1 & b_2 & b_3 \end{vmatrix} = 0. \tag{10.3}$$

Expanding the left hand side of (10.3) by the first row, and putting,

$$A = \begin{vmatrix} a_2 & a_3 \\ b_2 & b_3 \end{vmatrix}, \quad B = -\begin{vmatrix} a_1 & a_3 \\ b_1 & b_3 \end{vmatrix}, \quad C = \begin{vmatrix} a_1 & a_2 \\ b_1 & b_2 \end{vmatrix}, \tag{10.4}$$

gives

$$A(X - q_1) + B(Y - q_2) + C(Z - q_3) = 0. \tag{10.5}$$

Note that A, B, C are not all zero: this follows from the fact that $\mathbf{a}$ and $\mathbf{b}$ are linearly independent. If we put

$$D = -Aq_1 - Bq_2 - Cq_3, \tag{10.6}$$

then (10.5) can be rewritten in the form

$$AX + BY + CZ + D = 0. \tag{10.7}$$

Equation (10.7) is satisfied by the points $P(x,y,z) \in \mathcal{P}$, and only by these, and so it is the Cartesian equation of the plane $\mathcal{P}$.

If $\mathcal{P}$ is specified by giving three points which are not collinear, $P_0(x_0,y_0,z_0)$, $P_1(x_1,y_1,z_1)$ and $P_2(x_2,y_2,z_2)$, then its associated vector subspace is generated by the vectors $\overrightarrow{P_0P_1}$ and $\overrightarrow{P_0P_2}$, and (10.3) takes the form

$$\begin{vmatrix} X-x_0 & Y-y_0 & Z-z_0 \\ x_1-x_0 & y_1-y_0 & z_1-z_0 \\ x_2-x_0 & y_2-y_0 & z_2-z_0 \end{vmatrix} = 0. \tag{10.8}$$

Equation (10.8) is the Cartesian equation for the plane passing through the points P_0, P_1 and P_2.

A line $\ell \subset \mathbf{A}$ has parametric equations of the form,

$$\begin{aligned} x &= a + \ \, lt \\ y &= b + mt \\ z &= c + \ nt, \end{aligned} \tag{10.9}$$

where $Q(a,b,c) \in \ell$ and $\mathbf{v}(l,m,n)$ is its direction vector. The line ℓ can also be defined by two Cartesian equations, that is, as the intersection of two planes, because it has codimension 2:

$$\begin{aligned} AX + BY + CZ + D &= 0 \\ A'X + B'Y + C'Z + D' &= 0. \end{aligned} \tag{10.10}$$

The direction of ℓ is the 1-dimensional vector subspace of $\mathbf{V}$ given by the associated homogeneous system,

$$\begin{aligned} AX + BY + CZ &= 0 \\ A'X + B'Y + C'Z &= 0. \end{aligned} \tag{10.11}$$

It follows that the vector $\mathbf{v}(l,m,n)$ with coordinates

$$l = \begin{vmatrix} B & C \\ B' & C' \end{vmatrix}, \quad m = \begin{vmatrix} A & C \\ A' & C' \end{vmatrix}, \quad n = \begin{vmatrix} A & B \\ A' & B' \end{vmatrix}$$

is a direction vector for ℓ, because (l,m,n) is a non-zero solution of (10.11). This observation provides a practical method for finding the direction vector of a line which is given by Cartesian equations.

If, on the other hand, the line ℓ is specified by a point $Q(a,b,c)$ on it and a direction vector $\mathbf{v}(l,m,n)$, i.e. by (10.9), then the Cartesian equations for the line can be found by requiring that the order two minors of the matrix

$$\begin{pmatrix} X-a & Y-b & Z-c \\ l & m & n \end{pmatrix} \tag{10.12}$$

be zero. Indeed, this condition on the indeterminates X, Y, Z is satisfied by the coordinates of points $P(x,y,z) \in \mathbf{A}$ for which the vector $\overrightarrow{QP}$ is proportional to $\mathbf{v}$, that is, by points in ℓ.

Suppose that, for example, $l \neq 0$, then the condition above becomes

$$\begin{aligned} m(X-a) - l(Y-b) &= 0 \\ n(X-a) - l(Z-c) &= 0, \end{aligned} \tag{10.13}$$

or, equivalently,

$$\begin{aligned} mX - lY - (ma - lb) &= 0 \\ nX - lZ - \ (na - lc) &= 0, \end{aligned} \tag{10.14}$$

which are the equations of two *distinct* planes containing ℓ. That the planes are distinct can be seen because, since $l \neq 0$, the matrix of coefficients of (10.14) has rank 2. These planes therefore define the line ℓ. Note finally that (10.13) implies that the remaining minor in (10.12) also vanishes:

$$\begin{aligned} &n(Y-b) - m(Z-c) = \\ &\qquad l^{-1}\{m[n(X-a) - l(Z-c)] - n[m(X-a) - l(Y-b)]\} = 0. \end{aligned}$$

Proposition 10.1
Let $\mathcal{P}$ and $\mathcal{P}'$ be two planes in $\mathbf{A}$ with Cartesian equations

$$\begin{aligned} AX + BY + CZ + D &= 0 \\ A'X + B'Y + C'Z + D' &= 0 \end{aligned} \tag{10.15}$$

respectively. Then

1) *$\mathcal{P}$ and $\mathcal{P}'$ are parallel if and only if the matrix*

$$\begin{pmatrix} A & B & C \\ A' & B' & C' \end{pmatrix} \tag{10.16}$$

   has rank 1.

2) *If the matrix* (10.16) *has rank 1 then $\mathcal{P}$ and $\mathcal{P}'$ are disjoint if the augmented matrix*

$$\begin{pmatrix} A & B & C & D \\ A' & B' & C' & D' \end{pmatrix} \tag{10.17}$$

has rank 2; if its rank is 1 then they coincide.

3) *If $\mathcal{P}$ and $\mathcal{P}'$ are not parallel then they intersect, and $\mathcal{P} \cap \mathcal{P}'$ is a line; this occurs if and only if matrix* (10.16) *has rank 2.*

Proof

The cases considered in the proposition correspond to the different possibilities that can occur for the system of equations (10.15), for which (10.16) and (10.17) are the matrix of coefficients and the augmented matrix, respectively.

Since the vector subspaces associated to $\mathcal{P}$ and $\mathcal{P}'$ are given by the associated homogeneous equations

$$\begin{aligned} AX + BY + CZ &= 0 \\ A'X + B'Y + C'Z &= 0, \end{aligned} \tag{10.18}$$

$\mathcal{P}$ and $\mathcal{P}'$ are parallel if and only if the equations in (10.18) are proportional, which is in turn true if and only if matrix (10.16) has rank 1. In this case, matrix (10.17) has rank 1 or 2, according to whether system (10.15) is compatible or not, that is, according to whether $\mathcal{P} \cap \mathcal{P}' \neq \emptyset$ or $\mathcal{P} \cap \mathcal{P}' = \emptyset$. In the first case we must have that $\mathcal{P} = \mathcal{P}'$ since they are parallel. This proves (1) and (2).

By (1) matrix (10.16) has rank 2 if and only if $\mathcal{P}$ and $\mathcal{P}'$ are not parallel. In this case system (10.15) is compatible because also (10.17) has rank 2. System (10.15) therefore has a 1-parameter family of solutions and $\mathcal{P} \cap \mathcal{P}'$ is thus a line. □

Let us now consider the case of a line and a plane. The following proposition describes all possibilities for their relative positions.

Proposition 10.2

Let ℓ be a line with parametric equations (10.9), *and Cartesian equations* (10.10), *and let $\mathcal{P}''$ be a plane with equation*

$$A''X + B''Y + C''Z + D'' = 0. \tag{10.19}$$

1) *ℓ and $\mathcal{P}''$ are parallel if and only if*

$$\begin{vmatrix} A & B & C \\ A' & B' & C' \\ A'' & B'' & C'' \end{vmatrix} = 0 \tag{10.20}$$

or equivalently, if and only if

$$A''l + B''m + C''n = 0. \tag{10.21}$$

2) *If* (10.20) *is satisfied, then* $\ell \subset \mathcal{P}''$ *if and only if the matrix*

$$\begin{pmatrix} A & B & C & D \\ A' & B' & C' & D' \\ A'' & B'' & C'' & D'' \end{pmatrix} \tag{10.22}$$

has rank 2, otherwise they are disjoint (and the matrix has rank 3).

3) *If* ℓ *and* $\mathcal{P}''$ *are not parallel, then they are incident, and* $\ell \cap \mathcal{P}''$ *consists of only 1 point; this occurs if and only if*

$$\begin{vmatrix} A & B & C \\ A' & B' & C' \\ A'' & B'' & C'' \end{vmatrix} \neq 0 \tag{10.23}$$

or equivalently, if and only if,

$$A''l + B''m + C''n \neq 0. \tag{10.24}$$

Proof

Suppose that (10.20) is satisfied. Then the homogeneous system

$$\begin{aligned} AX + BY + CZ &= 0 \\ A'X + B'Y + C'Z &= 0 \\ A''X + B''Y + C''Z &= 0 \end{aligned} \tag{10.25}$$

is equivalent to the system consisting of only the first two equations, which defines the line ℓ. Thus every direction vector of ℓ satisfies the third equation in (10.25) which is the equation of the vector subspace associated to $\mathcal{P}''$; that is, $\mathcal{P}''$ and ℓ are parallel. Conversely, if $\mathcal{P}''$ and ℓ are parallel then the third equation in (10.25) is linearly dependent on the first two, and so (10.20) is satisfied. The fact that (10.20) and (10.21) are equivalent follows by expanding the determinant (10.20) by its first row. This proves (1).

If $\mathcal{P}''$ and ℓ are parallel, then either $\ell \subset \mathcal{P}''$ or they are disjoint accordingly as the system

$$\begin{aligned} AX + BY + CZ + D &= 0 \\ A'X + B'Y + C'Z + D' &= 0 \\ A''X + B''Y + C''Z + D'' &= 0 \end{aligned} \tag{10.26}$$

is or is not compatible. Part (2) is now proved using Theorem 5.7.

By (1), $\mathcal{P}''$ and ℓ are not parallel if and only if (10.23) and (10.24) (which, as already noted, are equivalent) are satisfied. In this case the system (10.26) is compatible and admits a unique solution, by Theorem 5.7. That is, $\mathcal{P}''$ and ℓ have a unique point in common. □

Equation (10.20), or equation (10.21), is the *condition for a line and a plane to be parallel.*

We now proceed to the case of the relative position of two lines.

Two lines ℓ and ℓ_1 are said to be *coplanar* if there is a plane which contains them both.

Proposition 10.3

Two lines ℓ and ℓ_1 in $\mathbf{A}$ *are coplanar if and only if one of the following conditions is satisfied:*

1) *ℓ and ℓ_1 are parallel;*
2) *ℓ and ℓ_1 are incident.*

In particular ℓ and ℓ_1 in $\mathbf{A}$ *are coplanar if and only if they are not skew.*

Proof

Since two lines in an affine plane which do not meet are parallel it follows that two coplanar lines are either incident or parallel.

Conversely, if $\ell = \ell_1$, then of course it is obvious that they are coplanar. If ℓ and ℓ_1 are parallel and distinct then they are contained in the plane passing through any point $Q \in \ell$ and with associated vector subspace $\langle \mathbf{v}, \overrightarrow{QQ_1} \rangle$, where $\mathbf{v}$ is a direction vector for ℓ and ℓ_1 and $Q_1 \in \ell_1$. If ℓ and ℓ_1 are incident and distinct, then they lie in the plane passing through the point $\ell \cap \ell_1$ and with associated vector subspace $\langle \mathbf{v}, \mathbf{v}_1 \rangle$, where $\mathbf{v}$ and $\mathbf{v}_1$ are direction vectors for ℓ and ℓ_1 respectively. □

The following result gives conditions under which two lines will be coplanar.

Proposition 10.4
Let ℓ and ℓ_1 be two lines in $\mathbf{A}$ *and suppose that ℓ has Cartesian equations* (10.10) *and ℓ_1 has equations*

$$\begin{aligned} A_1X + B_1Y + C_1Z + D_1 &= 0 \\ A_1'X + B_1'Y + C_1'Z + D_1' &= 0. \end{aligned} \tag{10.27}$$

Let $Q(a,b,c) \in \ell$ and $Q_1(a_1,b_1,c_1) \in \ell_1$, and let $\mathbf{v}(l,m,n)$ and $\mathbf{v}_1(l_1, m_1, n_1)$ be direction vectors for ℓ and ℓ_1 respectively. The following conditions are equivalent:
1) ℓ and ℓ_1 are coplanar;

2)
$$\begin{vmatrix} a-a_1 & b-b_1 & c-c_1 \\ l & m & n \\ l_1 & m_1 & n_1 \end{vmatrix} = 0;$$

3)
$$\begin{vmatrix} A & B & C & D \\ A' & B' & C' & D' \\ A_1 & B_1 & C_1 & D_1 \\ A_1' & B_1' & C_1' & D_1' \end{vmatrix} = 0.$$

Proof
(1) ⇒ (2) If ℓ and ℓ_1 are parallel, then $\mathbf{v}$ and $\mathbf{v}_1$ are proportional, and so (2) is satisfied. If ℓ and ℓ_1 are incident then they are contained in the same plane $\mathcal{P}$; the vector $\overrightarrow{QQ_1}$ belongs to the vector subspace assocciated to $\mathcal{P}$ which is generated by $\mathbf{v}$ and $\mathbf{v}_1$, and again (2) is satisfied.
(2) ⇒ (1) If (2) is satisfied, then either the last two rows of the matrix

$$\begin{pmatrix} a-a_1 & b-b_1 & c-c_1 \\ l & m & n \\ l' & m' & n' \end{pmatrix} \tag{10.28}$$

are proportional and the lines are parallel, or the first row is a linear combination of the other two, and the plane $\mathcal{P}$ with associated vector subspace $\langle \mathbf{v}, \mathbf{v}_1 \rangle$ passing through Q and containing ℓ contains also Q_1, and so contains ℓ_1. In either case the lines are coplanar.
(3) ⇒ (1) If the system comprised of (10.10) and (10.27) is compatible, then ℓ and ℓ_1 are coplanar because they have a point in common.

Suppose that this system is not compatible. Then, since the rank R of the matrix

$$\begin{pmatrix} A & B & C & D \\ A' & B' & C' & D' \\ A_1 & B_1 & C_1 & D_1 \\ A'_1 & B'_1 & C'_1 & D'_1 \end{pmatrix} \tag{10.29}$$

is at least 3, and the rank r of the matrix

$$\begin{pmatrix} A & B & C \\ A' & B' & C' \\ A_1 & B_1 & C_1 \\ A'_1 & B'_1 & C'_1 \end{pmatrix} \tag{10.30}$$

is at least two, we must have $R = 3$ and $r = 2$. Now, $r = 2$ implies that the last two rows of (10.30) are linear combinations of the first two. Thus ℓ and ℓ_1 have the same direction; that is, they are parallel, and hence coplanar.

(1) $\Rightarrow$ (3) If ℓ and ℓ_1 are incident, then the system consisting of (10.10) and (10.27) is compatible, and so (10.29) has rank less than 4. If instead ℓ and ℓ_1 are parallel, then (10.30) has rank 2 and so (10.29) has rank at most 3. In either case (3) holds. □

Let ℓ be a line in **A**. The set Φ of planes in **A** which contain ℓ is called a *proper pencil of planes*, and ℓ is called the *axis of the pencil* (Fig. 10.2).

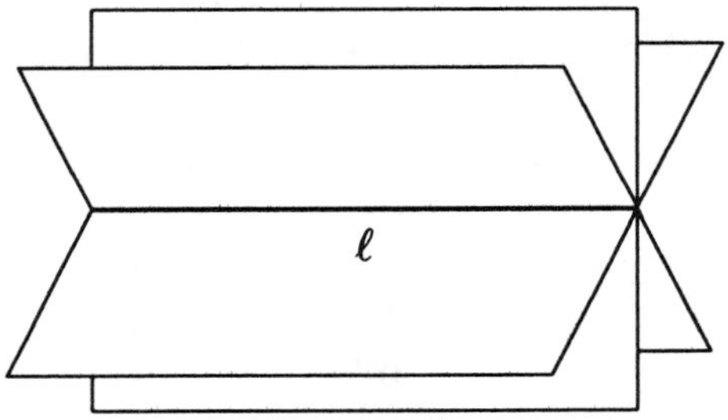

Fig. 10.2

Consider two distinct planes $\mathcal{P}$ and $\mathcal{P}_1$ in Φ, with respective equations

$$AX + BY + CZ + D = 0$$
$$A_1X + B_1Y + C_1Z + D_1 = 0.$$

As in the case of pencils of lines, one can show that every plane belonging to Φ has an equation of the form

$$\lambda(AX + BY + CZ + D) + \mu(A_1X + B_1Y + C_1Z + D_1) = 0$$

for some $\lambda, \mu \in \mathbf{K}$ not both zero.

In this case too, we can use a non-homogeneous parameter t to represent the planes of the pencil Φ in the form

$$AX + BY + CZ + D + t(A_1X + B_1Y + C_1Z + D_1) = 0,$$

remembering though, that the plane $\mathcal{P}_1$ is the one plane which cannot be put in this form.

Let Φ be the pencil of planes with axis ℓ and let $\mathcal{Q}$ be any plane not parallel to ℓ . The planes belonging to Φ intersect $\mathcal{Q}$ in the pencil of lines in $\mathcal{Q}$ with centre the point $\ell \cap \mathcal{Q}$ (Fig. 10.3).

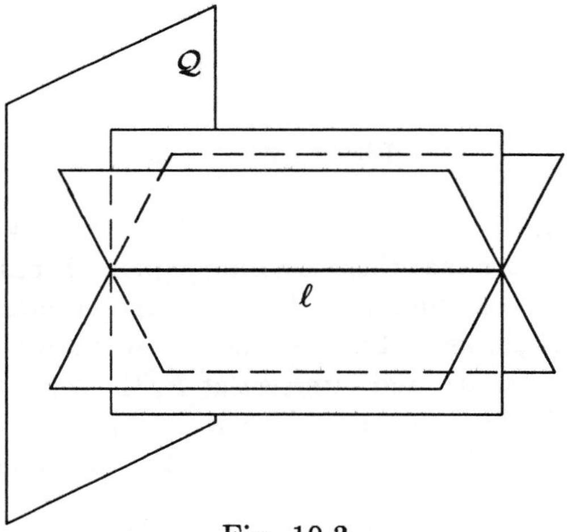

Fig. 10.3

Let $\mathbf{W}$ be a subspace of $\mathbf{V}$ of dimension 2. The set Ψ of planes in $\mathbf{A}$ having associated vector subspace $\mathbf{W}$ is called an *improper pencil of planes* and $\mathbf{W}$ is called the *vector subspace associated to the pencil.*

If $\mathcal{Q}$ is a plane in the improper pencil Ψ, with equation

$$AX + BY + CZ + D = 0,$$

then the other planes in the pencil have equations

$$AX + BY + CZ + t = 0$$

as t varies.

Example 10.5
Let ℓ and ℓ' be a pair of skew lines in $\mathbf{A}$, and let $P \in \mathbf{A}$ be a point not belonging to $\ell \cup \ell'$ (Fig. 10.4).

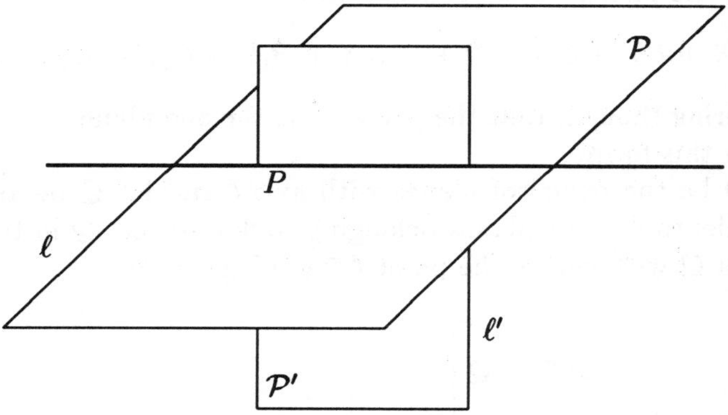

Fig. 10.4

There is a unique line through P which is coplanar with both of ℓ and ℓ'. To see this, note that there is a unique plane $\mathcal{P}$ passing through P and containing ℓ, and similarly, there is a unique plane $\mathcal{P}'$ passing through P and containing ℓ'. Since $\mathcal{P}$ and $\mathcal{P}'$ are distinct (otherwise ℓ and ℓ' would be coplanar), and intersect at P, the set $\mathcal{P} \cap \mathcal{P}'$ is a line passing through P which, by construction, is coplanar to both ℓ and ℓ'. The uniqueness of this line follows from the uniqueness of $\mathcal{P}$ and $\mathcal{P}'$.

Note that the line we have just constructed is not parallel to both ℓ and ℓ', for otherwise ℓ and ℓ' would be parallel to each other. Thus this line meets at least one of ℓ and ℓ'.

EXERCISES

10.1 Establish which of the following triples of points in $\mathbf{A}^3(\mathbf{C})$ are collinear:

a) $\{(2,1,-3),\ (1,-1,2),\ (3/2,0,-1/2)\}$

b) $\{(1,1,1),\ (2,-1,3),\ (2,1,-5)\}$

c) $\{(i,0,0), (1+i,2i,1), (1,2,-i)\}$
d) $\{(1,0,0), (2,-1,-1), (-2,-2,1)\}$
e) $\{(1,0,-1), (2,1,2), (-1,-1,3)\}$
f) $\{(1-i,i,2), (3,6i,-3), (2-i,3i,1)\}$.

10.2 In each of the following, find the value (if one exists) of the real parameter m for which the triple of points is collinear in $\mathbf{A}^3(\mathbf{R})$:
a) $\{(2,-1,2), (1,1,1), (4,-m+1,4)\}$;
b) $\{(3,0,0), (0,1,1), (m,m,m)\}$;
c) $\{(1,-m,0), (m,1,1), (1,-1,-3)\}$;
d) $\{(1,m,0), (2,\sqrt{2},1), (2,110,1)\}$.

10.3 After checking, for each of the following, that the points are not collinear, find parametric and Cartesian equations for the planes determined by the points
a) $\{(2,\sqrt{2},1), (1,1,\sqrt{2}), (0,0,1)\}$;
b) $\{(5,-1,0), (1,1,\sqrt{5}), (-3,1,\pi/2)\}$;
c) $\{(1,1,1), (-2,1,0), (2,2,2)\}$;
d) $\{(1,1000,0), (3.55,2\pi,0), (1,10^5,0)\}$.

10.4 In each of the following, find a Cartesian equation of the plane in $\mathbf{A}^3(\mathbf{C})$ passing through Q and parallel to the plane $\mathcal{P}$:
a) $Q=(-1,2,2), \quad \mathcal{P}: X+2Y+3Z+1=0$;
b) $Q=(i,i,i), \quad \mathcal{P}: 2X-Y=0$;
c) $Q=(1,i,i+1), \quad \mathcal{P}: iY-2Z+3i+10=0$;
d) $Q=(1-2i,1,\pi i), \quad \mathcal{P}: iY=3$.

10.5 For each of the following, determine whether or not the three planes belong to the same pencil.
a) $X-Y+Z=0, \quad -X+3Y-5Z+2=0, \quad Y-2Z+1=0$
b) $2X-3Y+3=0, \quad X-Y+6=0, \quad X-3Z=-1$
c) $X-5Y+1=0, \quad X-5Y=0, \quad 2X+Z=0$
d) $X-Y+Z+5=0, \quad 2X-2Y+2Z+77=0,$
$-X+Y-Z=0$.

10.6 For each of the following, find parametric and Cartesian equations for the line in $\mathbf{A}^3(\mathbf{R})$ passing through the point Q and parallel to the vector $\mathbf{v}$.
a) $Q=(1,1,0) \quad \mathbf{v}=(2,-1,\sqrt{2})$
b) $Q=(-2,2,-2) \quad \mathbf{v}=(1,1,0)$
c) $Q=(1,2,3) \quad \mathbf{v}=(1,2,3)$

d) $Q = (0, 0, 0)$ $\mathbf{v} = (1, 0, 0)$
e) $Q = (1, 1, 0)$ $\mathbf{v} = (1, 1, -1)$.

10.7 Find parametric equations for each of the following lines in $\mathbf{A}^3(\mathbf{C})$.
a) $X - iY = 0, \quad 2Y + Z + 1 = 0$
b) $3X + Z - 1 = 0, \quad Y + Z - 5 = 0$
c) $X - 1 = 0, \quad Z - 1 = 0$
d) $2iX - (i + 2)Y + Z - 3 + i = 0, \quad Z + iY = 2i.$

10.8 For each of the following, find parametric equations for the line in $\mathbf{A}^3(\mathbf{C})$ passing through the point Q parallel to the line ℓ.

a) $Q = (1, 1, 0)$, $\ell{:}X - iY = 0,\ Z + 1 = 0$
b) $Q = (1, 0, 0)$, $\ell{:}X + 2Y - 1 = 0,\ X = 2$
c) $Q = (2, 1, -5)$, $\ell{:}Y = 2,\ X = iZ + 7$
d) $Q = (3, 0, 0)$, $\ell{:}3X - Y - Z + 1 = 0,$
$X - 5Y + \sqrt{2}Z - 7000 = 0.$

10.9 In each of the following, find a Cartesian equation of the plane in $\mathbf{A}^3(\mathbf{R})$ passing through Q and parallel to the lines ℓ and ℓ':

a) $Q = (1, -1, -2)$, $\ell : X - Y = 1,\ X + Z = 5$
$\ell' : X = 1,\ Z = 2$
b) $Q = (0, 1, 3)$, $\ell : X + Y = -5,\ X - Y + 2Z = 0$
$\ell' : 2X - 2Y = 1,\ X - Y + 2Z = 1$
c) $Q = (3, 3, 3)$, $\ell : X - 2Y = -1,\ X + Z = -1$
$\ell' : 2X - 2Z = 1,\ X - 2Y = -1.$

10.10 In each of the following, determine whether the lines ℓ and ℓ' are skew or coplanar. If they are coplanar, find whether they are incident or parallel, and then, after checking that they are distinct, find a Cartesian equation for the plane containing them.

a) $\ell : x = 1 + t,\ y = -t,\ z = 2 + 2t,$
$\ell' : x = 2 - t,\ y = -1 + 3t,\ z = t$
b) $\ell : 2X + Y + 1 = 0,\ Y - Z = 2$
$\ell' : x = 2 - t,\ y = 3 + 2t,\ z = 1$
c) $\ell : 2X + 3Y - Z = 0,\ 5X + 2Z - 1 = 0,$
$\ell' : 3X - 3Y + 3Z - 1 = 0,\ 5X + 2Z + 1 = 0$

d) $\ell : 2X + Z - 1 = 0,\ Y - Z + 1 = 0,$
$\ell' : 2X - Y + 3Z = 0,\ 2X + Y - 3 = 0$

e) $\ell : X + 1 = 0,\ Z - 2 = 0,$
$\ell' : 2X + Y - 2Z + 6 = 0,\ Y + Z - 2 = 0.$

10.11 In each of the following, find the relative positions of the line ℓ and the plane $\mathcal{P}$ in $\mathbf{A}^3(\mathbf{R})$, and, if they are incident, determine the point of intersection.

a) $\ell : x = 1 + t,\ y = 2 - 2t,\ z = 1 - 4t,$
$\mathcal{P} : 2X - Y + Z - 1 = 0$

b) $\ell : x = 2 - t,\ y = 1 + 2t,\ z = -1 + 3t,$
$\mathcal{P} : 2X + 2Y - Z + 1 = 0$

c) $\ell : X + Z - 2 = 0,\ Y = 1,$
$\mathcal{P} : X - Y + 2Z - 5 = 0$

d) $\ell : X + Z + 1 = 0,\ X - Z = 0,$
$\mathcal{P} : X + Z - 1 = 0.$

10.12 In each of the following, find a Cartesian equation for the plane in $\mathbf{A}^3(\mathbf{R})$ containing the point Q and the line ℓ.

a) $Q = (3, 3, 1),\quad \ell : x = 2 + 3t,\ y = 5 + t,\ z = 1 + 7t$

b) $Q = (2, 1, 0),\quad \ell : X - Y + 1 = 0,\ 3X + 5Z - 7 = 0$

c) $Q = (1, 0, 2),\quad \ell : Y + 2Z - 5 = 0,\ Z = 1$

d) $Q = (2, 2, 2),\quad \ell : x = 5 + t,\ y = -3 - 3t,\ z = 3 + 3t.$

10.13 In each of the following, find Cartesian equations for the line ℓ in $\mathbf{A}^3(\mathbf{R})$ passing through Q, contained in the plane $\mathcal{P}$ and intersecting the line ℓ'.

a) $Q = (1, 1, 0),\qquad \mathcal{P} : 2X - Y + Z - 1 = 0,$
$\ell' : x = 2 - t,\ y = 2 + t,\ z = t$

b) $Q = (-1, -1, -1),\ \mathcal{P} : X + Y + Z + 3 = 0,$
$\ell' : X - 2Z + 4 = 0,\ 2Y - Z = 0$

c) $Q = (1, 2, 3),\qquad \mathcal{P} : 2X - Y = 0,$
$\ell' : X + Z + 1 = 0,\ 2X + 2Y - Z - 3 = 0.$

10.14 In each of the following, find Cartesian equations of the line ℓ in $\mathbf{A}^3(\mathbf{R})$ passing through the point Q and coplanar to the lines ℓ' and ℓ''. Furthermore, establish whether ℓ meets or is parallel to ℓ' and ℓ''.

a) $Q = (1, 1, 2)$ $\ell' : 3X - 5Y + Z = -1,\ 2X - 3Z = -9,$
$\ell'' : X + 5Y = 3,\ 2X + 2Y - 7Z = -7$

b) $Q = (2, 0, -2)$ $\ell' : -X + 3Y = 2,\ X + Y + Z = -1,$
$\ell'' : x = 2 - t,\ y = 3 + 5t,\ z = -t$

c) $Q = (1, -1, -1)$ $\ell' : 2X + Y = -1,\ -2X + 3Y + Z = 0,$
$\ell'' : Y = 2,\ Z = 1.$

10.15 In each of the following, find the value of the real parameter k for which the lines ℓ and ℓ' are coplanar. Find a Cartesian equation for the plane that contains them, and find the point of intersection whenever they meet.

a) $\ell : x = k + t,\ y = 1 + 2t,\ z = -1 + kt,$
$\ell' : x = 2 - 2t,\ y = 3 + 3t,\ z = 1 - t$

b) $\ell : x = 3 - t,\ y = 1 + 2t,\ z = k + t,$
$\ell' : x = 1 + t,\ y = 1 + 2t,\ z = 1 + 3t$

c) $\ell : X - kY + Z + 1 = 0,\ Y - k = 0,$
$\ell' : X - Z + k = 0,\ Y = 1.$

10.16 Find a Cartesian equation for the plane $\mathcal{P}$ in $\mathbf{A}^3(\mathbf{R})$ which contains the line of intersection of the two planes

$$X + Y = 3, \quad \text{and} \quad 2Y + 3Z = 4$$

and is parallel to the vector $\mathbf{v} = (3, -1, 2)$.

11

Linear maps

Let $\mathbf{V}$ and $\mathbf{W}$ be two $\mathbf{K}$-vector spaces. A *map*

$$F : \mathbf{V} \to \mathbf{W}$$

is said to be *linear* if for every $\mathbf{v}, \mathbf{v}' \in \mathbf{V}$ and $c \in \mathbf{K}$ one has

$$F(\mathbf{v} + \mathbf{v}') = F(\mathbf{v}) + F(\mathbf{v}') \tag{11.1}$$

$$F(c\mathbf{v}) = cF(\mathbf{v}). \tag{11.2}$$

The two properties (11.1) and (11.2) are equivalent to the single property

$$F(c\mathbf{v} + c'\mathbf{v}') = cF(\mathbf{v}) + c'F(\mathbf{v}') \tag{11.3}$$

for every $\mathbf{v}, \mathbf{v}' \in \mathbf{V}$ and $c, c' \in \mathbf{K}$. Indeed, (11.3) becomes (11.1) by putting $c = c' = 1$, while it becomes (11.2) if $c' = 0$; conversely, if F satisfies (11.1) and (11.2) then

$$F(c\mathbf{v} + c'\mathbf{v}') = F(c\mathbf{v}) + F(c'\mathbf{v}') = cF(\mathbf{v}) + c'F(\mathbf{v}');$$

the first equality following from (11.1) and the second from (11.2).

By applying (11.3) repeatedly, it is easy to see that if F is linear, then for every $\mathbf{v}_1, \mathbf{v}_2, \ldots, \mathbf{v}_n \in \mathbf{V}$ and $c_1, c_2, \ldots, c_n \in \mathbf{K}$ one has

$$F(c_1\mathbf{v}_1 + c_2\mathbf{v}_2 + \cdots + c_n\mathbf{v}_n) = c_1F(\mathbf{v}_1) + c_2F(\mathbf{v}_2) + \cdots + c_nF(\mathbf{v}_n). \tag{11.4}$$

Notice that (11.2) with $c = 0$ implies that for F linear

$$F(\mathbf{0}) = \mathbf{0}.$$

A linear map $F : \mathbf{V} \to \mathbf{V}$ is called a *linear operator* on $\mathbf{V}$, or an *endomorphism* of $\mathbf{V}$.

A linear map $F : \mathbf{V} \to \mathbf{K}$ is called a *linear functional* on $\mathbf{V}$.

If $G : \mathbf{U} \to \mathbf{V}$ and $F : \mathbf{V} \to \mathbf{W}$ are linear maps, then so is their composite $F \circ G : \mathbf{U} \to \mathbf{W}$. The proof is immediate, and is left to the reader.

A linear map $F : \mathbf{V} \to \mathbf{W}$ is called an *isomorphism* if it is bijective. The inverse $F^{-1} : \mathbf{W} \to \mathbf{V}$ of an isomorphism is also linear and so is itself an isomorphism. Indeed, let $\mathbf{w}, \mathbf{w}' \in \mathbf{W}$ and let $\mathbf{v} = F^{-1}(\mathbf{w})$, $\mathbf{v}' = F^{-1}(\mathbf{w}')$. Then

$$\begin{aligned} F^{-1}(\mathbf{w}+\mathbf{w}') &= F^{-1}\big(F(\mathbf{v})+F(\mathbf{v}')\big) = F^{-1}\big(F(\mathbf{v}+\mathbf{v}')\big) \\ &= \mathbf{v}+\mathbf{v}' = F^{-1}(\mathbf{w})+F^{-1}(\mathbf{w}'). \end{aligned}$$

Similarly, for $c \in \mathbf{K}$,

$$F^{-1}(c\mathbf{w}) = F^{-1}\big(cF(\mathbf{v})\big) = F^{-1}\big(F(c\mathbf{v})\big) = c\mathbf{v} = cF^{-1}(\mathbf{w}).$$

An isomorphism of a space $\mathbf{V}$ to itself is said to be an *automorphism* of $\mathbf{V}$.

In the following, $\mathrm{Hom}(\mathbf{V}, \mathbf{W})$ will denote the set of all linear maps from $\mathbf{V}$ to $\mathbf{W}$, $\mathrm{End}(\mathbf{V})$ will denote the set of all linear operators on $\mathbf{V}$ (endomorphisms of $\mathbf{V}$), and $\mathbf{V}^*$ will denote the set of all linear functionals on $\mathbf{V}$. Finally, $\mathrm{GL}(\mathbf{V})$ will denote the subset of $\mathrm{End}(\mathbf{V})$ consisting of all automorphisms of $\mathbf{V}$ (GL stands for 'General Linear').

Examples 11.1

1. For every $\mathbf{V}$ and $\mathbf{W}$ the zero map $\mathbf{0} : \mathbf{V} \to \mathbf{W}$ defined by

   $$\mathbf{0}(\mathbf{v}) = \mathbf{0} \in \mathbf{W}$$

   for all $\mathbf{v} \in \mathbf{V}$ is a linear map.

   In every vector space $\mathbf{V}$ the identity map $\mathbf{1}_\mathbf{V} : \mathbf{V} \to \mathbf{V}$ which sends every vector to itself, is an automorphism. If $F, G \in \mathrm{GL}(\mathbf{V})$ then F^{-1} and $F \circ G$ also belong to $\mathrm{GL}(\mathbf{V})$.

   Let $c \in \mathbf{K}$. The map $c\mathbf{1}_\mathbf{V}$ defined by

   $$(c\mathbf{1}_\mathbf{V})(\mathbf{v}) = c\mathbf{v}$$

   is linear. If $c = 0$ then $c\mathbf{1}_\mathbf{V} = \mathbf{0}$. If $c \neq 0$ then $c\mathbf{1}_\mathbf{V}$ is an automorphism whose inverse is $c^{-1}\mathbf{1}_\mathbf{V}$.

If $\dim(\mathbf{V}) = 1$ the only linear operators on $\mathbf{V}$ are of the form $c1_{\mathbf{V}}$. For suppose $F \in \mathrm{End}(\mathbf{V})$ and $\mathbf{e} \in \mathbf{V}$, $\mathbf{e} \neq \mathbf{0}$, then $\{\mathbf{e}\}$ is a basis of $\mathbf{V}$ so $F(\mathbf{e}) = c\mathbf{e}$ for some $c \in \mathbf{K}$, and so for every $\mathbf{v} = x\mathbf{e} \in \mathbf{V}$,

$$F(\mathbf{v}) = F(x\mathbf{e}) = xF(\mathbf{e}) = x(c\mathbf{e}) = c\mathbf{v}.$$

2. Let $\boldsymbol{e} = \{\mathbf{e}_1, \ldots, \mathbf{e}_n\}$ be a basis of the $\mathbf{K}$-vector space $\mathbf{V}$, and

$$\phi_{\boldsymbol{e}} : \mathbf{V} \to \mathbf{K}^n$$

be the map defined by

$$\phi_{\boldsymbol{e}}(x_1\mathbf{e}_1 + \cdots + x_n\mathbf{e}_n) = (x_1, \ldots, x_n).$$

That is, $\phi_{\boldsymbol{e}}$ is the map which associates to every vector $\mathbf{v} = x_1\mathbf{e}_1 + \cdots + x_n\mathbf{e}_n \in \mathbf{V}$ the n-tuple $(x_1, \ldots, x_n)$ of its coordinates with respect to the basis $\boldsymbol{e}$.

Now $\phi_{\boldsymbol{e}}$ is an *isomorphism*, since, for every $c \in \mathbf{K}$, $\mathbf{v} = x_1\mathbf{e}_1 + \cdots + x_n\mathbf{e}_n$, and $\mathbf{v}' = y_1\mathbf{e}_1 + \cdots + y_n\mathbf{e}_n \in \mathbf{V}$ one has

$$\begin{aligned}
\phi_{\boldsymbol{e}}(\mathbf{v} + \mathbf{v}') &= \phi_{\boldsymbol{e}}\big((x_1 + y_1)\mathbf{e}_1 + \cdots + (x_n + y_n)\mathbf{e}_n\big) \\
&= (x_1 + y_1, \ldots, x_n + y_n) \\
&= (x_1, \ldots, x_n) + (y_1, \ldots, y_n) = \phi_{\boldsymbol{e}}(\mathbf{v}) + \phi_{\boldsymbol{e}}(\mathbf{v}'), \\
\phi_{\boldsymbol{e}}(c\mathbf{v}) &= (cx_1, \ldots, cx_n) = c(x_1, \ldots, x_n) = c\phi_{\boldsymbol{e}}(\mathbf{v}),
\end{aligned}$$

and so $\phi_{\boldsymbol{e}}$ is linear. Moreover, by the properties of the coordinates of a vector, $\phi_{\boldsymbol{e}}$ is bijective.

$\phi_{\boldsymbol{e}}$ is the *isomorphism defined by the basis* $\boldsymbol{e}$.

In the case that $\mathbf{V} = \mathbf{K}^n$ and $\boldsymbol{e}$ is the canonical basis, then $\phi_{\boldsymbol{e}}$ is the identity on $\mathbf{K}^n$.

3. Let $\mathbf{V}$ be a $\mathbf{K}$-vector space and let $\mathbf{U}$ and $\mathbf{W}$ be complementary subspaces of $\mathbf{V}$, that is suppose

$$\mathbf{V} = \mathbf{U} \oplus \mathbf{W}. \tag{11.5}$$

Since every $\mathbf{v} \in \mathbf{V}$ can be expressed in a unique way as $\mathbf{v} = \mathbf{u} + \mathbf{w}$, with $\mathbf{u} \in \mathbf{U}$ and $\mathbf{w} \in \mathbf{W}$, we can define a map

$$p : \mathbf{V} \to \mathbf{W}$$

by putting

$$p(\mathbf{u} + \mathbf{w}) = \mathbf{w}.$$

p is called the *projection of* $\mathbf{V}$ *on to* $\mathbf{W}$ *defined by the decomposition* (11.5).

The projection p is a linear map. Indeed, if $\mathbf{v} = \mathbf{u} + \mathbf{w}$, $\mathbf{v}' = \mathbf{u}' + \mathbf{w}'$ and $c, c' \in \mathbf{K}$ then

$$\begin{aligned} p(c\mathbf{v} + c'\mathbf{v}') &= p(c\mathbf{u} + c\mathbf{w} + c'\mathbf{u}' + c'\mathbf{w}') \\ &= p(c\mathbf{u} + c'\mathbf{u}' + c\mathbf{w} + c'\mathbf{w}') \\ &= c\mathbf{w} + c'\mathbf{w}' = cp(\mathbf{v}) + c'p(\mathbf{v}'). \end{aligned}$$

In particular, if $\mathbf{W}$ is a hyperplane then $\mathbf{U} = \langle \mathbf{u} \rangle$ for some $u \in \mathbf{V} \setminus \mathbf{W}$, and in this case p is called the *projection of* $\mathbf{V}$ *on to* $\mathbf{W}$ *in the direction* $\langle \mathbf{u} \rangle$.

If $\{\mathbf{e}_1, \ldots, \mathbf{e}_n\}$ is a basis of $\mathbf{V}$ and $1 \leq k < n$, then the projection of $\mathbf{V}$ on to $\langle \mathbf{e}_{k+1}, \ldots, \mathbf{e}_n \rangle$ defined by the decomposition $\mathbf{V} = \langle \mathbf{e}_1, \ldots, \mathbf{e}_k \rangle \oplus \langle \mathbf{e}_{k+1}, \ldots, \mathbf{e}_n \rangle$ is given by

$$p(c_1\mathbf{e}_1 + \cdots + c_n\mathbf{e}_n) = c_{k+1}\mathbf{e}_{k+1} + \cdots + c_n\mathbf{e}_n.$$

If π is the ordinary plane, ℓ a line in π and $\mathbf{V}$ and $\mathbf{W}$ are the real vector spaces of geometric vectors of π and of ℓ respectively, and $\mathbf{u} \in \mathbf{V} \setminus \mathbf{W}$ is a vector not parallel to ℓ, then the projection of $\mathbf{V}$ on to $\mathbf{W}$ in the direction $\mathbf{u}$ is the map illustrated in Fig. 11.1

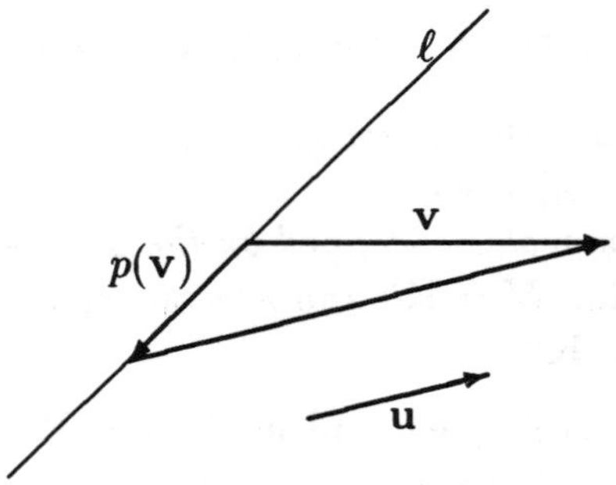

Fig. 11.1

In a similar way, given any plane π in ordinary space Σ, letting $\mathbf{V}$ and $\mathbf{W}$ be the spaces of geometric vectors of Σ and π respectively, and given any $\mathbf{u} \in \mathbf{V} \setminus \mathbf{W}$ the projection of $\mathbf{V}$ on to $\mathbf{W}$ in the direction $\langle \mathbf{u} \rangle$ is depicted in Fig. 11.2.

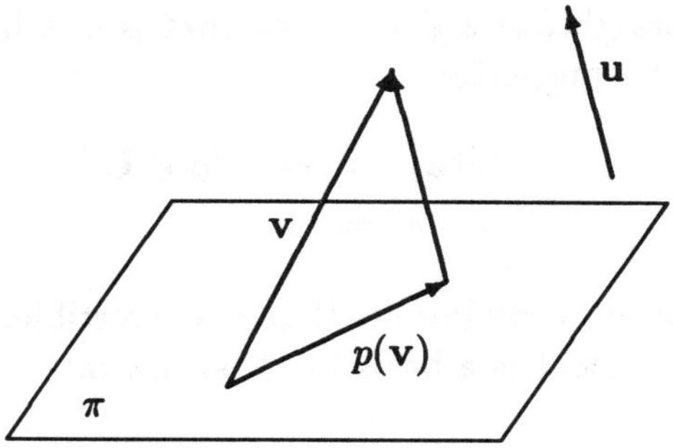

Fig. 11.2

The projections described here for vectors in the ordinary plane and space are based on the same geometrical principal as the projection of an affine space on to a subspace defined at the end of Chapter 8.

4. Let $\mathbf{V}$ be a $\mathbf{K}$-vector space and $\boldsymbol{e} = \{\mathbf{e}_1, \ldots, \mathbf{e}_n\}$ be a basis for $\mathbf{V}$. For each $i = 1, \ldots, n$ define

$$\eta_i : \mathbf{V} \to \mathbf{K}$$

by

$$\eta_i(c_1\mathbf{e}_1 + \cdots + c_n\mathbf{e}_n) = c_i,$$

that is, associating to each vector its i-th coordinate with respect to $\boldsymbol{e}$. It is easy to see that η_i is a linear functional on $\mathbf{V}$. Indeed, given any

$$\mathbf{v} = c_1\mathbf{e}_1 + \cdots + c_n\mathbf{e}_n, \qquad \mathbf{v}' = d_1\mathbf{e}_1 + \cdots + d_n\mathbf{e}_n$$

in $\mathbf{V}$ and $k, k' \in \mathbf{K}$, one has

$$\begin{aligned} \eta_i(k\mathbf{v} + k'\mathbf{v}') &= \eta_i\big((kc_1 + k'd_1)\mathbf{e}_1 + \cdots + (kc_n + k'd_n)\mathbf{e}_n\big) \\ &= kc_i + k'd_i = k\eta_i(\mathbf{v}) + k'\eta_i(\mathbf{v}'). \end{aligned}$$

5. Let $\mathbf{V}$ be a vector space over $\mathbf{K}$ and $\mathbf{U}, \mathbf{W}$ be supplementary subspaces of $\mathbf{V}$. Define the map $\rho : \mathbf{V} \to \mathbf{V}$ by

$$\rho(\mathbf{u} + \mathbf{w}) = \mathbf{u} - \mathbf{w}$$

for every $\mathbf{v} = \mathbf{u} + \mathbf{w}$, with $\mathbf{u} \in \mathbf{U}$, $\mathbf{w} \in \mathbf{W}$.

It is straightforward to check that ρ is a linear operator with the following properties

$$\begin{aligned} \rho(\mathbf{u}) &= \mathbf{u} \quad \forall \mathbf{u} \in \mathbf{U}, \\ \rho \circ \rho &= \mathbf{1}_{\mathbf{V}}. \end{aligned}$$

The second property implies that ρ is invertible, and so $\rho \in \mathrm{GL}(\mathbf{V})$.

If $\{\mathbf{e}_1, \ldots, \mathbf{e}_n\}$ is a basis for $\mathbf{V}$ such that

$$\mathbf{U} = \langle \mathbf{e}_1, \ldots, \mathbf{e}_k \rangle \qquad \mathbf{W} = \langle \mathbf{e}_{k+1}, \ldots, \mathbf{e}_n \rangle,$$

for some k with $1 \leq k < n$ then

$$\rho(c_1\mathbf{e}_1 + \cdots + c_n\mathbf{e}_n) = c_1\mathbf{e}_1 + \ldots + c_k\mathbf{e}_k - c_{k+1}\mathbf{e}_{k+1} - \cdots - c_n\mathbf{e}_n.$$

6. Let $\mathbf{V}$ be a vector space over $\mathbf{K}$, $\mathbf{U}$ a subspace and consider the quotient space $\mathbf{V}/\mathbf{U}$. The map

$$p : \mathbf{V} \to \mathbf{V}/\mathbf{U}$$

defined by

$$p(\mathbf{v}) = \mathbf{v} + \mathbf{U}$$

is linear. It is called the *natural projection* of $\mathbf{V}$ on to $\mathbf{V}/\mathbf{U}$. It is left to the reader to check that p is indeed linear.

7. Let $\mathbf{U}$ be a subspace of the $\mathbf{K}$-vector space $\mathbf{V}$. The inclusion i of $\mathbf{U}$ into $\mathbf{V}$ is a linear map. This too is left to the reader to check.

Proposition 11.2
Let $\mathbf{V}$ and $\mathbf{W}$ be $\mathbf{K}$-vector spaces, and $F : \mathbf{V} \to \mathbf{W}$ a linear map. Let $\mathbf{v}_1, \ldots, \mathbf{v}_n \in \mathbf{V}$, and $\mathbf{w}_i = F(\mathbf{v}_i)$ for $i = 1, \ldots, n$. If $\mathbf{v}_1, \ldots, \mathbf{v}_n$ are linearly dependent then $\mathbf{w}_1, \ldots, \mathbf{w}_n$ are also linearly dependent.

Equivalently, if $\mathbf{w}_1, \ldots, \mathbf{w}_n$ are linearly independent then so too are $\mathbf{v}_1, \ldots, \mathbf{v}_n$.

Proof
We prove the first statement. Let $c_1, \ldots, c_n \in \mathbf{K}$ be scalars which are not all 0, satisfying

$$c_1\mathbf{v}_1 + \cdots + c_n\mathbf{v}_n = \mathbf{0}.$$

Then

$$\begin{aligned} c_1\mathbf{w}_1 + \cdots + c_n\mathbf{w}_n &= c_1F(\mathbf{v}_1) + \cdots + c_nF(\mathbf{v}_n) \\ &= F(c_1\mathbf{v}_1 + \cdots + c_n\mathbf{v}_n) \\ &= F(\mathbf{0}) = \mathbf{0}, \end{aligned}$$

and so $\mathbf{w}_1, \ldots, \mathbf{w}_n$ are linearly dependent. □

Theorem 11.3
Let $\mathbf{V}$ *and* $\mathbf{W}$ *be two* $\mathbf{K}$*-vector spaces,* $\boldsymbol{e} = \{\mathbf{e}_1, \mathbf{e}_2, \ldots, \mathbf{e}_n\}$ *a basis for* $\mathbf{V}$ *and* $\mathbf{w}_1, \mathbf{w}_2, \ldots, \mathbf{w}_n$ *arbitrary vectors of* $\mathbf{W}$. *Then there is a linear map* $F : \mathbf{V} \to \mathbf{W}$ *such that*

$$F(\mathbf{e}_i) = \mathbf{w}_i, \quad i = 1, 2, \ldots, n.$$

Proof
If F exists it is unique, because for every $\mathbf{v} = x_1\mathbf{e}_1 + \cdots + x_n\mathbf{e}_n \in \mathbf{V}$ one has, by (11.4), that

$$F(\mathbf{v}) = x_1F(\mathbf{e}_1) + \cdots + x_nF(\mathbf{e}_n) = x_1\mathbf{w}_1 + \cdots + x_n\mathbf{w}_n \qquad (11.6)$$

and the coefficients $x_1, \ldots, x_n$ are uniquely determined as $\boldsymbol{e}$ is a basis.

It thus suffices to show that the map defined by (11.6) is linear. First we check condition (11.1). If

$$\begin{aligned} \mathbf{v} &= x_1\mathbf{e}_1 + \cdots + x_n\mathbf{e}_n \\ \mathbf{v}' &= y_1\mathbf{e}_1 + \cdots + y_n\mathbf{e}_n \end{aligned}$$

are elements of $\mathbf{V}$ then

$$\begin{aligned} F(\mathbf{v} + \mathbf{v}') &= (x_1 + y_1)\mathbf{w}_1 + \cdots + (x_n + y_n)\mathbf{w}_n \\ &= (x_1\mathbf{w}_1 + \cdots + x_n\mathbf{w}_n) + (y_1\mathbf{w}_1 + \cdots + y_n\mathbf{w}_n) \\ &= F(\mathbf{v}) + F(\mathbf{v}'). \end{aligned}$$

Now check (11.2). Let $c \in \mathbf{K}$ and $\mathbf{v} = x_1\mathbf{e}_1 + \cdots + x_n\mathbf{e}_n$. Then

$$\begin{aligned} F(c\mathbf{v}) &= cx_1\mathbf{w}_1 + \cdots + cx_n\mathbf{w}_n \\ &= c(x_1\mathbf{w}_1 + \cdots + x_n\mathbf{w}_n) = cF(\mathbf{v}). \end{aligned}$$

Thus F is linear. □

Definition 11.4
Let $F : \mathbf{V} \to \mathbf{W}$ be a linear map. The *kernel* of F is

$$\ker(F) = \{\mathbf{v} \in \mathbf{V} \mid F(\mathbf{v}) = \mathbf{0}\},$$

that is it is the subset of $\mathbf{V}$ consisting of all vectors which are mapped to $\mathbf{0} \in \mathbf{W}$ by F.

The *image* of F is the subset of $\mathbf{W}$

$$\mathrm{Im}(F) = \{\mathbf{w} = F(\mathbf{v}) \mid \mathbf{v} \in \mathbf{V}\}.$$

The sets $\ker(F)$ and $\mathrm{Im}(F)$ are vector subspaces of $\mathbf{V}$ and $\mathbf{W}$ respectively (the proof of this is left to the reader). If they are finite dimensional then the dimension of $\mathrm{Im}(F)$ is called the *rank* of F, denoted $\mathrm{rk}(F)$, and the dimension of $\ker(F)$ is called the *nullity* of F.

From (11.4) it follows that if $\{\mathbf{e}_1, \dots, \mathbf{e}_n\}$ is a basis for $\mathbf{V}$ then

$$\mathrm{Im}(F) = \langle F(\mathbf{e}_1), \dots, F(\mathbf{e}_n)\rangle.$$

Proposition 11.5
A linear map $F : \mathbf{V} \to \mathbf{W}$ *is injective if and only if* $\ker(F) = \langle \mathbf{0} \rangle$.

Proof
If F is injective then $\mathbf{0}$ is the only element of $\ker(F)$, so $\ker(F) = \langle \mathbf{0} \rangle$.
Conversely, suppose that $\ker(F) = \langle \mathbf{0} \rangle$. If $\mathbf{v}, \mathbf{v}'$ are such that $F(\mathbf{v}) = F(\mathbf{v}')$ then

$$F(\mathbf{v} - \mathbf{v}') = F(\mathbf{v}) - F(\mathbf{v}') = \mathbf{0}$$

and so $\mathbf{v} - \mathbf{v}' \in \ker(F)$. Since $\mathbf{0}$ is the only element of $\ker(F)$ it follows that $\mathbf{v} - \mathbf{v}' = \mathbf{0}$, or $\mathbf{v} = \mathbf{v}'$. Thus F is injective. □

We now have a particularly important theorem.

Theorem 11.6
Let $F : \mathbf{V} \to \mathbf{W}$ *be a linear map of* $\mathbf{K}$*-vector spaces, with* $\dim(\mathbf{V}) = n$. *Then* $\ker(F)$ *and* $\mathrm{Im}(F)$ *are finite dimensional, and*

$$\dim\big(\ker(F)\big) + \mathrm{rk}(F) = n.$$

Proof

Since $\mathbf{V}$ is finite dimensional and $\ker(F)$ is a subspace of $\mathbf{V}$ it follows that $\ker(F)$ is also finite dimensional. Let $s = \dim(\ker(F))$, and fix a basis $\{\mathbf{n}_1, \ldots, \mathbf{n}_s\}$ of $\ker(F)$. Let $\mathbf{v}_{s+1}, \ldots, \mathbf{v}_n \in \mathbf{V}$ be such that $\{\mathbf{n}_1, \ldots, \mathbf{n}_s, \mathbf{v}_{s+1}, \ldots, \mathbf{v}_n\}$ is a basis of $\mathbf{V}$. To complete the proof, it is sufficient to show that $\{F(\mathbf{v}_{s+1}), \ldots, F(\mathbf{v}_n)\}$ forms a basis for $\mathrm{Im}(F)$.

Every vector $\mathbf{w} \in \mathrm{Im}(F)$ is of the form

$$\begin{aligned} \mathbf{w} &= F(a_1\mathbf{n}_1 + \cdots + a_s\mathbf{n}_s + b_{s+1}\mathbf{v}_{s+1} + \cdots + b_n\mathbf{v}_n) \\ &= a_1F(\mathbf{n}_1) + \cdots + a_sF(\mathbf{n}_s) + b_{s+1}F(\mathbf{v}_{s+1}) + \cdots + b_nF(\mathbf{v}_n) \\ &= b_{s+1}F(\mathbf{v}_{s+1}) + \cdots + b_nF(\mathbf{v}_n), \end{aligned}$$

for some scalars $a_1, \ldots, a_s, b_{s+1}, \ldots, b_n$. Thus $F(\mathbf{v}_{s+1}), \ldots, F(\mathbf{v}_n)$ generate $\mathrm{Im}(F)$.

Suppose $c_{s+1}, \ldots, c_n \in \mathbf{K}$ are such that

$$c_{s+1}F(\mathbf{v}_{s+1}) + \cdots + c_nF(\mathbf{v}_n) = \mathbf{0}$$

so that $c_{s+1}\mathbf{v}_{s+1} + \cdots + c_n\mathbf{v}_n \in \ker(F)$. Since $\{\mathbf{n}_1, \ldots, \mathbf{n}_s\}$ is a basis for $\ker(F)$, there are $d_1, \ldots, d_s \in \mathbf{K}$ such that

$$c_{s+1}\mathbf{v}_{s+1} + \cdots + c_n\mathbf{v}_n = d_1\mathbf{n}_1 + \cdots + d_s\mathbf{n}_s$$

i.e. such that

$$d_1\mathbf{n}_1 + \cdots + d_s\mathbf{n}_s - c_{s+1}\mathbf{v}_{s+1} - \cdots - c_n\mathbf{v}_n = \mathbf{0}.$$

However, $\mathbf{n}_1, \ldots, \mathbf{n}_s, \mathbf{v}_{s+1}, \ldots, \mathbf{v}_n$ are linearly independent and so all the coefficients are zero; in particular $c_{s+1} = \ldots = c_s = 0$ and $F(\mathbf{v}_{s+1}), \ldots, F(\mathbf{v}_n)$ are indeed linearly independent. □

Notice that the preceding theorem does not require that $\mathbf{W}$ be finite dimensional.

Corollary 11.7

If $\mathbf{U}$ *is a subspace of a finite dimensional* $\mathbf{K}$*-vector space* $\mathbf{V}$*, then*

$$\dim(\mathbf{V}/\mathbf{U}) = \dim(\mathbf{V}) - \dim(\mathbf{U}). \tag{11.7}$$

Proof
The natural projection $p : \mathbf{V} \to \mathbf{V}/\mathbf{U}$ is surjective and has kernel $\mathbf{U}$. □

The following corollary provides a simple characterization of isomorphisms, and follows immediately from Theorem 11.6. Its proof is left to the reader.

Corollary 11.8
If $\mathbf{V}$ and $\mathbf{W}$ are $\mathbf{K}$-vector spaces of the same finite dimension, and $F : \mathbf{V} \to \mathbf{W}$ is a linear map then the following conditions are equivalent:
1) $\ker(F) = \langle \mathbf{0} \rangle$;
2) $\mathrm{Im}(F) = \mathbf{W}$;
3) *F is an isomorphism.*

Theorem 11.9 (Homomorphism theorem for vector spaces)
Let $F : \mathbf{V} \to \mathbf{W}$ be a linear map of $\mathbf{K}$-vector spaces. F defines an isomorphism

$$F' : \frac{\mathbf{V}}{\ker(F)} \longrightarrow \mathrm{Im}(F)$$

such that

$$F = i \circ F' \circ p, \tag{11.8}$$

where $p : \mathbf{V} \to \mathbf{V}/\ker(F)$ is the natural projection, and $i : \mathrm{Im}(F) \to \mathbf{W}$ is the inclusion.

Proof
Let $\mathbf{v}_1, \mathbf{v}_2 \in \mathbf{V}$. Now, $F(\mathbf{v}_1) = F(\mathbf{v}_2)$ if and only if $F(\mathbf{v}_2 - \mathbf{v}_1) = \mathbf{0}$, that is, if and only if $\mathbf{v}_2 - \mathbf{v}_1 \in \ker(F)$. In other words,

$$F(\mathbf{v}_1) = F(\mathbf{v}_2) \Leftrightarrow \mathbf{v}_2 \in [\mathbf{v}_1 + \ker(F)]. \tag{11.9}$$

We can therefore define a map $F' : \mathbf{V}/\ker(F) \to \mathrm{Im}(F)$ by

$$F'(\mathbf{v} + \ker(F)) = F(\mathbf{v}).$$

F' is bijective because it is obviously surjective, and by (11.9) it is also injective. The proof that F' is linear and of (11.8) is left to the reader. □

Definition 11.10
Two $\mathbf{K}$-vector spaces $\mathbf{V}$ and $\mathbf{W}$ are said to be *isomorphic* if there is an isomorphism $\mathbf{V} \to \mathbf{W}$. One also says that $\mathbf{V}$ *is isomorphic to* $\mathbf{W}$.

Every vector space is isomorphic to itself, because $\mathbf{1}_V$ is an isomorphism. If $\mathbf{V}$ is isomorphic to $\mathbf{W}$ then $\mathbf{W}$ is isomorphic to $\mathbf{V}$ because the inverse of an isomorphism is an isomorphism. Finally, since the composite of two isomorphisms is an isomorphism, if $\mathbf{V}$ is isomorphic to $\mathbf{W}$ and $\mathbf{W}$ is isomorphic to $\mathbf{U}$ then $\mathbf{V}$ is isomorphic to $\mathbf{U}$. Thus isomorphism is an equivalence relation between vector spaces.

Theorem 11.11
Two finite dimensional $\mathbf{K}$*-vector spaces are isomorphic if and only if they have the same dimension.*

Proof
Suppose $\mathbf{V}$ and $\mathbf{W}$ are isomorphic, and let $F : \mathbf{V} \to \mathbf{W}$ be an isomorphism. Since $\ker(F) = \langle \mathbf{0} \rangle$, by Theorem 11.6 one has

$$\dim(\mathbf{W}) = \mathrm{rk}(F) = \dim(\mathbf{V}).$$

Conversely, suppose $\dim(\mathbf{V}) = n = \dim(\mathbf{W})$, and let $\{\mathbf{v}_1, \ldots, \mathbf{v}_n\}$ and $\{\mathbf{w}_1, \ldots, \mathbf{w}_n\}$ be bases for $\mathbf{V}$ and $\mathbf{W}$ respectively. The linear map $F : \mathbf{V} \to \mathbf{W}$ defined by

$$F(\mathbf{v}_i) = \mathbf{w}_i, \quad i = 1, \ldots, n,$$

is surjective because $\mathbf{w}_1, \ldots, \mathbf{w}_n$ generate $\mathbf{W}$. It now follows from Theorem 11.6 that $\dim(\ker(F)) = 0$ and so F is also injective. Thus F is an isomorphism. □

We conclude this chapter with a discussion of a few properties of the space of linear functionals on a vector space.

Let $\mathbf{V}$ be a $\mathbf{K}$-vector space. Recall that a linear functional on $\mathbf{V}$ is a linear map $L : \mathbf{V} \to \mathbf{K}$; the set of all linear functionals is denoted $\mathbf{V}^*$. One can define on $\mathbf{V}^*$ the structure of a $\mathbf{K}$-vector space by defining operations in the following manner.

For $L_1, L_2 \in \mathbf{V}^*$, define $L_1 + L_2 \in \mathbf{V}^*$ by

$$(L_1 + L_2)(\mathbf{v}) = L_1(\mathbf{v}) + L_2(\mathbf{v}), \quad \forall \mathbf{v} \in \mathbf{V}.$$

To see that $L_1 + L_2$ is linear, let $\mathbf{v}, \mathbf{v}' \in \mathbf{V}$ and $c \in \mathbf{K}$. Then

$$\begin{aligned}(L_1 + L_2)(\mathbf{v} + \mathbf{v}') &= L_1(\mathbf{v} + \mathbf{v}') + L_2(\mathbf{v} + \mathbf{v}') \\ &= L_1(\mathbf{v}) + L_1(\mathbf{v}') + L_2(\mathbf{v}) + L_2(\mathbf{v}') \\ &= L_1(\mathbf{v}) + L_2(\mathbf{v}) + L_1(\mathbf{v}') + L_2(\mathbf{v}') \\ &= (L_1 + L_2)(\mathbf{v}) + (L_1 + L_2)(\mathbf{v}').\end{aligned}$$

$$\begin{aligned}(L_1 + L_2)(c\mathbf{v}) &= L_1(c\mathbf{v}) + L_2(c\mathbf{v}) \\ &= cL_1(\mathbf{v}) + cL_2(\mathbf{v}) \\ &= c\big(L_1(\mathbf{v}) + L_2(\mathbf{v})\big) \\ &= c(L_1 + L_2)(\mathbf{v}).\end{aligned}$$

Thus $L_1 + L_2$ is linear.

For $L \in \mathbf{V}^*$ and $c \in \mathbf{K}$, define cL by

$$(cL)(\mathbf{v}) = cL(\mathbf{v}).$$

The proof that cL is linear is left to the reader.

The zero functional $\mathbf{0} \in \mathbf{V}^*$, defined by $\mathbf{0}(\mathbf{v}) = \mathbf{0}$ for every $\mathbf{v} \in \mathbf{V}$ obviously satisfies

$$L + \mathbf{0} = L, \quad \forall L \in \mathbf{V}^*.$$

The task of checking that these operations do indeed make $\mathbf{V}^*$ into a $\mathbf{K}$-vector space is left to the reader. The vector space $\mathbf{V}^*$ is called *the dual vector space of* $\mathbf{V}$.

Suppose that $\mathbf{V}$ is finite dimensional, and let $\mathbf{e} = \{\mathbf{e}_1, \ldots, \mathbf{e}_n\}$ be a basis for $\mathbf{V}$. Let $1 \leq i \leq n$. By Theorem 11.3 there is a unique $\eta_i \in \mathbf{V}^*$ satisfying

$$\eta_i(\mathbf{e}_j) = \delta_{ij}. \tag{11.10}$$

(The Kronecker symbol δ_{ij} is defined in (2.3).) We thus obtain n linear functionals $\eta_1, \ldots, \eta_n$ that are the same as those considered in Example 11.1(4).

Theorem 11.12

Let $\mathbf{V}$ *be a finite dimensional* $\mathbf{K}$*-vector space, and let* $\mathbf{e} = \{\mathbf{e}_1, \ldots, \mathbf{e}_n\}$ *be a basis for* $\mathbf{V}$. *The set* $\{\eta_1, \ldots, \eta_n\}$ *of linear functionals defined in* (11.10) *is a basis for* $\mathbf{V}^*$; *in particular*

$$\dim(\mathbf{V}) = \dim(\mathbf{V}^*)$$

so $\mathbf{V}$ *and* $\mathbf{V}^*$ *are isomorphic.*

Proof

Let $L \in \mathbf{V}^*$, and $a_1 = L(\mathbf{e}_1), a_2 = L(\mathbf{e}_2), \dots, a_n = L(\mathbf{e}_n)$. The functional $a_1\eta_1 + a_2\eta_2 + \cdots + a_n\eta_n$ satisfies the identity

$$\begin{aligned}(a_1\eta_1 + \cdots + a_n\eta_n)(\mathbf{e}_i) &= (a_1\eta_1)(\mathbf{e}_i) + \cdots + (a_i\eta_i)(\mathbf{e}_i) + \cdots \\ &\quad +(a_n\eta_n)(\mathbf{e}_i) \\ &= a_1\eta_1(\mathbf{e}_i) + \cdots + a_i\eta_i(\mathbf{e}_i) + \cdots + a_n\eta_n(\mathbf{e}_i) \\ &= a_i\eta_i(\mathbf{e}_i) = a_i\end{aligned}$$

and so $a_1\eta_1 + \cdots + a_n\eta_n$ has the same values as L on the basis $\{\mathbf{e}_1, \dots, \mathbf{e}_n\}$. It follows from Theorem 11.3 that

$$a_1\eta_1 + \cdots + a_n\eta_n = L.$$

Thus $\eta_1, \dots, \eta_n$ generate $\mathbf{V}^*$. Suppose now that $a_1, \dots, a_n$ are such that

$$a_1\eta_1 + \cdots + a_n\eta_n = \mathbf{0}.$$

Then for every $i = 1, \dots, n$,

$$\begin{aligned} 0 &= \mathbf{0}(\mathbf{e}_i) = (a_1\eta_1 + \cdots + a_n\eta_n)(\mathbf{e}_i) \\ &= a_1\eta_1(\mathbf{e}_i) + \cdots + a_n\eta_n(\mathbf{e}_i) = a_i\eta_i(\mathbf{e}_i) = a_i.\end{aligned}$$

Thus $a_1 = \cdots = a_n = 0$, and so $\eta_1, \dots, \eta_n$ are linearly independent. □

The set of linear functionals $\{\eta_1, \dots, \eta_n\}$ is called the *dual basis* to the basis $\{\mathbf{e}_1, \dots, \mathbf{e}_n\}$ of $\mathbf{V}$.

If $L = a_1\eta_1 + \cdots + a_n\eta_n \in \mathbf{V}^*$ then

$$\begin{aligned} L(x_1\mathbf{e}_1 + \cdots + x_n\mathbf{e}_n) &= (a_1\eta_1 + \cdots + a_n\eta_n)(x_1\mathbf{e}_1 + \cdots + x_n\mathbf{e}_n) \\ &= (a_1\eta_1)(x_1\mathbf{e}_1 + \cdots + x_n\mathbf{e}_n) + \cdots \\ &\quad +(a_n\eta_n)(x_1\mathbf{e}_1 + \cdots + x_n\mathbf{e}_n) \\ &= a_1x_1 + \cdots + a_nx_n,\end{aligned}$$

since $\eta_i(x_1\mathbf{e}_1 + \cdots + x_n\mathbf{e}_n) = x_i$ for $i = 1, \dots, n$. Thus, *every linear functional on* $\mathbf{V}$ *can be expressed in a unique way as a homogeneous polynomial of degree 1 in the coordinates of* $\mathbf{v}$ *with respect to the basis* $\{\mathbf{e}_1, \dots, \mathbf{e}_n\}$. *The coefficients of the polynomial are the coordinates of the functional with respect to the dual basis* $\{\eta_1, \dots, \eta_n\}$.

In particular, for $\mathbf{V} = \mathbf{K}^n$ and $\{\mathbf{E}_1, \ldots, \mathbf{E}_n\}$ the canonical basis, every linear functional L on $\mathbf{K}^n$ can be written uniquely as a homogeneous polynomial of degree 1:

$$L(x_1, x_2, \ldots, x_n) = a_1x_1 + a_2x_2 + \cdots + a_nx_n. \tag{11.11}$$

Note that if a functional $L : \mathbf{V} \to \mathbf{K}$ is non-zero then $\mathrm{Im}(L) = \mathbf{K}$ since the image is a subspace of $\mathbf{K}$ which is not $\langle \mathbf{0} \rangle$. It then follows from Theorem 11.6 that $\dim(\ker(F)) = n - 1$.

Proposition 11.13
Let $\mathbf{V}$ be a $\mathbf{K}$-vector space and suppose that $f, g \in \mathbf{V}^*$ are such that $f(\mathbf{v}) = 0$ for every $\mathbf{v} \in \ker(g)$, that is, $\ker(g) \subset \ker(f)$. Then $f = cg$ for some $c \in \mathbf{K}$.

Proof
If $g = \mathbf{0}$ then $\ker(g) = \mathbf{V}$ and so $\ker(f) = \mathbf{V}$, whence $f = \mathbf{0} = g$. Conversely suppose that $g \neq \mathbf{0}$. In this case $\ker(g)$ is a hyperplane. Choose a vector $\mathbf{e} \in \mathbf{V} \setminus \ker(g)$ and let c satisfy $f(\mathbf{e}) = cg(\mathbf{e})$. Now, $\mathbf{V} = \langle \mathbf{e} \rangle \oplus \ker(g)$ and so every $\mathbf{v} \in \mathbf{V}$ is of the form $\mathbf{v} = x\mathbf{e} + \mathbf{w}$ with $\mathbf{w} \in \ker(g)$ and $x \in \mathbf{K}$. Comparing

$$f(\mathbf{v}) = f(x\mathbf{e} + \mathbf{w}) = f(x\mathbf{e}) + f(\mathbf{w}) = xf(\mathbf{e}) = xcg(\mathbf{e}),$$

and

$$cg(\mathbf{v}) = cg(x\mathbf{e} + \mathbf{w}) = cg(x\mathbf{e}) + cg(\mathbf{w}) = cxg(\mathbf{e})$$

shows that $f = cg$. □

Given two $\mathbf{K}$-vector spaces $\mathbf{V}$ and $\mathbf{W}$ the set $\mathrm{Hom}(\mathbf{V}, \mathbf{W})$ can be given the structure of a $\mathbf{K}$-vector space by defining $F + G$ and cF for $F, G \in \mathrm{Hom}(\mathbf{V}, \mathbf{W})$ and $c \in \mathbf{K}$, in the following manner:

$$\begin{aligned} (F + G)(\mathbf{v}) &= F(\mathbf{v}) + G(\mathbf{v}) \\ (cF)(\mathbf{v}) &= cF(\mathbf{v}) \end{aligned}$$

for every $\mathbf{v} \in \mathbf{V}$.

The proof of the fact that this does indeed define a $\mathbf{K}$-vector space structure, with zero being the zero map $\mathbf{0}$, is very similar to the proof for $\mathbf{V}^* = \mathrm{Hom}(\mathbf{V}, \mathbf{K})$ and the details are left to the reader.

In the particular case that $\mathbf{V} = \mathbf{K}$, the space $\mathrm{Hom}(\mathbf{K}, \mathbf{W})$ is isomorphic to $\mathbf{W}$. An isomorphism $\lambda : \mathbf{W} \to \mathrm{Hom}(\mathbf{K}, \mathbf{W})$ can be defined as follows: for each $\mathbf{w} \in \mathbf{W}$ let $\lambda(\mathbf{w}) \in \mathrm{Hom}(\mathbf{K}, \mathbf{W})$ be the linear map which satisfies $\lambda(\mathbf{w})(1) = \mathbf{w}$. It is left to the reader to check that λ is an isomorphism, it is called the *canonical isomorphism* of $\mathbf{W}$ with $\mathrm{Hom}(\mathbf{K}, \mathbf{W})$.

Complements 11.14

1. Let $\mathbf{V}$ be a vector space over $\mathbf{K}$ and $\mathbf{V}^*$ its dual. The dual of $\mathbf{V}^*$, that is the space $\mathrm{Hom}(\mathbf{V}^*, \mathbf{K})$, is called the *double dual* of $\mathbf{V}$ and denoted $\mathbf{V}^{**}$. If $\mathbf{V}$ is finite dimensional then by Theorem 11.12 we have that

$$\dim(\mathbf{V}^{**}) = \dim(\mathbf{V}^*) = \dim(\mathbf{V}),$$

and the three spaces $\mathbf{V}$, $\mathbf{V}^*$, and $\mathbf{V}^{**}$ are isomorphic.

If a basis $\{\mathbf{e}_1, \ldots, \mathbf{e}_n\}$ for $\mathbf{V}$ is chosen, then we have the dual basis $\{\eta_1, \ldots, \eta_n\}$ for $\mathbf{V}^*$, and, as described above, there is an isomorphism $\phi : \mathbf{V} \to \mathbf{V}^*$ defined by $\phi(\mathbf{e}_i) = \eta_i$ for $i = 1, \ldots, n$. This isomorphism depends on the choice of basis, in the sense that if a different basis of $\mathbf{V}$ was chosen initially then the resulting isomorphism would be different (see Exercise 11.6).

For the space $\mathbf{V}^{**}$ the situation is different: it is possible to define an isomorphism $\beta : \mathbf{V} \to \mathbf{V}^{**}$ in a totally intrinsic manner; that is, β does not depend on any choice of basis. For this reason, it is called the *canonical isomorphism of* $\mathbf{V}$ *and* $\mathbf{V}^{**}$.

This canonical isomorphism β is defined by associating to any $\mathbf{v} \in \mathbf{V}$ the functional $\beta(\mathbf{v}) : \mathbf{V}^* \to \mathbf{K}$ given by, for any $L \in \mathbf{V}^*$

$$\beta(\mathbf{v})(L) = L(\mathbf{v}).$$

Let us check that $\beta(\mathbf{v})$ is linear:

$$\begin{aligned} \beta(\mathbf{v})(cL + c'L') &= (cL + c'L')(\mathbf{v}) \\ &= (cL)(\mathbf{v}) + (c'L')(\mathbf{v}) \\ &= cL(\mathbf{v}) + c'L'(\mathbf{v}) \\ &= c\beta(\mathbf{v})(L) + c'\beta(\mathbf{v})(L'), \end{aligned}$$

for every $L, L' \in \mathbf{V}^*$ and $c, c' \in \mathbf{K}$. Thus $\beta(\mathbf{v})$ is a linear functional on $\mathbf{V}^*$.

We now show that $\beta : \mathbf{V} \to \mathbf{V}^{**}$ is linear. For every $\mathbf{v}, \mathbf{v}' \in \mathbf{V}$ and every $c, c' \in \mathbf{K}$ we have

$$\begin{aligned} \beta(c\mathbf{v} + c'\mathbf{v}')(L) &= L(c\mathbf{v} + c'\mathbf{v}') \\ &= cL(\mathbf{v}) + c'L(\mathbf{v}') \\ &= c\beta(\mathbf{v})(L) + c'\beta(\mathbf{v}')(L) \\ &= \big(c\beta(\mathbf{v}) + c'\beta(\mathbf{v}')\big)(L), \end{aligned}$$

for every $L \in \mathbf{V}^*$, and so $\beta(c\mathbf{v} + c'\mathbf{v}') = c\beta(\mathbf{v}) + c'\beta(\mathbf{v}')$, that is β is linear.

To show that β is an isomorphism it is enough to show that $\ker(\beta) = \langle \mathbf{0} \rangle$, since $\dim(\mathbf{V}) = \dim(\mathbf{V}^{**})$. Suppose, on the contrary, that there is a $\mathbf{v} \in \mathbf{V}$, $\mathbf{v} \neq \mathbf{0}$, such that $\beta(\mathbf{v}) = \mathbf{0} \in \mathbf{V}^{**}$. Then

$$\beta(\mathbf{v})(L) = L(\mathbf{v}) = \mathbf{0}$$

for all $L \in \mathbf{V}^*$. Let $\mathbf{v}_2, \dots, \mathbf{v}_n \in \mathbf{V}$ be such that $\{\mathbf{v}, \mathbf{v}_2, \dots, \mathbf{v}_n\}$ is a basis for $\mathbf{V}$. The functional $L \in \mathbf{V}^*$ defined by

$$L(\mathbf{v}) = 1, \quad L(\mathbf{v}_i) = 0, \quad i = 2, 3, \dots, n,$$

gives a contradiction.

2. Let $\mathbf{V}$ be a vector space over $\mathbf{K}$ of dimension n. For $L \in \mathbf{V}^*$, the vectors in the kernel of L are sometimes said to be *orthogonal to* L. If Φ is a subset of $\mathbf{V}^*$, a vector $\mathbf{v} \in \mathbf{V}$ is said to be *orthogonal to* Φ if $\mathbf{v}$ is orthogonal to every $L \in \Phi$, that is if $L(\mathbf{v}) = 0$ for all $L \in \Phi$. The set

$$\Phi^\perp = \{\mathbf{v} \in \mathbf{V} \mid \mathbf{v} \text{ is orthogonal to } \Phi\} = \bigcap_{L \in \Phi} \ker(L) \subset \mathbf{V}$$

is a vetor subspace of $\mathbf{V}$ as it is the intersection of a family of vector subspaces of $\mathbf{V}$. We will call $\Phi^\perp$ the *subspace of* $\mathbf{V}$ *orthogonal to* Φ.

Suppose that Φ is a vector subspace of $\mathbf{V}^*$ of dimension t and let $\{L_1, \dots, L_t\}$ be a basis for Φ. Then

$$\Phi^\perp = \ker(L_1) \cap \dots \cap \ker(L_t) \tag{11.12}$$

and

$$\dim(\Phi^\perp) = n - t. \tag{11.13}$$

It is obvious that $\Phi^\perp \subset \ker(L_1) \cap \ldots \cap \ker(L_t)$. To see the converse, let $\mathbf{v} \in \ker(L_1) \cap \ldots \cap \ker(L_t)$. Then for every $L = a_1L_1 + \cdots + a_tL_t$ one has

$$L(\mathbf{v}) = a_1L_1(\mathbf{v}) + \cdots + a_tL_t(\mathbf{v}) = 0,$$

that is, $\mathbf{v} \in \ker(L)$. Thus $\ker(L_1) \cap \ldots \cap \ker(L_t) \subset \Phi^\perp$ and (11.12) follows.

Let $m = \dim(\Phi^\perp)$. Observe that $\Phi^\perp$ coincides with the kernel of the map

$$\begin{aligned} \phi : \mathbf{V} &\rightarrow \mathbf{K}^t \\ \mathbf{v} &\mapsto (L_1(\mathbf{v}), \ldots, L_t(\mathbf{v})) \end{aligned}$$

It follows from Theorem 11.6 that $m \geq n - t$. Suppose, for a contradiction, that $m > n - t$. Choose a basis $\{\mathbf{e}_1, \ldots, \mathbf{e}_n\}$ of $\mathbf{V}$ such that $\{\mathbf{e}_1, \ldots, \mathbf{e}_m\}$ is a basis for $\Phi^\perp$. The $t \times n$ matrix

$$\begin{pmatrix} 0 & \cdots & 0 & L_1(\mathbf{e}_{m+1}) & \cdots & L_1(\mathbf{e}_n) \\ 0 & \cdots & 0 & L_2(\mathbf{e}_{m+1}) & \cdots & L_2(\mathbf{e}_n) \\ \vdots & & \vdots & \vdots & & \vdots \\ 0 & \cdots & 0 & L_t(\mathbf{e}_{m+1}) & \cdots & L_t(\mathbf{e}_n) \end{pmatrix}$$

has rank at most $t - 1$, since it has $m > n - t$ columns which are zero. On the other hand, its t rows are linearly independent, because they are the vectors with the coordinates of $L_1, \ldots, L_t$ with respect to the basis $\{\eta_1, \ldots, \eta_n\}$, dual to $\{\mathbf{e}_1, \ldots, \mathbf{e}_n\}$. This is a contradiction; consequently $m = n - t$.

3. Let $\mathbf{V}$ and $\mathbf{W}$ be two complex vector spaces. A map $F : \mathbf{V} \rightarrow \mathbf{W}$ is said to be *antilinear* if it satisfies:

$$F(\mathbf{v} + \mathbf{v}') = F(\mathbf{v}) + F(\mathbf{v}'), \tag{11.14}$$

$$F(c\mathbf{v}) = \bar{c}F(\mathbf{v}), \tag{11.15}$$

for every $\mathbf{v}, \mathbf{v}' \in \mathbf{V}$ and every $c \in \mathbf{C}$. Equivalently, it should satisfy

$$F(c\mathbf{v} + c'\mathbf{v}') = \bar{c}F(\mathbf{v}) + \bar{c}'F(\mathbf{v}')$$

for every $\mathbf{v}, \mathbf{v}' \in \mathbf{V}$ and every $c, c' \in \mathbf{C}$. Showing that these conditions are equivalent is left to the reader.

EXERCISES

11.1 Let $g : \mathbf{U} \to \mathbf{V}$, $f : \mathbf{V} \to \mathbf{W}$ be linear maps. Prove that

$$\ker(g) \subset \ker(f \circ g)$$

$$\mathrm{Im}(f) \supset \mathrm{Im}(f \circ g).$$

11.2 Prove that, if $F : \mathbf{V} \to \mathbf{W}$ is a linear map with $\ker(F) = \langle \mathbf{0} \rangle$, and $\mathbf{v}_1, \ldots, \mathbf{v}_n$ are linearly independent, then $F(\mathbf{v}_1), \ldots, F(\mathbf{v}_n)$ are linearly independent.

11.3 Let $H \subset \mathbf{R}^3$ be the plane with equation $X_1 + X_2 + X_3 = 0$, and let $\mathbf{u} = (0, 1, 1)$. First show that $\mathbf{R}^3 = H \oplus \langle \mathbf{u} \rangle$, and then find the analytic expression for the projection $p : \mathbf{R}^3 \to H$ in the direction $\langle \mathbf{u} \rangle$.

11.4 Let $\mathbf{W}$ be the subspace of $\mathbf{R}^4$ given by

$$2X_1 + X_3 = 0, \quad X_2 - 3X_4 = 0,$$

and let $\mathbf{U} = \langle (1,0,0,0), (0,1,0,0) \rangle$. First show that $\mathbf{R}^4 = \mathbf{U} \oplus \mathbf{W}$, and then find the analytic expression for the projection $p : \mathbf{R}^4 \to \mathbf{W}$ defined by the above direct sum decomposition of $\mathbf{R}^4$.

11.5 Expressing linear functionals on $\mathbf{R}^3$ as homogeneous polynomials in X_1, X_2, X_3, with real coefficients, find the bases of $(\mathbf{R}^3)^*$ dual to each of the following bases of $\mathbf{R}^3$.

a) $\{(1,0,0)\,(0,1,0),\ (0,0,1)\}$
b) $\{(2,0,0),\ (0,1/\sqrt{2},0),\ (0,0,-1/6)\}$
c) $\{(1,-1,0),\ (0,1,1)\,(1,0,2)\}$
d) $\{(1,0,0),\ (1,1,0),\ (1,1,1)\}$.

11.6 Let $\mathbf{V}$ be a $\mathbf{K}$-vector space of dimension 1 and let $\mathbf{e} \in \mathbf{V} \setminus \{\mathbf{0}\}$. Let $\eta \in \mathbf{V}^*$ be the functional dual to $\mathbf{e}$, (i.e. defined by the condition $\eta(\mathbf{e}) = 1$). Prove that for every $a \in \mathbf{K}^*$:
a) the linear functional dual to $a\mathbf{e}$ is $\zeta = a^{-1}\eta$;
b) if $\phi, \psi : \mathbf{V} \to \mathbf{V}^*$ are the isomorphisms defined by $\phi(\mathbf{e}) = \eta$ and $\psi(a\mathbf{e}) = \zeta$, then $\psi = a^{-2}\phi$.

12

Linear maps and matrices; affine changes of coordinates

Let $\mathbf{V}$ and $\mathbf{W}$ be two finite dimensional $\mathbf{K}$-vector spaces, with bases $\boldsymbol{v} = \{\mathbf{v}_1, \ldots, \mathbf{v}_n\}$ and $\boldsymbol{w} = \{\mathbf{w}_1, \ldots, \mathbf{w}_m\}$ respectively.

Let $F : \mathbf{V} \to \mathbf{W}$ be a linear map. The $m \times n$ matrix whose j-th column ($j = 1, \ldots, n$) consists of the coordinates of the vector $F(\mathbf{v}_j) \in \mathbf{W}$ with respect to the basis $\boldsymbol{w}$ is called *the matrix of F with respect to the the bases $\boldsymbol{v}$ and $\boldsymbol{w}$*, and is denoted $M_{\boldsymbol{w}, \boldsymbol{v}}(F)$. Explicitly,

$$M_{\boldsymbol{w}, \boldsymbol{v}}(F) = \begin{pmatrix} m_{11} & m_{12} & \cdots & m_{1n} \\ m_{21} & m_{22} & \cdots & m_{2n} \\ \vdots & \vdots & & \vdots \\ m_{m1} & m_{m2} & \cdots & m_{mn} \end{pmatrix},$$

where

$$F(\mathbf{v}_j) = m_{1j}\mathbf{w}_1 + m_{2j}\mathbf{w}_2 + \cdots + m_{mj}\mathbf{w}_m.$$

Obviously, $M_{\boldsymbol{w}, \boldsymbol{v}}(F)$ depends not only on F but also on the bases $\boldsymbol{v}$ and $\boldsymbol{w}$. The importance of the matrix $M_{\boldsymbol{w}, \boldsymbol{v}}(F)$ lies in the fact that, once the bases $\boldsymbol{v}$ and $\boldsymbol{w}$ are fixed, the map can be reconstructed from the matrix, as shown in the following proposition.

Proposition 12.1
Let $\mathbf{V}$ and $\mathbf{W}$ be two K-vector spaces with bases $\boldsymbol{v} = \{\mathbf{v}_1, \ldots, \mathbf{v}_n\}$ and $\boldsymbol{w} = \{\mathbf{w}_1, \ldots, \mathbf{w}_m\}$ respectively, and let $F : \mathbf{V} \to \mathbf{W}$ be a linear map. For every $\mathbf{v} = x_1\mathbf{v}_1 + \cdots + x_n\mathbf{v}_n \in \mathbf{V}$ one has

$$F(\mathbf{v}) = y_1\mathbf{w}_1 + \cdots + y_m\mathbf{w}_m$$

where

$$\begin{pmatrix} y_1 \\ \vdots \\ y_n \end{pmatrix} = M_{\boldsymbol{w},\boldsymbol{v}}(F) \begin{pmatrix} x_1 \\ \vdots \\ x_n \end{pmatrix}.$$

Proof

One has,

$$\begin{aligned} F(\mathbf{v}) &= F(x_1\mathbf{v}_1 + \cdots + x_n\mathbf{v}_n) \\ &= x_1F(\mathbf{v}_1) + \cdots + x_nF(\mathbf{v}_n) \\ &= x_1(m_{11}\mathbf{w}_1 + \cdots + m_{m1}\mathbf{w}_n) + \cdots \\ &\qquad + x_n(m_{1n}\mathbf{w}_1 + \cdots + m_{mn}\mathbf{w}_n) \\ &= (\sum_{j=1}^{n} m_{1j}x_j)\mathbf{w}_1 + \cdots + (\sum_{j=1}^{n} m_{mj}x_j)\mathbf{w}_m. \end{aligned}$$

The column vector with the coordinates of $F(\mathbf{v})$ is therefore

$$\begin{pmatrix} y_1 \\ \vdots \\ y_n \end{pmatrix} = \begin{pmatrix} \sum_{j=1}^{n} m_{1j}x_j \\ \vdots \\ \sum_{j=1}^{n} m_{mj}x_j \end{pmatrix} = M_{\boldsymbol{w},\boldsymbol{v}}(F) \begin{pmatrix} x_1 \\ \vdots \\ x_n \end{pmatrix}.$$

□

Theorem 12.2

Let $\mathbf{V}$ *and* $\mathbf{W}$ *be two* K*-vector spaces with bases* $\boldsymbol{v} = \{\mathbf{v}_1, \ldots, \mathbf{v}_n\}$ *and* $\boldsymbol{w} = \{\mathbf{w}_1, \ldots, \mathbf{w}_m\}$ *respectively. The map*

$$\begin{aligned} M_{\boldsymbol{w},\boldsymbol{v}} : \mathrm{Hom}(\mathbf{V}, \mathbf{W}) &\to M_{m,n}(\mathbf{K}) \\ F &\mapsto M_{\boldsymbol{w},\boldsymbol{v}}(F) \end{aligned}$$

is an isomorphism of $\mathbf{K}$*-vector spaces. In particular,*

$$\dim(\mathrm{Hom}(\mathbf{V}, \mathbf{W})) = mn.$$

Proof

Let $F, G \in \mathrm{Hom}(\mathbf{V}, \mathbf{W})$, $M = M_{\boldsymbol{w},\boldsymbol{v}}(F)$, $N = M_{\boldsymbol{w},\boldsymbol{v}}(G)$, and $c \in \mathbf{K}$.

If $\mathbf{v} = x_1\mathbf{v}_1 + \cdots + x_n\mathbf{v}_n$, let $\mathbf{x} = (x_1 \ \ldots \ x_n)^t$, the column n-vector with the coordinates of $\mathbf{v}$ with respect to the basis $\boldsymbol{v}$. The column m-vectors with the coordinates of $F(\mathbf{v})$ and $G(\mathbf{v})$ with respect to $\boldsymbol{w}$ are

$M\mathbf{x}$ and $N\mathbf{x}$ respectively, while the column vector with the coordinates of $(F+G)(\mathbf{v}) = F(\mathbf{v}) + G(\mathbf{v})$ is $M\mathbf{x} + N\mathbf{x}$. Thus,

$$M_{\boldsymbol{w},\boldsymbol{v}}(F+G) = M_{\boldsymbol{w},\boldsymbol{v}}(F) + M_{\boldsymbol{w},\boldsymbol{v}}(G).$$

Furthermore, the column m-vector with the coordinates of $(cF)(\mathbf{v}) = cF(\mathbf{v})$ is $c(M\mathbf{x}) = cM\mathbf{x}$, and so

$$M_{\boldsymbol{w},\boldsymbol{v}}(cF) = cM_{\boldsymbol{w},\boldsymbol{v}}(F).$$

Therefore $M_{\boldsymbol{w},\boldsymbol{v}}$ is a linear map.

Now let $A \in M_{m,n}(\mathbf{K})$, and define $F_A : \mathbf{V} \to \mathbf{W}$ by

$$F_A(x_1\mathbf{v}_1 + \cdots + x_n\mathbf{v}_n) = (A^{(1)}\mathbf{x})\mathbf{w}_1 + \cdots + (A^{(m)}\mathbf{x})\mathbf{w}_m,$$

where $\mathbf{x} = (x_1 \ldots x_n)^t$ and $A^{(1)}, \ldots A^{(m)}$ are the rows of A. In other words, $A\mathbf{x}$ is the column vector with the coordinates of $F_A(\mathbf{v})$.

First we check that F_A is a linear map. Consider two arbitrary vectors $\mathbf{v} = x_1\mathbf{v}_1 + \cdots + x_n\mathbf{v}_n$, and $\mathbf{v}' = x_1'\mathbf{v}_1 + \cdots + x_n'\mathbf{v}_n$ in $\mathbf{V}$. Then

$$\begin{aligned}
F_A(\mathbf{v}+\mathbf{v}') &= A^{(1)}(\mathbf{x}+\mathbf{x}')\mathbf{w}_1 + \cdots + A^{(m)}(\mathbf{x}+\mathbf{x}')\mathbf{w}_m \\
&= (A^{(1)}\mathbf{x} + A^{(1)}\mathbf{x}')\mathbf{w}_1 + \cdots + (A^{(m)}\mathbf{x} + A^{(m)}\mathbf{x}')\mathbf{w}_m \\
&= (A^{(1)}\mathbf{x})\mathbf{w}_1 + \cdots + (A^{(m)}\mathbf{x})\mathbf{w}_m + \\
&\qquad (A^{(1)}\mathbf{x}')\mathbf{w}_1 + \cdots + (A^{(m)}\mathbf{x}')\mathbf{w}_m \\
&= F_A(\mathbf{v}) + F_A(\mathbf{v}').
\end{aligned}$$

And for $c \in \mathbf{K}$,

$$\begin{aligned}
F_A(c\mathbf{v}) &= (A^{(1)}c\mathbf{x})\mathbf{w}_1 + \cdots + (A^{(m)}c\mathbf{x})\mathbf{w}_m \\
&= c(A^{(1)}\mathbf{x})\mathbf{w}_1 + \cdots + c(A^{(m)}\mathbf{x})\mathbf{w}_m \\
&= c[(A^{(1)}\mathbf{x})\mathbf{w}_1 + \cdots + (A^{(m)}\mathbf{x})\mathbf{w}_m] \\
&= cF_A(\mathbf{v}).
\end{aligned}$$

Thus F_A is linear.

By definition,

$$M_{\boldsymbol{w},\boldsymbol{v}}(F_A) = A.$$

On the other hand, if $A = M_{\boldsymbol{w},\boldsymbol{v}}(F)$ it is obvious that $F_A = F$. Thus the map

$$\begin{aligned}
M_{m,n}(\mathbf{K}) &\to \mathrm{Hom}(\mathbf{V},\mathbf{W}) \\
A &\mapsto F_A
\end{aligned}$$

is the inverse of $M_{\boldsymbol{w},\boldsymbol{v}}$. The last assertion of the theorem follows from Example 4.15(7). □

The map F_A introduced during the proof of the preceding theorem is called the *linear map associated to the matrix A with respect to the bases* $\boldsymbol{v}$ *and* $\boldsymbol{w}$. It is defined, for each $A \in M_{m,n}(\mathbf{K})$, by

$$F_A(x_1\mathbf{v}_1 + \cdots + x_n\mathbf{v}_n) = y_1\mathbf{w}_1 + \cdots + y_m\mathbf{w}_m,$$

where $\mathbf{y} = A\mathbf{x}$, with $\mathbf{x} = (x_1 \ldots x_n)^t$, $\mathbf{y} = (y_1 \ldots y_m)^t$.

Note that $M_{\boldsymbol{w},\boldsymbol{v}}$ maps the linear map $\mathbf{0}$ to the zero $m \times n$ matrix. From the definition, it follows moreover that the coordinates of the vector $F_A(\mathbf{v})$ are homogeneous polynomials of degree 1 in the coordinates of $\mathbf{v}$.

Proposition 12.3

Let $\mathbf{U}, \mathbf{V}, \mathbf{W}$ *be* $\mathbf{K}$*-vector spaces of dimensions* s, n *and* m, *and let* $\boldsymbol{u} = \{\mathbf{u}_1, \ldots, \mathbf{u}_s\}$, $\boldsymbol{v} = \{\mathbf{v}_1, \ldots, \mathbf{v}_n\}$ *and* $\boldsymbol{w} = \{\mathbf{w}_1, \ldots, \mathbf{w}_m\}$ *their respective bases. Let* $G : \mathbf{U} \to \mathbf{V}$ *and* $F : \mathbf{V} \to \mathbf{W}$ *be linear maps. Then*

$$M_{\boldsymbol{w},\boldsymbol{v}}(F \circ G) = M_{\boldsymbol{w},\boldsymbol{v}}(F)M_{\boldsymbol{v},\boldsymbol{u}}(G).$$

Proof

For $\mathbf{u} = z_1\mathbf{u}_1 + \cdots + z_s\mathbf{u}_s \in \mathbf{U}$, let

$$\begin{aligned} G(\mathbf{u}) &= x_1\mathbf{v}_1 + \cdots + x_n\mathbf{v}_n \\ F(G(\mathbf{u})) = (F \circ G)(\mathbf{u}) &= y_1\mathbf{w}_1 + \cdots + y_m\mathbf{w}_m. \end{aligned}$$

Putting $\mathbf{z} = (z_1\, z_2 \ldots z_s)^t$, $\mathbf{x} = (x_1\, x_2 \ldots x_n)^t$, and $\mathbf{y} = (y_1\, y_2 \ldots y_m)^t$ gives

$$\begin{aligned} \mathbf{x} &= M_{\boldsymbol{v},\boldsymbol{u}}(G)(\mathbf{z}) \\ \mathbf{y} &= M_{\boldsymbol{w},\boldsymbol{v}}(F)\mathbf{x} = M_{\boldsymbol{w},\boldsymbol{v}}(F)[M_{\boldsymbol{v},\boldsymbol{u}}(G)\mathbf{z}] \\ &= [M_{\boldsymbol{w},\boldsymbol{v}}(F)M_{\boldsymbol{v},\boldsymbol{u}}(G)]\mathbf{z}. \end{aligned}$$

□

If $\mathbf{V} = \mathbf{W}$ and if $\boldsymbol{v} = \{\mathbf{v}_1, \ldots, \mathbf{v}_n\}$ and $\boldsymbol{w} = \{\mathbf{w}_1, \ldots, \mathbf{w}_n\}$ are two bases of $\mathbf{V}$, then to each linear operator F on $\mathbf{V}$ there is associated a square matrix $M_{\boldsymbol{w},\boldsymbol{v}}(F) \in M_n(\mathbf{K})$.

If $\boldsymbol{v} = \boldsymbol{w}$ it follows from Proposition 12.3 that $F \in \mathrm{GL}(\mathbf{V})$ if and only if $M_{\boldsymbol{v},\boldsymbol{v}} \in \mathrm{GL}_n(\mathbf{K})$, in which case one has

$$M_{\boldsymbol{v},\boldsymbol{v}}(F^{-1}) = M_{\boldsymbol{v},\boldsymbol{v}}(F)^{-1}.$$

Furthermore, $M_{\boldsymbol{v},\boldsymbol{v}}(F) = \mathbf{I}_n$ if and only if $F = \mathbf{1}_{\mathbf{V}}$.

A particularly important case arises when $\boldsymbol{v}$ and $\boldsymbol{w}$ are two distinct bases of $\mathbf{V}$ and $F = \mathbf{1}_{\mathbf{V}}$, the identity map. In this case, $M_{\boldsymbol{w},\boldsymbol{v}}(\mathbf{1}_{\mathbf{V}})$ is called the *change of basis matrix from the basis* $\boldsymbol{v}$ *to the basis* $\boldsymbol{w}$.

By definition, the j-th column of $M_{\boldsymbol{w},\boldsymbol{v}}(\mathbf{1}_{\mathbf{V}})$ consists of the coordinates of $\mathbf{v}_j$ with respect to the basis $\boldsymbol{w}$. For every vector $\mathbf{v} \in \mathbf{V}$ one has

$$\mathbf{v} = x_1\mathbf{v}_1 + \cdots + x_n\mathbf{v}_n = y_1\mathbf{w}_1 + \cdots + y_n\mathbf{w}_n$$

and, putting $\mathbf{x} = (x_1 \ldots x_n)^t$, $\mathbf{y} = (y_1 \ldots y_n)^t$, one has

$$\mathbf{y} = M_{\boldsymbol{w},\boldsymbol{v}}(\mathbf{1}_{\mathbf{V}})\mathbf{x}.$$

Thus, from the matrix $M_{\boldsymbol{w},\boldsymbol{v}}(\mathbf{1}_{\mathbf{V}})$ one can find the coordinates $\mathbf{y}$ of any vector $\mathbf{v}$ with respect to the basis $\boldsymbol{w}$ given the coordinates $\mathbf{x}$ of $\mathbf{v}$ with respect to the base $\boldsymbol{v}$.

Note that by Proposition 12.3,

$$M_{\boldsymbol{v},\boldsymbol{w}}(\mathbf{1}_{\mathbf{V}})M_{\boldsymbol{w},\boldsymbol{v}}(\mathbf{1}_{\mathbf{V}}) = M_{\boldsymbol{v},\boldsymbol{v}}(\mathbf{1}_{\mathbf{V}}) = \mathbf{I}_n$$

and so

$$M_{\boldsymbol{w},\boldsymbol{v}}(\mathbf{1}_{\mathbf{V}}) = M_{\boldsymbol{w},\boldsymbol{v}}(\mathbf{1}_{\mathbf{V}})^{-1}. \tag{12.1}$$

Suppose now that $\mathbf{V}$ is a *real* vector space. Two bases

$$\boldsymbol{e} = \{\mathbf{e}_1, \ldots, \mathbf{e}_n\} \quad \text{and} \quad \boldsymbol{f} = \{\mathbf{f}_1, \ldots, \mathbf{f}_n\}$$

of $\mathbf{V}$ are said to *have the same orienatation* if $\det(M_{\boldsymbol{e},\boldsymbol{f}}(\mathbf{1}_{\mathbf{V}})) > 0$, and one writes $\boldsymbol{e} \sim_{\mathrm{or}} \boldsymbol{f}$. Otherwise the bases have *different orientations.*

Clearly, every base has the same orientation as itself since

$$M_{\boldsymbol{e},\boldsymbol{e}}(\mathbf{1}_{\mathbf{V}}) = \mathbf{I}_n.$$

Moreover, since $M_{\boldsymbol{f},\boldsymbol{e}}(\mathbf{1}_{\mathbf{V}}) = M_{\boldsymbol{e},\boldsymbol{f}}(\mathbf{1}_{\mathbf{V}})^{-1}$, whether two bases have the same or different orientation does not depend on the order in which they are taken. Finally, if $\boldsymbol{e}$ and $\boldsymbol{f}$ have the same orientation and $\boldsymbol{f}$

and $\boldsymbol{g} = \{\mathbf{g}_1, \ldots, \mathbf{g}_n\}$ have the same orientation, then $\boldsymbol{e}$ and $\boldsymbol{g}$ have the same orientation. For,

$$\begin{aligned}\det(M_{\boldsymbol{e},\boldsymbol{g}}(\mathbf{1_V})) &= \det(M_{\boldsymbol{e},\boldsymbol{f}}(\mathbf{1_V})M_{\boldsymbol{f},\boldsymbol{g}}(\mathbf{1_V})) \\ &= \det(M_{\boldsymbol{e},\boldsymbol{f}}(\mathbf{1_V}))\det(M_{\boldsymbol{f},\boldsymbol{g}}(\mathbf{1_V})) > 0.\end{aligned}$$

We see therefore that $\sim_{\text{or}}$ is an equivalence relation on the set $\mathcal{B}$ of all bases of $\mathbf{V}$. Each equivalence class is called an orientation of $\mathbf{V}$.

How many orientations are there of $\mathbf{V}$? Clearly there are at most two, since if there were three distinct bases $\boldsymbol{e}, \boldsymbol{f}$ and $\boldsymbol{g}$ defining three different orientations one could deduce the absurd statement:

$$0 > \det(M_{\boldsymbol{e},\boldsymbol{g}}(\mathbf{1_V})) = \det(M_{\boldsymbol{e},\boldsymbol{f}}(\mathbf{1_V}))\det(M_{\boldsymbol{f},\boldsymbol{g}}(\mathbf{1_V})) > 0.$$

On the other hand, if $\boldsymbol{e} = \{\mathbf{e}_1, \ldots, \mathbf{e}_n\}$ is a basis then $\boldsymbol{e}$ and the basis $\boldsymbol{f} = \{-\mathbf{e}_1, \ldots, \mathbf{e}_n\}$ have different orientations since

$$M_{\boldsymbol{e},\boldsymbol{f}}(\mathbf{1_V}) = \begin{pmatrix} -1 & 0 & \cdots & 0 \\ 0 & 1 & \cdots & 0 \\ \vdots & \vdots & & \vdots \\ 0 & 0 & \cdots & 1 \end{pmatrix}.$$

Thus any real vector space $\mathbf{V}$ has *exactly two orientations*, that is the set $\mathcal{B}$ consists of two equivalence classes for the relation $\sim_{\text{or}}$. The orientation of $\mathbf{V}$ to which $\boldsymbol{e} \in \mathcal{B}$ belongs is called the *orientation of* $\mathbf{V}$ *defined by* $\boldsymbol{e}$.

Examples 12.4

1. If $\mathbf{V} = \mathbf{K}^n, \mathbf{W} = \mathbf{K}^m$ and $\boldsymbol{v}$ and $\boldsymbol{w}$ are the canonical bases of $\mathbf{K}^n$ and $\mathbf{K}^m$ respectively, then the linear map F_A associated to a matrix $A \in M_{m,n}(\mathbf{K})$ is given by

$$F_A(\mathbf{x}) = A\mathbf{x} \qquad \text{for every } x \in \mathbf{K}^n,$$

where the elements of $\mathbf{K}^n$ and $\mathbf{K}^m$ are viewed as column vectors. Since the columns of A are the vectors $F_A(\mathbf{E}_1), F_A(\mathbf{E}_2), \ldots, F_A(\mathbf{E}_n)$, one has

$$\text{Im}(F_A) = \langle F_A(\mathbf{E}_1), \ldots, F_A(\mathbf{E}_n)\rangle.$$

In particular $\text{rk}(F_A) = \text{rk}(A)$.

Note that the coordinates of the vector $A\mathbf{x}$ are m homogeneous linear polynomials in $x_1, \ldots, x_n$. Conversely, any linear map $F : \mathbf{K}^n \to \mathbf{K}^m$ is of the form

$$F(x_1, \ldots, x_n) = (F_1(x_1, \ldots, x_n), \ldots, F_n(x_1, \ldots, x_n))$$

in which each $F_j(x_1, \ldots, x_n)$ is a linear functional on $\mathbf{K}^n$: indeed, F_j is equal to the composite

$$\eta_j \circ F : \mathbf{K}^n \to \mathbf{K}^m \to \mathbf{K},$$

which is linear. By (11.11), we see that each of the $F_j(x_1, \ldots, x_n)$ is a homogeneous linear polynomial in $x_1, \ldots, x_n$.

Thus, *every linear map $F : \mathbf{K}^n \to \mathbf{K}^m$ is determined by m homogeneous linear polynomials in $x_1, \ldots, x_n$, and the rows of the corresponding matrix are the coefficients in*

$$F_1(x_1, \ldots, x_n), F_2(x_1, \ldots, x_n), \ldots, F_m(x_1, \ldots, x_n).$$

For example, the matrix

$$A = \begin{pmatrix} 1 & 2 & \sqrt{2} \\ 3 & 1 & -1/2 \end{pmatrix} \in M_{2,3}(\mathbf{R})$$

is associated to the linear map $F_A : \mathbf{R}^3 \to \mathbf{R}^2$ defined by

$$F_A(x_1, x_2, x_3) = (x_1 + 2x_2 + \sqrt{2}x_3,\ 3x_1 + x_2 - \tfrac{1}{2}x_3).$$

On the other hand, the linear map $F_A : \mathbf{R}^4 \to \mathbf{R}^3$ defined by

$$F_A(x_1, x_2, x_3, x_4) =$$
$$(2x_1 - x_3 + x_4,\ x_2 - \sqrt{3}x_3 + \tfrac{3}{2}x_4,\ x_1 - x_2 + x_3 + 5x_4)$$

has associated matrix

$$\begin{pmatrix} 2 & 0 & -1 & 1 \\ 0 & 1 & -\sqrt{3} & 3/2 \\ 1 & -1 & 1 & 5 \end{pmatrix}.$$

2. Let $A \in M_{m,n}(\mathbf{K})$ and $\mathbf{b} = (b_1 \ldots b_m)^t \in \mathbf{K}^m$, and consider the system of m equations in n unknowns

$$A\mathbf{X} = \mathbf{b} \tag{12.2}$$

where $\mathbf{X} = (X_1 \dots X_n)^t$. A vector $\mathbf{x} = (x_1 \dots x_n)^t \in \mathbf{K}^n$ is a solution of (12.2) if and only if $F_A(\mathbf{x}) = \mathbf{b}$, where $F_A : \mathbf{K}^n \to \mathbf{K}^m$ is the linear map associated to A. For such an $\mathbf{x}$ to exist it is necessary and sufficient that $\mathbf{b} \in \text{Im}(F_A)$. On the other hand, since $\text{Im}(F_A)$ is generated by the columns of A, in order that $\mathbf{b} \in \text{Im}(F_A)$, that is for system (12.2) to be compatible, it is necessary and sufficient that $\text{rk}(A) = \text{rk}(A\,\mathbf{b})$. This argument gives a second proof of the theorem of Kronecker-Rouché-Capelli.

If system (12.2) is compatible, we know that the space of solutions has dimension $n - r$, where $r = \text{rk}(A)$. This can be proved by noting that the space of solutions of system (12.2) has the same dimension as the space of solutions of the associated homogeneous system

$$A\mathbf{X} = \mathbf{0},$$

and that this space is just $\ker(F_A)$. By Theorem 11.6, we have that

$$\dim(\ker(F_A)) = n - \text{rk}(A) = n - r.$$

3. Let $\mathbf{V}$ be a real vector space of dimension 3, and let $\boldsymbol{e} = \{\mathbf{e}_1, \mathbf{e}_2, \mathbf{e}_3\}$ be a basis of $\mathbf{V}$. Consider the following bases, whose vectors are given in coordinates with respect to $\boldsymbol{e}$:

$$\begin{aligned} \boldsymbol{v} &= \{\mathbf{v}_1(1,1,0),\ \mathbf{v}_2(2,1,1),\ \mathbf{v}_3(0,-2,1)\} \\ \boldsymbol{w} &= \{\mathbf{w}_1(-1,0,1),\ \mathbf{w}_2(1,-2,-3),\ \mathbf{w}_3(1,1,1)\}. \end{aligned}$$

Then

$$\begin{aligned} M_{\boldsymbol{e},\boldsymbol{v}}(\mathbf{1_V}) &= \begin{pmatrix} 1 & 2 & 0 \\ 1 & 1 & -2 \\ 0 & 1 & 1 \end{pmatrix} \\ M_{\boldsymbol{v},\boldsymbol{e}}(\mathbf{1_V}) &= M_{\boldsymbol{e},\boldsymbol{v}}(\mathbf{1_V})^{-1} = \begin{pmatrix} 3 & -2 & -4 \\ -1 & 1 & 2 \\ 1 & -1 & -1 \end{pmatrix}. \end{aligned}$$

Similarly,

$$\begin{aligned} M_{\boldsymbol{e},\boldsymbol{w}}(\mathbf{1_V}) &= \begin{pmatrix} -1 & 1 & 1 \\ 0 & -2 & 1 \\ 1 & -3 & 1 \end{pmatrix} \\ M_{\boldsymbol{w},\boldsymbol{e}}(\mathbf{1_V}) &= M_{\boldsymbol{e},\boldsymbol{w}}(\mathbf{1_V})^{-1} = \begin{pmatrix} 1/2 & -2 & 3/2 \\ 1/2 & -1 & 1/2 \\ 1 & -1 & 1 \end{pmatrix}. \end{aligned}$$

To find $M_{\boldsymbol{w},\boldsymbol{v}}(\mathbf{1_V})$ one can use Proposition 12.3. Thus

$$\begin{aligned} M_{\boldsymbol{w},\boldsymbol{v}}(\mathbf{1_V}) &= M_{\boldsymbol{w},\boldsymbol{e}}(\mathbf{1_V})M_{\boldsymbol{e},\boldsymbol{v}}(\mathbf{1_V}) \\ &= \begin{pmatrix} 1/2 & -2 & 3/2 \\ 1/2 & -1 & 1/2 \\ 1 & -1 & 1 \end{pmatrix} \begin{pmatrix} 1 & 2 & 0 \\ 1 & 1 & -2 \\ 0 & 1 & 1 \end{pmatrix} \\ &= \begin{pmatrix} -3/2 & 1/2 & 11/2 \\ -1/2 & 1/2 & 5/2 \\ 0 & 2 & 3 \end{pmatrix}. \end{aligned}$$

Finally, to find $M_{\boldsymbol{v},\boldsymbol{w}}(\mathbf{1_V})$ we can use the identity

$$M_{\boldsymbol{v},\boldsymbol{w}}(\mathbf{1_V}) = M_{\boldsymbol{w},\boldsymbol{v}}(\mathbf{1_V})^{-1}$$

and calculate the inverse of $M_{\boldsymbol{v},\boldsymbol{w}}(\mathbf{1_V})$, or else write

$$\begin{aligned} M_{\boldsymbol{v},\boldsymbol{w}}(\mathbf{1_V}) &= M_{\boldsymbol{v},\boldsymbol{e}}(\mathbf{1_V})M_{\boldsymbol{e},\boldsymbol{w}}(\mathbf{1_V}) \\ &= \begin{pmatrix} 3 & -2 & -4 \\ -1 & 1 & 2 \\ 1 & -1 & -1 \end{pmatrix} \begin{pmatrix} -1 & 1 & 1 \\ 0 & -2 & 1 \\ 1 & -3 & 1 \end{pmatrix} \\ &= \begin{pmatrix} -7 & 19 & -3 \\ 3 & -9 & 2 \\ -2 & 6 & -1 \end{pmatrix}. \end{aligned}$$

4. Let $\mathbf{V}$ and $\mathbf{W}$ be real vector spaces with $\dim(\mathbf{V}) = 4$ and $\dim(\mathbf{W}) = 2$, and let $\boldsymbol{v} = \{\mathbf{v}_1, \mathbf{v}_2, \mathbf{v}_3, \mathbf{v}_4\}$ and $\boldsymbol{w} = \{\mathbf{w}_1, \mathbf{w}_2\}$ be bases of $\mathbf{V}$ and $\mathbf{W}$. Let $F : \mathbf{V} \to \mathbf{W}$ be the linear map for which

$$M_{\boldsymbol{w},\boldsymbol{v}}(F) = \begin{pmatrix} 1 & 3 & -2 & 1/2 \\ 2 & 0 & 1 & 0 \end{pmatrix}.$$

Let

$$\begin{aligned} \boldsymbol{e} &= \{\mathbf{e}_1(1,1,1,2),\ \mathbf{e}_2(2,-1,3,0),\ \mathbf{e}_3(\sqrt{2},1,0,0),\ \mathbf{e}_4(1,-1/2,1,5)\} \\ \boldsymbol{f} &= \{\mathbf{f}_1(2,1),\ \mathbf{f}_2(-1,1)\} \end{aligned}$$

be new bases of $\mathbf{V}$ and $\mathbf{W}$ respectively, given by their coordinates in the bases $\boldsymbol{v}$ and $\boldsymbol{w}$. By Proposition 12.3 one has

$$M_{\boldsymbol{f},\boldsymbol{e}}(F) = M_{\boldsymbol{f},\boldsymbol{w}}(\mathbf{1_W})M_{\boldsymbol{w},\boldsymbol{v}}(F)M_{\boldsymbol{v},\boldsymbol{e}}(\mathbf{1_V}).$$

Since

$$M_{\boldsymbol{w},\boldsymbol{f}}(\mathbf{1w}) = \begin{pmatrix} 2 & -1 \\ 1 & 1 \end{pmatrix},$$

$$M_{\boldsymbol{f},\boldsymbol{w}}(\mathbf{1w}) = M_{\boldsymbol{w},\boldsymbol{f}}(\mathbf{1w})^{-1} = \begin{pmatrix} 1/3 & 1/3 \\ -1/3 & 2/3 \end{pmatrix},$$

and

$$M_{\boldsymbol{v},\boldsymbol{e}}(\mathbf{1v}) = \begin{pmatrix} 1 & 2 & \sqrt{2} & 1 \\ 1 & -1 & 1 & -1/2 \\ 1 & 3 & 0 & 1 \\ 2 & 0 & 0 & 5 \end{pmatrix},$$

one gets

$$M_{\boldsymbol{f},\boldsymbol{e}}(F) =$$

$$= \begin{pmatrix} 1/3 & 1/3 \\ -1/3 & 2/3 \end{pmatrix} \begin{pmatrix} 1 & 3 & -2 & 1/2 \\ 2 & 0 & 1 & 0 \end{pmatrix} \begin{pmatrix} 1 & 2 & \sqrt{2} & 1 \\ 1 & -1 & 1 & -1/2 \\ 1 & 3 & 0 & 1 \\ 2 & 0 & 0 & 5 \end{pmatrix}$$

$$= \begin{pmatrix} 1 & 1 & -1/3 & 1/6 \\ 1 & -1 & 4/3 & -1/6 \end{pmatrix} \begin{pmatrix} 1 & 2 & \sqrt{2} & 1 \\ 1 & -1 & 1 & -1/2 \\ 1 & 3 & 0 & 1 \\ 2 & 0 & 0 & 5 \end{pmatrix}$$

$$= \begin{pmatrix} 2 & 0 & \sqrt{2}+1 & 1 \\ 1 & 7 & \sqrt{2}-1 & 2 \end{pmatrix}.$$

5. In the vector space $M_2(\mathbf{C})$ of 2×2 matrices with complex entries, consider the bases

$$\begin{aligned} \boldsymbol{e} &= \{\mathbf{1}_{11}, \mathbf{1}_{12}, \mathbf{1}_{21}, \mathbf{1}_{22}\} \\ \boldsymbol{p} &= \{\mathbf{I}_2, \sigma_1, \sigma_2, \sigma_3\}, \end{aligned}$$

where

$$\mathbf{1}_{11} = \begin{pmatrix} 1 & 0 \\ 0 & 0 \end{pmatrix}, \quad \mathbf{1}_{12} = \begin{pmatrix} 0 & 1 \\ 0 & 0 \end{pmatrix}$$

$$\mathbf{1}_{21} = \begin{pmatrix} 0 & 0 \\ 1 & 0 \end{pmatrix}, \quad \mathbf{1}_{22} = \begin{pmatrix} 0 & 0 \\ 0 & 1 \end{pmatrix}$$

and

$$\sigma_1 = \begin{pmatrix} 0 & 1 \\ 1 & 0 \end{pmatrix}, \quad \sigma_2 = \begin{pmatrix} 0 & -i \\ i & 0 \end{pmatrix}, \quad \sigma_3 = \begin{pmatrix} 1 & 0 \\ 0 & -1 \end{pmatrix}.$$

(The matrices σ_1, σ_2, and σ_3 are the Pauli matrices, see Exercise 4.13.) Since

$$\mathbf{I}_2 = \mathbf{1}_{11} + \mathbf{1}_{22}, \quad \sigma_1 = \mathbf{1}_{12} + \mathbf{1}_{21},$$

$$\sigma_2 = -i\mathbf{1}_{12} + i\mathbf{1}_{21}, \quad \sigma_3 = \mathbf{1}_{11} - \mathbf{1}_{22},$$

one has

$$M_{\boldsymbol{e},\boldsymbol{p}}(\mathbf{1}) = \begin{pmatrix} 1 & 0 & 0 & 1 \\ 0 & 1 & -i & 0 \\ 0 & 1 & i & 0 \\ 1 & 0 & 0 & -1 \end{pmatrix},$$

and

$$M_{\boldsymbol{p},\boldsymbol{e}}(\mathbf{1}) = M_{\boldsymbol{e},\boldsymbol{p}}(\mathbf{1})^{-1} = \begin{pmatrix} 1/2 & 0 & 0 & 1/2 \\ 0 & 1/2 & 1/2 & 0 \\ 0 & i/2 & -i/2 & 0 \\ 1/2 & 0 & 0 & -1/2 \end{pmatrix}.$$

In an affine space $\mathbf{A}$ with associated vector space $\mathbf{V}$ consider two affine frames $E\mathbf{e}_1 \ldots \mathbf{e}_n$ and $F\mathbf{f}_1 \ldots \mathbf{f}_n$, and any given point $P \in \mathbf{A}$.

Let $\mathbf{x} = (x_1, \ldots, x_n)^t$ be the column vector of the coordinates of P with respect to the frame $E\mathbf{e}_1 \ldots \mathbf{e}_n$, and let $\mathbf{y} = (y_1, \ldots, y_n)^t$ be the column vector of P with respect to $F\mathbf{f}_1 \ldots \mathbf{f}_n$. Then

$$\begin{aligned} \overrightarrow{EP} &= x_1\mathbf{e}_1 + \cdots + x_n\mathbf{e}_n \\ \overrightarrow{FP} &= y_1\mathbf{f}_1 + \cdots + y_n\mathbf{f}_n. \end{aligned}$$

Suppose that we are given $\mathbf{x}$ and we wish to find $\mathbf{y}$. Denote by $\boldsymbol{e} = \{\mathbf{e}_1, \ldots, \mathbf{e}_n\}$ and $\boldsymbol{f} = \{\mathbf{f}_1, \ldots, \mathbf{f}_n\}$ the two bases of $\mathbf{V}$, with $A = (a_{ij}) = M_{\boldsymbol{f},\boldsymbol{e}}(\mathbf{1_V})$ and $\mathbf{c} = (c_1, \ldots, c_n)^t$ the vector of E with respect to the frame $F\mathbf{f}_1 \ldots \mathbf{f}_n$. The vector identity

$$\overrightarrow{FP} = \overrightarrow{FE} + \overrightarrow{EP},$$

written with respect to the basis $\boldsymbol{f}$ yields the equation

$$\mathbf{y} = A\mathbf{x} + \mathbf{c}. \tag{12.3}$$

Equation (12.3) is the formula for the *change of affine coordinates from the frame* $E\mathbf{e}_1 \ldots \mathbf{e}_n$ *to the frame* $F\mathbf{f}_1 \ldots \mathbf{f}_n$.

As should be expected, this formula depends only on the bases $\boldsymbol{e}$ and $\boldsymbol{f}$ and on the points E and F, the origins of the two reference frames.

Suppose $\mathbf{A}$ is a *real* affine space. Two affine frames $E\mathbf{e}_1 \ldots \mathbf{e}_n$ and $F\mathbf{f}_1 \ldots \mathbf{f}_n$ are said to have the *same orientation* if the bases $\boldsymbol{e}$ and $\boldsymbol{f}$ of $\mathbf{V}$ have the same orientation; otherwise the frames are said to have *different orientations.*

Consider the set $\mathcal{R}$ of all affine reference frames of $\mathbf{A}$, and define an equivalence relation on $\mathcal{R}$ by saying that two frames are equivalent if they have the same orientation. In exactly the same way as with vector spaces, we can see that this is indeed an equivalence relation, and that there are precisely two equivalence classes, which are called *orientations of* $\mathbf{A}$. The orientation to which a given frame $E\mathbf{e}_1 \ldots \mathbf{e}_n$ belongs is called the *orientation of* $\mathbf{A}$ *defined by* $E\mathbf{e}_1 \ldots \mathbf{e}_n$.

Examples 12.5

1. If $E\mathbf{e}_1 \ldots \mathbf{e}_n$ is a reference frame for the affine space $\mathbf{A}$, and we are given a second frame $F\mathbf{e}_1 \ldots \mathbf{e}_n$ (obtained from the first by changing only the origin) then formula (12.3) becomes simply

$$\mathbf{y} = \mathbf{x} + \mathbf{c}. \tag{12.4}$$

If instead the second frame is $E\mathbf{f}_1 \ldots \mathbf{f}_n$ (changing only the basis but not the origin) then (12.3) becomes

$$\mathbf{y} = A\mathbf{x}. \tag{12.5}$$

Any change of affine frame can be obtained by composing one of type (12.4) and one of type (12.5), or vice versa. The proof is left to the reader.

2. Let $E\mathbf{e}_1 \ldots \mathbf{e}_n$, $F\mathbf{f}_1 \ldots \mathbf{f}_n$, $G\mathbf{g}_1 \ldots \mathbf{g}_n$ be three affine frames in $\mathbf{A}$, and let $\mathbf{x} = (x_1, \ldots, x_n)^t$, $\mathbf{y} = (y_1, \ldots, y_n)^t$ and $\mathbf{z} = (z_1, \ldots, z_n)^t$ be the coordinates of a point $P \in \mathbf{A}$ with respect to each of the three frames. Suppose that the change of coordinates from $\mathbf{x}$ to $\mathbf{y}$ is given by (12.3), and that from $\mathbf{y}$ to $\mathbf{z}$ is given by

$$\mathbf{y} = B\mathbf{y} + \mathbf{d} \tag{12.6}$$

with $B = (b_{ij})$, and $\mathbf{d} = (d_1, \ldots, d_n)^t$. Then the formula giving the change of coordinates from $\mathbf{x}$ to $\mathbf{z}$ is obtained by substituting

(12.3) into (12.6):

$$\mathbf{z} = BA\mathbf{x} + (\mathbf{d} + B\mathbf{c}).$$

3. With the notation of (12.3), the formula expressing the change of coordinates from $\mathbf{y}$ to $\mathbf{x}$, i.e. the one given by the inverse of formula (12.3), is
$$\mathbf{x} = A^{-1}\mathbf{y} - A^{-1}\mathbf{c}. \tag{12.7}$$
For, substituting (12.3) in the right hand side of (12.7) gives the identity $\mathbf{x} = \mathbf{x}$.

4. Let $\mathbf{A}$ be an affine space of dimension 3, and let $\mathbf{V}$ be its associated vector space. Let $E\mathbf{e}_1\,\mathbf{e}_2\,\mathbf{e}_3$ and $F\mathbf{f}_1\,\mathbf{f}_2\,\mathbf{f}_3$ be two affine frames in $\mathbf{A}$. Suppose that
$$\mathbf{f}_1 = \mathbf{e}_1 + \mathbf{e}_2, \quad \mathbf{f}_2 = -\mathbf{e}_1 + \mathbf{e}_3, \quad \mathbf{f}_3 = \mathbf{e}_1 + \mathbf{e}_2 + \mathbf{e}_3,$$
and that F has coordinates $(5, -3/2, 1/2)$ in the frame $E\mathbf{e}_1\,\mathbf{e}_2\,\mathbf{e}_3$.

To find the formula for changing coordinates from $E\mathbf{e}_1\,\mathbf{e}_2\,\mathbf{e}_3$ to $F\mathbf{f}_1\,\mathbf{f}_2\,\mathbf{f}_3$ we need the matrix $M_{\boldsymbol{f},\boldsymbol{e}}(\mathbf{1_V})$ as well as the coordinates c_1, c_2, c_3 of E with respect to $F\mathbf{f}_1\,\mathbf{f}_2\,\mathbf{f}_3$. One has
$$M_{\boldsymbol{f},\boldsymbol{e}}(\mathbf{1_V}) = \begin{pmatrix} 1 & -1 & 1 \\ 1 & 0 & 1 \\ 0 & 1 & 1 \end{pmatrix}$$
whence
$$M_{\boldsymbol{f},\boldsymbol{e}}(\mathbf{1_V}) = M_{\boldsymbol{e},\boldsymbol{f}}(\mathbf{1_V})^{-1} = \begin{pmatrix} -1 & 2 & -1 \\ -1 & 1 & 0 \\ 1 & -1 & 1 \end{pmatrix}.$$

To find (c_1, c_2, c_3) we use the matrix $M_{\boldsymbol{f},\boldsymbol{e}}(\mathbf{1_V})$ that we have just calculated. Thus
$$c_1\mathbf{f}_1 + c_2\mathbf{f}_2 + c_3\mathbf{f}_3 = \overrightarrow{FE} = -\overrightarrow{EF} = -(5\mathbf{e}_1 - \frac{3}{2}\mathbf{e}_2 + \frac{1}{2}\mathbf{e}_3).$$
Thus,
$$\begin{pmatrix} c_1 \\ c_2 \\ c_3 \end{pmatrix} = M_{\boldsymbol{f},\boldsymbol{e}}(\mathbf{1_V}) \begin{pmatrix} -5 \\ 3/2 \\ -1/2 \end{pmatrix} = \begin{pmatrix} -1 & 2 & -1 \\ -1 & 1 & 0 \\ 1 & -1 & 1 \end{pmatrix} \begin{pmatrix} -5 \\ 3/2 \\ -1/2 \end{pmatrix} = \begin{pmatrix} 17/2 \\ 13/2 \\ -7 \end{pmatrix}.$$

We conclude that the formula for changing coordinates from $E\mathbf{e}_1\,\mathbf{e}_2\,\mathbf{e}_3$ to $F\mathbf{f}_1\,\mathbf{f}_2\,\mathbf{f}_3$ is

$$\begin{pmatrix} y_1 \\ y_2 \\ y_3 \end{pmatrix} = \begin{pmatrix} -1 & 2 & -1 \\ -1 & 1 & 0 \\ 1 & -1 & 1 \end{pmatrix} \begin{pmatrix} x_1 \\ x_2 \\ x_3 \end{pmatrix} + \begin{pmatrix} 17/2 \\ 13/2 \\ -7 \end{pmatrix}$$

or,

$$\begin{aligned} y_1 &= -x_1 + 2x_2 - x_3 + \frac{17}{2} \\ y_1 &= -x_1 + x_2 + \frac{13}{2} \\ y_1 &= x_1 - x_2 + x_3 - 7. \end{aligned}$$

EXERCISES

12.1 Let $F : \mathbf{R}^2 \to \mathbf{R}^3$ be the linear map

$$F(x_1, x_2) = (x_1 + x_2,\ x_1 - 2x_2,\ x_1).$$

Find $M_{\boldsymbol{b}', \boldsymbol{b}}(F)$, where

$$\boldsymbol{b} = \{(1,1),\ (0,-1)\}, \quad \boldsymbol{b}' = \{(1,1,1),\ (1,-2,0),\ (0,0,1)\}.$$

12.2 Let $F : \mathbf{R}^2 \to \mathbf{R}^3$ be defined by

$$F(x_1, x_2) = (-\frac{x_1}{2} + \frac{3x_2}{2},\ \frac{3x_1}{2} - \frac{x_2}{2},\ 2x_1).$$

Find $M_{\boldsymbol{b}', \boldsymbol{b}}(F)$, where

$$\boldsymbol{b} = \{(1,1),\ (1,-1)\}, \quad \boldsymbol{b}' = \{(1,0,1),\ (0,1,1),\ (2,-1,-1)\}.$$

12.3 Let $F : \mathbf{C}^3 \to \mathbf{C}^2$ be the linear map defined by the matrix

$$A = \begin{pmatrix} 2 & 1 & -1 \\ i & i & 1+i \end{pmatrix}$$

with respect to the bases $\boldsymbol{b} = \{(1,i,i),\ (i,i,1),\ (0,i,0)\}$ of $\mathbf{C}^3$ and $\boldsymbol{b}' = \{(1,1),\ (i,-i)\}$ of $\mathbf{C}^2$. Find the matrix that represents F with respect to the canonical bases.

12.4 Let $\mathbf{v}_1 = (-1,1,1)$, $\mathbf{v}_2 = (1,1,0)$, $\mathbf{v}_3 = (0,2,1) \in \mathbf{R}^3$. Prove that there is no linear map $F : \mathbf{R}^3 \to \mathbf{R}^3$ satisfying

$$F(\mathbf{v}_1) = (1,0,0), \quad F(\mathbf{v}_2) = (0,1,0), \quad F(\mathbf{v}_3) = (0,0,1).$$

12.5 For each of the following pairs of bases $\boldsymbol{b}$ and $\boldsymbol{b}'$ of $\mathbf{R}^2$ find $M_{\boldsymbol{b},\boldsymbol{b}'}(\mathbf{1})$:

a) $\boldsymbol{b} = \{(1,0),\ (0,1)\}$ $\quad \boldsymbol{b}' = \{(1,\sqrt{3}),\ (\sqrt{3},1)\}$
b) $\boldsymbol{b} = \{(1,-1),\ (1,1)\}$ $\quad \boldsymbol{b}' = \{(1,0),\ (1,1)\}$
c) $\boldsymbol{b} = \{(2,1),\ (2,2)\}$ $\quad \boldsymbol{b}' = \{(\sqrt{5},-\sqrt{5}),\ (\sqrt{5},\sqrt{5})\}$.

12.6 For each of the following pairs of bases $\boldsymbol{b}$ and $\boldsymbol{b}'$ of $\mathbf{C}^2$ find $M_{\boldsymbol{b},\boldsymbol{b}'}(\mathbf{1})$:

a) $\boldsymbol{b} = \{(1,i),\ (i,1)\}$ $\quad \boldsymbol{b}' = \{(2,1),\ (1,2)\}$
b) $\boldsymbol{b} = \{(i,i),\ (-1,1)\}$ $\quad \boldsymbol{b}' = \{(i,0),\ (0,i)\}$.

12.7 For each of the following pairs of bases $\boldsymbol{b}$ and $\boldsymbol{b}'$ of $\mathbf{Q}^3$ find $M_{\boldsymbol{b},\boldsymbol{b}'}(\mathbf{1})$:

a) $\boldsymbol{b} = \{(1,0,1),\ (1,1,0),\ (0,1,1)\}$
$\boldsymbol{b}' = \{(1,1,1),\ (0,1,1),\ (0,0,1)\}$
b) $\boldsymbol{b} = \{(1,-1,1),\ (-1,1,1),\ (1,1,1)\}$
$\boldsymbol{b}' = \{(13,5,-6),\ (8,-10,-4),\ (-17,0,-7)\}$.

12.8 Let $O\mathbf{ij}$ be a fixed frame for a real affine plane $\mathbf{A}$. Find the formula for changing coordinates from the frame $O\mathbf{ij}$ to the frame $O'\mathbf{i}'\mathbf{j}'$ where $O' = O'(1,2)$, $\mathbf{i}' = \mathbf{i} + 3\mathbf{j}$ and $\mathbf{j}' = \mathbf{i} + \mathbf{j}$.

12.9 Let $\mathbf{A}$ be a real affine plane in which there is a given affine frame $O\mathbf{ij}$. Let ℓ, ℓ' and ℓ'' be the lines with equations

$$\ell : X + Y = 0, \quad \ell' : X - Y - 1 = 0, \quad \ell'' : 2X + Y + 2 = 0.$$

Putting $O' = \ell \cap \ell'$, $U = \ell \cap \ell''$ and $U' = \ell' \cap \ell''$, let $\mathbf{i}' = \overrightarrow{O'U}$, $\mathbf{j}' = \overrightarrow{O'U'}$. First show that the vectors $\mathbf{i}'$ and $\mathbf{j}'$ are linearly independent, and then find the formula for changing coordinates from the frame $O\mathbf{ij}$ to the frame $O'\mathbf{i}'\mathbf{j}'$.

12.10 Let **A** be a 3-dimensional real affine space with a fixed frame $O\mathbf{ijk}$. Find the formula for changing coordinates from the frame $O\mathbf{ijk}$ to the frame $O'\mathbf{i}'\mathbf{j}'\mathbf{k}'$, where

$$O' = O'\left(\frac{\pi}{3}, -\pi, \frac{\pi}{3}\right), \quad \mathbf{i}' = \mathbf{i} + \mathbf{k},$$

$$\mathbf{j}' = \mathbf{j} - \mathbf{k}, \quad \mathbf{k}' = \mathbf{i} + \mathbf{j} + \mathbf{k}.$$

12.11 Let $\boldsymbol{e} = \{\mathbf{e}_1, \ldots, \mathbf{e}_n\}$, $\boldsymbol{b} = \{\mathbf{b}_1, \ldots, \mathbf{b}_n\}$ be two bases of the $\mathbf{K}$-vector space $\mathbf{V}$, and let $\boldsymbol{\eta} = \{\eta_1, \ldots, \eta_n\}$ and $\boldsymbol{\beta} = \{\beta_1, \ldots, \beta_n\}$ be the bases of $\mathbf{V}^*$ dual to $\boldsymbol{e}$ and $\boldsymbol{b}$ respectively. Show that

$$M_{\boldsymbol{\beta}, \boldsymbol{\eta}}(1_{\mathbf{V}^*}) = [M_{\boldsymbol{e}, \boldsymbol{b}}(1_{\mathbf{V}})]^t.$$

13

Linear operators

Let $\mathbf{V}$ be a finite dimensional vector space over K, and let $\boldsymbol{e} = \{\mathbf{e}_1, \ldots, \mathbf{e}_n\}$ be a basis of $\mathbf{V}$. For every operator $F \in \mathrm{End}(\mathbf{V})$ we will write $M_{\boldsymbol{e}}(F)$ rather than $M_{\boldsymbol{e},\boldsymbol{e}}(F)$, and $M_{\boldsymbol{e}}(F)$ is called the *matrix of F with respect to the basis $\boldsymbol{e}$.*

From Theorem 12.2 it follows that the map

$$\begin{array}{rcl} M_{\boldsymbol{e}} : \mathrm{End}(\mathbf{V}) & \rightarrow & M_n(\mathbf{K}) \\ F & \mapsto & M_{\boldsymbol{e}}(F) \end{array}$$

is an isomorphism of $\mathbf{K}$-vector spaces. In particular,

$$M_{\boldsymbol{e}}(\mathbf{1_V}) = \mathbf{I}_n,$$

and $M_{\boldsymbol{e}}(F) \in \mathrm{GL}_n(\mathbf{K})$ if and only if $F \in \mathrm{GL}(\mathbf{V})$, that is F is an automorphism if and only if $M_{\boldsymbol{e}}(F)$ is invertible. Thus $M_{\boldsymbol{e}}$ induces a bijection, which we denote by the same symbol,

$$M_{\boldsymbol{e}} : \mathrm{GL}(\mathbf{V}) \rightarrow \mathrm{GL}_n(\mathbf{K}).$$

Let $\boldsymbol{f} = \{\mathbf{f}_1, \ldots, \mathbf{f}_n\}$ be a second basis of $\mathbf{V}$. By Proposition 12.3 it follows that

$$M_{\boldsymbol{f}}(F) = M_{\boldsymbol{f},\boldsymbol{e}}(\mathbf{1_V}) M_{\boldsymbol{e}}(F) M_{\boldsymbol{e},\boldsymbol{f}}(\mathbf{1_V}).$$

Since $M_{\boldsymbol{f},\boldsymbol{e}}(\mathbf{1_V}) = M_{\boldsymbol{e},\boldsymbol{f}}(\mathbf{1_V})^{-1}$, one has

$$M_{\boldsymbol{f}}(F) = M_{\boldsymbol{e},\boldsymbol{f}}(\mathbf{1_V})^{-1} M_{\boldsymbol{e}}(F) M_{\boldsymbol{e},\boldsymbol{f}}(\mathbf{1_V}) \qquad (13.1)$$

from which it follows immediately that

$$\det(M_{\boldsymbol{f}}(F)) = \det(M_{\boldsymbol{e}}(F)),$$

that is $\det(M_{\boldsymbol{e}}(F))$ does not depend on the basis $\boldsymbol{e}$ but only on the operator F. We therefore call $\det(M_{\boldsymbol{f}}(F))$ the *determinant of the operator* F and we denote it by $\det(F)$, without having to specify the matrix $M_{\boldsymbol{f}}(F)$ used to compute it.

Definition 13.1
Two matrices $A, B \in M_n(\mathbf{K})$ are said to be *similar* if there is a matrix $M \in \mathrm{GL}_n(\mathbf{K})$ such that $B = M^{-1}AM$.

Similarity is an equivalence relation in $M_n(\mathbf{K})$. Indeed, every matrix is similar to itself: $A = \mathbf{I}_n^{-1}A\mathbf{I}_n$; secondly, if $B = M^{-1}AM$ then multiplying by $M = (M^{-1})^{-1}$ on the left gives $MB = MM^{-1}AM = AM$, and then multiplying by M^{-1} on the right gives $MBM^{-1} = AMM^{-1} = A$, i.e. $(M^{-1})^{-1}BM^{-1} = A$. Finally, if $C = N^{-1}BN$ and $B = M^{-1}AM$ then

$$C = N^{-1}(M^{-1}AM)N = (MN)^{-1}A(MN)$$

and the relation is transitive.

Proposition 13.2
Let $\mathbf{V}$ *vector space over* $\mathbf{K}$ *with* $\dim(\mathbf{V}) = n$, *and let* $A, B \in M_n(\mathbf{K})$. *Then* A *and* B *are similar if and only if there is a linear operator* $F \in \mathrm{End}(\mathbf{V})$ *and bases* $\boldsymbol{e}$ *and* $\boldsymbol{f}$ *of* $\mathbf{V}$ *such that* $M_{\boldsymbol{e}}(F) = A$ *and* $M_{\boldsymbol{f}}(F) = B$.

Proof
If such an F, $\boldsymbol{e}$ and $\boldsymbol{f}$ exist, then then it follows from (13.1) that A and B are similar. Conversely, suppose that

$$B = M^{-1}AM. \tag{13.2}$$

Let $\boldsymbol{e}$ be any basis of $\mathbf{V}$ and let $F = F_A$ be the operator associated to the matrix A. For each $j = 1, \dots, n$ let $\mathbf{f}_j$ be the vector whose coordinates, with respect to $\mathbf{e}$, are the elements of the j-th column of M, that is $\mathbf{f}_j = m_{1j}\mathbf{e}_1 + \cdots + m_{nj}\mathbf{e}_n$. Since M has rank n the vectors

$\mathbf{f}_1, \ldots, \mathbf{f}_n$ are linearly independent and so form a basis of $\mathbf{V}$ which we call $\boldsymbol{f}$. Moreover,

$$M = M_{\boldsymbol{e}, \boldsymbol{f}}(\mathbf{1_V}).$$

From (13.2) it follows that $B = M_{\boldsymbol{f}}(F)$. □

Definition 13.3
Let $\mathbf{V}$ be a $\mathbf{K}$-vector space of dimension n. An operator $F \in \mathrm{End}(\mathbf{V})$ is said to be *diagonalizable* if there is a basis $\boldsymbol{e}$ of $\mathbf{V}$ such that $M_{\boldsymbol{e}}(F)$ is a diagonal matrix, i.e. is of the form

$$\begin{pmatrix} \lambda_1 & 0 & \cdots & 0 \\ 0 & \lambda_2 & \cdots & 0 \\ \vdots & \vdots & & \vdots \\ 0 & 0 & \cdots & \lambda_n \end{pmatrix}$$

for some $\lambda_1, \ldots, \lambda_n \in \mathbf{K}$.

In this case $\boldsymbol{e}$ is said to be a diagonalizing basis for F.

A matrix $A \in M_n(\mathbf{K})$ is said to be *diagonalizable* if it is similar to a diagonal matrix.

Clearly, if $F \in \mathrm{End}(\mathbf{V})$ and $\boldsymbol{e}$ is a basis $\mathbf{V}$ then F is diagonalizable if and only if $M_{\boldsymbol{e}}(F)$ is a diagonalizable matrix. In particular, $A \in M_n(\mathbf{K})$ is diagonalizable if and only if the operator $F_A : \mathbf{K}^n \to \mathbf{K}^n$ defined by A is diagonalizable.

If $F : \mathbf{V} \to \mathbf{V}$ is a diagonalizable linear operator and $\boldsymbol{e}$ is a diagonalizing basis for F then,

$$F(\mathbf{e}_i) = \lambda_i \mathbf{e}_i, \quad \text{for each } i = 1, \ldots, n. \tag{13.3}$$

Conversely, if there exists a basis $\boldsymbol{e}$ satisfying (13.3) then the matrix $M_{\boldsymbol{e}}(F)$ is diagonal, and so F is diagonalizable and $\boldsymbol{e}$ is a diagonalizing basis for F.

Note that, if $\dim(\mathbf{V}) = 1$ then every $F \in \mathrm{End}(\mathbf{V})$ is diagonalizable, and every basis is diagonalizing for F. If $\dim(\mathbf{V}) > 1$ then there are some operators $F \in \mathrm{End}(\mathbf{V})$ which are not diagonalizable. Similarly, not all matrices in $M_n(\mathbf{K})$ are diagonalizable if $n > 1$ — see Example 13.15(2).

The notions of 'eigenvector' and 'eigenvalue' arise naturally when considering the problem of existence of diagonalizing bases.

Definition 13.4
Let $\mathbf{V}$ be a $\mathbf{K}$-vector space and let $F \in \text{End}(\mathbf{V})$. A non-zero vector $\mathbf{v} \in \mathbf{V}$ is called an *eigenvector of* F if there is a scalar $\lambda \in \mathbf{K}$ such that $F(\mathbf{v}) = \lambda\mathbf{v}$. The scalar λ is then called *the eigenvalue associated to the eigenvector* $\mathbf{v}$.

The set of eigenvalues of an operator F is called the *spectrum of* F.

For $A \in M_n(\mathbf{K})$, an *eigenvector of* A is an eigenvector $\mathbf{x} \in \mathbf{K}^n$ of the operator $F_A : \mathbf{K}^n \to \mathbf{K}^n$ defined by A, and an *eigenvalue of* A is an eigenvalue of F_A.

For example, if $F = \mathbf{1_V}$, then every vector $\mathbf{v} \neq \mathbf{0}$ is an eigenvector of F with eigenvalue $\lambda = 1$.

If F is an operator with $\ker(F) \neq \langle 0 \rangle$, then every $\mathbf{v} \in \ker(F) \setminus \{\mathbf{0}\}$ is an eigenvector of F with eigenvalue $\lambda = 0$.

In the remainder of this chapter we describe some of the simple properties of eigenvectors and eigenvalues of an operator $F \in \text{End}(\mathbf{V})$. We suppose that $\dim(\mathbf{V}) = n \geq 1$.

Proposition 13.5
The eigenvalue associated to an eigenvector is uniquely determined.

Proof
If $\lambda\mathbf{v} = F(\mathbf{v}) = \mu\mathbf{v}$ for some $\lambda, \mu \in \mathbf{K}$ then $(\lambda - \mu)\mathbf{v} = \mathbf{0}$ and, since $\mathbf{v} \neq \mathbf{0}$, this implies $\lambda - \mu = 0$, i.e. $\lambda = \mu$. □

Proposition 13.6
If $\mathbf{v}_1, \mathbf{v}_2 \in \mathbf{V}$ *are eigenvectors with the same eigenvalue* λ, *then for every* $c_1, c_2 \in \mathbf{K}$ *the vector* $c_1\mathbf{v}_1 + c_2\mathbf{v}_2$, *if it is non-zero, is also an eigenvector with eigenvalue* λ.

Proof
One has

$$F(c_1\mathbf{v}_1 + c_2\mathbf{v}_2) = c_1F(\mathbf{v}_1) + c_2F(\mathbf{v}_2) = c_1\lambda\mathbf{v}_1 + c_2\lambda\mathbf{v}_2 = \lambda(c_1\mathbf{v}_1 + c_2\mathbf{v}_2).$$

□

From Proposition 13.6, it follows that for each $\lambda \in \mathbf{K}$,

$$\mathbf{V}_\lambda(F) = \{\mathbf{v} \in \mathbf{V} \mid \mathbf{v} \text{ is an eigenvector of } F \text{ with eigenvalue } \lambda\} \cup \{\mathbf{0}\}$$

is a vector subspace of $\mathbf{V}$, called *the eigenspace for the eigenvalue* λ.

For a matrix $A \in M_n(\mathbf{K})$ the *eigenspace for the eigenvalue* λ is defined to be the subspace $\mathbf{V}_\lambda(A) = \mathbf{V}_\lambda(F_A)$ in $\mathbf{K}^n$.

Proposition 13.7
If $\mathbf{v}_1, \ldots, \mathbf{v}_k \in \mathbf{V}$ are eigenvectors with eigenvalues $\lambda_1, \ldots, \lambda_k$, and these λ_i are pairwise distinct, then $\mathbf{v}_1, \ldots, \mathbf{v}_k$ are linearly independent.

Proof
The assertion is trivial if $k = 1$, as $\mathbf{v}_1 \neq \mathbf{0}$. We now proceed by induction on k, and suppose that $k \geq 2$. If

$$c_1\mathbf{v}_1 + c_2\mathbf{v}_2 + \cdots + c_k\mathbf{v}_k = \mathbf{0}, \tag{13.4}$$

then, applying F to both sides, gives

$$c_1F(\mathbf{v}_1) + c_2F(\mathbf{v}_2) + \cdots + c_kF(\mathbf{v}_k) = \mathbf{0},$$

that is,

$$c_1\lambda_1\mathbf{v}_1 + c_2\lambda_2\mathbf{v}_2 + \cdots + c_k\lambda_k\mathbf{v}_k = \mathbf{0}. \tag{13.5}$$

On the other hand, multiplying (13.4) by λ_1 gives

$$c_1\lambda_1\mathbf{v}_1 + c_2\lambda_1\mathbf{v}_2 + \cdots + c_k\lambda_1\mathbf{v}_k = \mathbf{0}, \tag{13.6}$$

and subtracting (13.6) from (13.5) gives

$$c_1(\lambda_2 - \lambda_1)\mathbf{v}_2 + \cdots + c_k(\lambda_k - \lambda_1)\mathbf{v}_k = \mathbf{0}. \tag{13.7}$$

By the inductive hypothesis, $\mathbf{v}_2, \ldots, \mathbf{v}_k$ are linearly independent, whence $c_2(\lambda_2 - \lambda_1) = \cdots = c_k(\lambda_k - \lambda_1) = 0$. Since $\lambda_j - \lambda_1 \neq 0$ for $j = 2, \ldots, k$ it follows that $c_2 = \cdots = c_k = 0$. Thus (13.4) reduces to $c_1\mathbf{v}_1 = \mathbf{0}$ which implies that $c_1 = 0$ since $\mathbf{v}_1 \neq \mathbf{0}$. □

Proposition 13.8
If every $\mathbf{v} \in \mathbf{V} \setminus \{\mathbf{0}\}$ is an eigenvector of F then there exists $\lambda \in \mathbf{K}$ such that $F = \lambda\mathbf{1}_\mathbf{V}$.

Proof

If $\dim(\mathbf{V}) = 1$ the assertion is obvious, and we can therefore suppose that $\dim(\mathbf{V}) > 1$. Let $\{\mathbf{e}_1, \ldots, \mathbf{e}_n\}$ be a basis of $\mathbf{V}$. From the hypothesis it follows that there are $\lambda_1, \ldots, \lambda_n \in \mathbf{K}$ such that $F(\mathbf{e}_i) = \lambda_i \mathbf{e}_i$, for $i = 1, \ldots, n$. Let i, j be distinct integers with $1 \leq i, j \leq n$, and let $\mathbf{v}_{ij} = \mathbf{e}_i + \mathbf{e}_j$. By hypothesis there is a scalar $\lambda_{ij} \in \mathbf{K}$ for which

$$F(\mathbf{v}_{ij}) = \lambda_{ij}\mathbf{v}_{ij} = \lambda_{ij}\mathbf{e}_i + \lambda_{ij}\mathbf{e}_j.$$

On the other hand,

$$F(\mathbf{v}_{ij}) = F(\mathbf{e}_i + \mathbf{e}_j) = F(\mathbf{e}_i) + F(\mathbf{e}_j) = \lambda_i\mathbf{e}_i + \lambda_j\mathbf{e}_j,$$

and from the independence of $\mathbf{e}_i$ and $\mathbf{e}_j$ it follows that $\lambda_i = \lambda_j = \lambda_{ij}$. In conclusion, we have $\lambda_1 = \cdots = \lambda_k$, and the proposition is proved. □

In order to find the eigenvalues of a linear operator or a matrix one uses the so-called 'characteristic polynomial', whose definition relies on the following simple result.

Proposition 13.9

Let $\mathbf{V}$ be a finite dimensional vector space and let $F \in \mathrm{End}(\mathbf{V})$. A scalar $\lambda \in \mathbf{K}$ is an eigenvalue of F if and only if the operator

$$F - \lambda \mathbf{1}_{\mathbf{V}} : \mathbf{V} \to \mathbf{V}$$

which is defined by

$$(F - \lambda \mathbf{1}_{\mathbf{V}})(\mathbf{v}) = F(\mathbf{v}) - \lambda\mathbf{v},$$

is not an isomorphism, that is, if and only if $\det(F - \lambda \mathbf{1}_{\mathbf{V}}) = 0$.

Proof

$(F - \lambda \mathbf{1}_{\mathbf{V}})$ fails to be an isomorphism if and only if $\ker(F - \lambda \mathbf{1}_{\mathbf{V}}) \neq \langle \mathbf{0} \rangle$, that is if and only if there is a $\mathbf{v} \in \mathbf{V}$ with $\mathbf{v} \neq \mathbf{0}$, for which $(F - \lambda \mathbf{1}_{\mathbf{V}})(\mathbf{v}) = \mathbf{0}$, i.e. for which

$$F(\mathbf{v}) = \lambda\mathbf{v}. \tag{13.8}$$

(13.8) states that F has an eigenvector $\mathbf{v}$ with eigenvalue λ. □

Let $\boldsymbol{e} = \{\mathbf{e}_1, \ldots, \mathbf{e}_n\}$ be a basis of $\mathbf{V}$. The matrix associated to the operator $\lambda \mathbf{1}_{\mathbf{V}}$ is

$$\begin{pmatrix} \lambda & 0 & \cdots & 0 \\ 0 & \lambda & \cdots & 0 \\ \vdots & \vdots & & \vdots \\ 0 & 0 & \cdots & \lambda \end{pmatrix}$$

and, if $A = (a_{ij}) = M_{\boldsymbol{e}}(F)$ then

$$M_{\boldsymbol{e}}(F - \lambda \mathbf{1}_{\mathbf{V}}) = \begin{pmatrix} a_{11} - \lambda & a_{12} & \cdots & a_{1n} \\ a_{21} & a_{22} - \lambda & \cdots & a_{2n} \\ \vdots & \vdots & & \vdots \\ a_{n1} & a_{n2} & \cdots & a_{nn} - \lambda \end{pmatrix}.$$

Definition 13.10
Let $A \in M_n(\mathbf{K})$ and let T be an indeterminate. The determinant

$$P_A(T) = |A - T\mathbf{I}_n| = \begin{vmatrix} a_{11} - T & a_{12} & \cdots & a_{1n} \\ a_{21} & a_{22} - T & \cdots & a_{2n} \\ \vdots & \vdots & & \vdots \\ a_{n1} & a_{n2} & \cdots & a_{nn} - T \end{vmatrix}$$

is a polynomial of degree n in T, called *the characteristic polynomial of A.*

If $F \in \mathrm{End}(\mathbf{V})$, $\boldsymbol{e} = \{\mathbf{e}_1, \ldots, \mathbf{e}_n\}$ is a basis of $\mathbf{V}$ and $A = M_{\boldsymbol{e}}(F)$ then $P_A(T)$ is the *characteristic polynomial of F*, and is denoted $P_F(T)$.

The definition of $P_F(T)$ is independent of the basis $\boldsymbol{e}$ because *two similar matrices*, such as the representatives of F in two different bases, *have the same characteristic polynomial.*

To see this, let A and B be similar matrices, so $B = M^{-1}AM$ for some $M \in \mathrm{GL}_n(\mathbf{K})$. Then

$$B - T\mathbf{I}_n = M^{-1}AM - T\mathbf{I}_n = M^{-1}(A - T\mathbf{I}_n)M.$$

Therefore $|B - T\mathbf{I}_n| = |M^{-1}|\,|A - T\mathbf{I}_n|\,|M| = |A - T\mathbf{I}_n|$.

Note that the coefficient of T^n in $P_A(T)$ is $(-1)^n$, consequently $(-1)^n P_A(T)$ is a monic polynomial.

The following corollary follows from Proposition 13.9.

Corollary 13.11
Let $\mathbf{V}$ *be a vector space of dimension* n*, and let* $F \in \mathrm{End}(F)$*. Then* $\lambda \in \mathbf{K}$ *is an eigenvalue of* F *if and only if* λ *is a root of the polynomial* $P_F(T)$*. In particular,* F *has at most* n *eigenvalues.*

Proof
The first assertion is a reformulation of Proposition 13.9. Since $P_A(T)$ has degree equal to $\dim(\mathbf{V})$, the last assertion follows from the fact that a polynomial of degree n has at most n roots. □

Corollary 13.11 provides a practical method for finding the eigenvalues of an operator. Illustrations of this method are left to the examples at the end of the chapter.

The question of whether or not a given operator is diagonalizable is a problem about finding eigenvalues and eigenvectors. Indeed one has the following result.

Proposition 13.12
Let $\mathbf{V}$ *be a finite dimensional vector space. An operator* $F \in \mathrm{End}(\mathbf{V})$ *is diagonalizable if and only if there is a basis of* $\mathbf{V}$ *consisting entirely of eigenvectors of* F*.*

Proof
This follows immediately from (13.3). □

The following result provides a necessary and sufficient condition for an operator to be diagonalizable.

Theorem 13.13
Let $\mathbf{V}$ *be a* $\mathbf{K}$*-vector space of dimension* n*, and let* $F \in \mathrm{End}(\mathbf{V})$*. If* $\{\lambda_1, \ldots, \lambda_k\} \subset \mathbf{K}$ *is the spectrum of* F*, then*

$$\dim(\mathbf{V}_{\lambda_1}(F)) + \cdots + \dim(\mathbf{V}_{\lambda_k}(F)) \leq n \tag{13.9}$$

with equality if and only if F *is diagonalizable.*

Proof
For each $i = 1, \ldots, k$ put $d(i) = \dim(\mathbf{V}_{\lambda_i}(F))$, and let $\{\mathbf{e}_{i1}, \ldots, \mathbf{e}_{id(i)}\}$ be a basis of $\mathbf{V}_{\lambda_i}(F)$. By virtue of Proposition 13.12 it will be enough

to prove that the vectors

$$\mathbf{e}_{11}, \ldots, \mathbf{e}_{1d(1)}, \mathbf{e}_{21}, \ldots, \mathbf{e}_{2d(2)}, \ldots, \mathbf{e}_{k1}, \ldots, \mathbf{e}_{kd(k)}$$

are linearly independent.

Suppose instead that

$$\begin{aligned} \mathbf{0} = c_{11}\mathbf{e}_{11} + \cdots + c_{1d(1)}\mathbf{e}_{1d(1)} + c_{21}\mathbf{e}_{21} + \cdots + c_{2d(2)}\mathbf{e}_{2d(2)} \\ + \cdots + c_{k1}\mathbf{e}_{k1} + \cdots + c_{kd(k)}\mathbf{e}_{kd(k)} \end{aligned} \tag{13.10}$$

for some scalars c_{ij}. Putting $\mathbf{v}_i = c_{i1}\mathbf{e}_{i1} + \cdots + c_{id(i)}\mathbf{e}_{id(i)}$, then $\mathbf{v}_i \in \mathbf{V}_{\lambda_i}(F)$ and (13.10) becomes simply

$$\mathbf{0} = \mathbf{v}_1 + \mathbf{v}_2 + \cdots + \mathbf{v}_k. \tag{13.11}$$

Since $\mathbf{v}_i = \mathbf{0}$ if and only if $c_{i1} = c_{i2} = \cdots = c_{id(i)} = 0$, it is enough to show that $\mathbf{v}_1 = \mathbf{v}_2 = \cdots = \mathbf{v}_k = \mathbf{0}$. If $\mathbf{v}_i \neq \mathbf{0}$ then $\mathbf{v}_i$ is an eigenvector with eigenvalue λ_i, and it follows from (13.7) that $\{\mathbf{v}_1, \ldots, \mathbf{v}_k\} \setminus \{\mathbf{0}\}$ is a linearly independent set of vectors. Thus the right hand side of (13.11) is equal to $\mathbf{0}$ if and only if all the summands are $\mathbf{0}$. □

A particularly important case of Theorem 13.13 is the following immediate corollary.

Corollary 13.14
If $\dim(\mathbf{V}) = n$, *and* $F \in \mathrm{End}(\mathbf{V})$ *has* n *distinct eigenvalues then it is diagonalizable.*

Note that the sufficient condition for diagonalizability given in this corollary is not necessary. For example, the operator $\mathbf{1_V}$ is diagonalizable for any $n = \dim(\mathbf{V})$, though it has only one eigenvalue, namely $\lambda = 1$.

Observations and examples 13.15

1. Corollary 13.11 provides a practical method for finding eigenvalues and eigenvectors of an operator or a matrix. By choosing a basis for $\mathbf{V}$ an operator becomes a matrix, so we consider only the case of matrices.

   Suppose we are given a matrix $A \in M_n(\mathbf{K})$. One begins by finding the eigenvalues of A by calculating the characteristic polynomial $P_A(T)$ of A and its roots in $\mathbf{K}$. For each eigenvalue $\lambda \in$

K, the homogeneous system of n equations in n unknowns $\mathbf{X} = (X_1, \ldots, X_n)^t$:

$$(A - \lambda \mathbf{I}_n)\mathbf{X} = \mathbf{0}$$

has rank $r < n$ and therefore has nontrivial solutions. The space of solutions is the eigenspace $\mathbf{V}_\lambda(A)$. If the sum of the dimensions of the eigenspaces found by letting λ vary over the roots of $P_A(T)$ is equal to n then A is diagonalizable, by Theorem 13.13. A diagonalizing basis can be found by taking the union of bases of the eigenspaces.

In principle, this provides a method of finding all vectors of a diagonalizing basis and the corresponding change of basis matrix.

Note, however, that an operator need not have any eigenvalues, nor eigenvectors, because the polynomial $P_A(T)$ might not have any roots in **K**. If $\mathbf{K} = \mathbf{C}$, then by the fundamental theorem of algebra it follows that $P_A(T)$ has roots in **C**, and so *every operator on a finite dimensional complex vector space has at least one eigenvalue*, and so *at least one eigenvector*. This does not of course mean that the operator is diagonalizable.

If $\mathbf{K} = \mathbf{R}$ then it can happen that an operator on the space **V** have no eigenvectors — see Example 4 below. If, however, $\dim(\mathbf{V})$ is odd, then the characteristic polynomial has odd degree and so has at least one real root. In conclusion, *every operator on an odd dimensional real vector space has at least one eigenvalue, and so it has at least one eigenvector.*

2. The characteristic polynomial of the $n \times n$ identity matrix is

$$P_{\mathbf{I}_n}(T) = (1 - T)^n,$$

while the characteristic polynomial of the zero matrix is

$$P_{\mathbf{0}}(T) = (-1)^n T^n.$$

More generally, the characteristic polynomial of the matrix $a\mathbf{I}_n$ is

$$P_{a\mathbf{I}_n}(T) = (a - T)^n,$$

In all these cases, the only eigenspace is $\mathbf{K}^n$.

3. If $A = (a_{ij}) \in M_n(\mathbf{K})$ is an upper or lower triangular matrix, then

$$P_A(T) = (a_{11} - T)(a_{22} - T) \cdots (a_{nn} - T).$$

If $a_{11}, a_{22}, \ldots, a_{nn}$ are distinct, then by Corollary 13.14 A is diagonalizable since it has n distinct eigenvalues; otherwise it need not be diagonalizable.

For example the $n \times n$ matrix $(n \geq 2)$

$$A = \begin{pmatrix} 0 & 1 & 0 & \cdots & 0 \\ 0 & 0 & 1 & \cdots & 0 \\ \vdots & \vdots & \vdots & & \vdots \\ 0 & 0 & 0 & \cdots & 1 \\ 0 & 0 & 0 & \cdots & 0 \end{pmatrix} \in M_n(\mathbf{K})$$

has characteristic polynomial

$$P_A(T) = (-1)^n T^n$$

and so it has the unique eigenvalue $\lambda = 0$. From this it follows that if A were diagonalizable then it would be similar to the zero matrix. But $\mathbf{0}$ is similar only to itself since for any $M \in \mathrm{GL}_n(\mathbf{K})$ one has $M^{-1}\mathbf{0}M = \mathbf{0}$. Thus A is not diagonalizable.

The matrix

$$B = A + \mathbf{I}_n = \begin{pmatrix} 1 & 1 & 0 & \cdots & 0 & 0 \\ 0 & 1 & 1 & \cdots & 0 & 0 \\ \vdots & \vdots & \vdots & & & \vdots \\ 0 & 0 & 0 & \cdots & 1 & 1 \\ 0 & 0 & 0 & \cdots & 0 & 1 \end{pmatrix} \in M_n(\mathbf{K})$$

has characteristic polynomial $P_B(T) = (1-T)^n$, equal to that of $\mathbf{I}_n$. But as for A above, B is not diagonalizable for if it were it would have to be similar to the identity matrix, which is impossible since $M^{-1}\mathbf{I}_n M = \mathbf{I}_n$ for all $M \in \mathrm{GL}_n(\mathbf{K})$.

4. The matrix

$$A = \begin{pmatrix} 0 & 1 \\ -1 & 0 \end{pmatrix} \in M_2(\mathbf{R})$$

has characteristic polynomial $1 + T^2$. Since this polynomial does not have real roots, A can have neither eigenvalues nor eigenvectors. However, if A is considered as a matrix with complex entries then it has two distinct eigenvalues $\lambda = \pm i$. It therefore follows from Corollary 13.14 that A is diagonalizable in $M_2(\mathbf{C})$, and it is similar to the diagonal matrix

$$B = \begin{pmatrix} i & 0 \\ 0 & -i \end{pmatrix}.$$

To find the matrix M which diagonalizes A, i.e. for which $B = M^{-1}AM$, we need to find a basis of $\mathbf{C}^2$ consisting of eigenvectors, which is equivalent to finding an eigenvector for each of the eigevalues. This can be done as follows.

By Proposition 13.9, the eigenvectors with eigenvalue i are the elements of the kernel of $A - i\mathbf{I}_2$; that is, they are the solutions of the homogeneous system,

$$\begin{aligned} -iX + Y &= 0 \\ -X - iY &= 0, \end{aligned}$$

which has rank 1, and is therefore equivalent to just the first equation. The solutions are the vectors of the form (t, it), which, as t varies, form the subsace $\mathbf{C}^2_i(A)$. Taking $t = 1$ for example, gives the eigenvector $(1, i)$ with eigenvalue $\lambda = i$. Analogously, considering the homogeneous system corresponding to $A + i\mathbf{I}_2$, one gets vectors of the form $(t, -it)$, and so $(i, 1)$ is an eigenvector of A with eigenvalue $\lambda = -i$. The basis $\boldsymbol{b} = \{(1, i), (i, 1)\}$ is thus diagonalizing. Let $\boldsymbol{e}$ be the canonical basis of $\mathbf{C}^2$, then

$$M = M_{\boldsymbol{e},\boldsymbol{b}}(\mathbf{1}) = \begin{pmatrix} 1 & i \\ i & 1 \end{pmatrix}$$

and

$$M^{-1}AM = \begin{pmatrix} 1/2 & -i/2 \\ -i/2 & 1/2 \end{pmatrix} \begin{pmatrix} 0 & 1 \\ -1 & 0 \end{pmatrix} \begin{pmatrix} 1 & i \\ i & 1 \end{pmatrix} = \begin{pmatrix} i & 0 \\ 0 & -i \end{pmatrix}.$$

5. Let $\mathbf{V}$ be a finite dimensional $\mathbf{K}$-vector space and let $F \in \mathrm{End}(\mathbf{V})$ with $\lambda \in \mathbf{K}$ an eigenvalue of F. The number $\dim(\mathbf{V}_\lambda(F))$ is called the *geometric multiplicity of* λ *for* F. The *algebraic multiplicity* of λ for F is instead the multiplicity of λ as a root of the characteristic polynomial $P_F(T)$; this is denoted $h(\lambda)$.

   In general, the geometric and algebraic multiplicities do not coincide. This occurs for the matrices A and B in Example 3 above. Clearly, for A one has $h(0) = n$ and for B, $h(1) = n$. On the other hand A and B are not diagonalizable, and so $\dim(\mathbf{V}_0(A)) < n$. Similarly for B.

   For any operator $F \in \mathrm{End}(\mathbf{V})$ and $\lambda \in \mathbf{K}$ one has

$$\dim(\mathbf{V}_\lambda(F)) \leq h(\lambda), \tag{13.12}$$

that is, the geometric multiplicity is not larger than the algebraic multiplicity.

To prove this, suppose that $d = \dim(\mathbf{V}_\lambda(F)) \geq 1$, and let $\{\mathbf{e}_1, \ldots, \mathbf{e}_n\}$ be a basis of $\mathbf{V}$ such that $\{\mathbf{e}_1, \ldots, \mathbf{e}_d\}$ is a basis of $\mathbf{V}_\lambda(F)$. Then

$$A = M_{\mathbf{e}} = \begin{pmatrix} \lambda\mathbf{I}_d & B \\ \mathbf{0} & C \end{pmatrix}$$

where $B \in M_{d,n-d}(\mathbf{K})$, $\mathbf{0} \in M_{n-d,d}(\mathbf{K})$ is the zero matrix, and $C \in M_{n-d,n-d}(\mathbf{K})$. Expanding the determinant $|A - T\mathbf{I}_n|$ using Laplace's rule with repect to the first d rows gives

$$P_F(T) = (\lambda - T)^d p(T),$$

where $p(T) = P_C(T)$ is a polynomial of degree $n - d$. Thus $h(\lambda) \geq d$.

If the field $\mathbf{K}$ is algebraically closed and $\lambda_1, \ldots, \lambda_k$ are the eigenvalues of F, then

$$h(\lambda_1) + \cdots + h(\lambda_k) = n.$$

It therefore follows from (13.12) and Theorem 13.13 that *if the field K is algebraically closed the operator F is diagonalizable if and only if*

$$\dim(\mathbf{V}_\lambda(F)) = h(\lambda)$$

for every eigenvalue λ. That is *if and only if the geometric multiplicity of λ is equal to the algebraic multiplicity of λ for every eigenvalue λ of F.*

6. If $A = (a_{ij}) \in M_2(\mathbf{K})$, then

$$P_A(T) = T^2 - \mathrm{tr}(A)T + \det(A),$$

where $\mathrm{tr}(A) = a_{11} + a_{22}$ is the *trace* of A. The proof of this is left to the reader.

7. Let

$$P(T) = g_0 + g_1 T + \cdots + g_{n-1}T^{n-1} + g_n T^n$$

be a monic polynomial of degree n with coefficients in $\mathbf{K}$, and let

$$M_P = \begin{pmatrix} 0 & 0 & \cdots & 0 & -g_0 \\ 1 & 0 & \cdots & 0 & -g_1 \\ 0 & 1 & \cdots & 0 & -g_2 \\ \vdots & \vdots & & \vdots & \vdots \\ 0 & 0 & \cdots & 1 & -g_{n-1} \end{pmatrix}.$$

The characteristic polynomial of M_P *is equal to* $(-1)^n P(T)$. If $n = 1$ this is obvious: $M_P - T = -g_0 - T$. Now suppose $n \geq 2$ and proceed by induction on n. One has

$$
\begin{aligned}
|M_P - T\mathbf{I}_n| &= \begin{vmatrix} -T & 0 & \cdots & 0 & -g_0 \\ 1 & -T & \cdots & 0 & -g_1 \\ 0 & 1 & \cdots & 0 & -g_2 \\ \vdots & \vdots & & \vdots & \vdots \\ 0 & 0 & \cdots & 1 & -g_{n-1} - T \end{vmatrix} \\
&= -T \begin{vmatrix} -T & \cdots & 0 & -g_1 \\ 1 & \cdots & 0 & -g_2 \\ \vdots & & \vdots & \vdots \\ 0 & \cdots & 1 & -g_{n-1} - T \end{vmatrix} \\
&\qquad + (-1)^n g_0 \begin{vmatrix} 1 & -T & \cdots & 0 \\ 0 & 1 & \cdots & 0 \\ \vdots & \vdots & & \vdots \\ 0 & 0 & \cdots & 1 \end{vmatrix} \\
&= -T(-1)^{n-1}(g_1 + g_2 T + \cdots + g_{n-1}T^{n-2} + T^{n-1}) \\
&\qquad + (-1)^n g_0 \\
&= (-1)^n P(T).
\end{aligned}
$$

Here the value of the first determinant is deduced from the inductive hypothesis.

This example shows for that any monic polynomial of degree $n \geq 1$ in $\mathbf{K}[T]$ there is a square matrix of order n for which it is the charactristic polynomial.

EXERCISES

13.1 Let $F : \mathbf{R}^3 \to \mathbf{R}^3$ be the linear operator

$$F(x, y, z) = (x + y - z,\ y + z,\ 2x).$$

Find the matrix $M_{\boldsymbol{b}}(F)$, where

$$\boldsymbol{b} = \{(1,1,0),\ (-1,0,1),\ (1,1,1)\}.$$

13.2 Let $F : \mathbf{R}^3 \to \mathbf{R}^3$ be the linear operator

$$F(x,y,z) = (2x,\ x-y,\ y-z).$$

Find the matrix $M_{\boldsymbol{b}}(F^2)$, where

$$\boldsymbol{b} = \{(1,1,0),\ (2,-1,1),\ (0,1,-1)\}.$$

13.3 Let $F : \mathbf{C}^3 \to \mathbf{C}^3$ be the linear operator defined with respect to the canonical basis by the matrix

$$\begin{pmatrix} 1 & 1 & 0 \\ 1 & -1 & -2 \\ 2 & 1 & -3 \end{pmatrix}.$$

Find the matrix $M_{\boldsymbol{b}}(F)$ representing F with respect to the basis $\boldsymbol{b} = \{(-2i,i,i),\ (-1,-1,1),\ (1,0,-1)\}$.

13.4 Let $F : \mathbf{R}^3 \to \mathbf{R}^3$ be the linear operator defined with respect to the canonical basis by the matrix

$$\begin{pmatrix} 2 & 1 & 1 \\ 0 & 1 & 2 \\ 0 & 0 & 3 \end{pmatrix}.$$

Find the matrix $M_{\boldsymbol{b}}(F)$, where

$$\boldsymbol{b} = \{(1,1,0),\ (-1,0,1),\ (1,1,1)\}.$$

13.5 Let $F : \mathbf{R}^3 \to \mathbf{R}^3$ be the linear operator defined with respect to the canonical basis by the matrix

$$\begin{pmatrix} 1 & 2 & 1 \\ 1 & -1 & 3 \\ 1 & 0 & 2 \end{pmatrix}.$$

Find the matrix $M_{\boldsymbol{b}}(F)$, where

$$\boldsymbol{b} = \{(-1,0,-1),\ (1,1,1),\ (1,-1,0)\}.$$

13.6 Let $F : \mathbf{C}^3 \to \mathbf{C}^3$ be the linear operator defined with respect to the canonical basis by the matrix

$$\begin{pmatrix} -1 & -1 & 1 \\ 2 & 1 & 2i \\ 1+i & 0 & 0 \end{pmatrix}.$$

Find the matrix $M_{\boldsymbol{b}}(F)$, where

$$\boldsymbol{b} = \{(i,1,-1),\ (-2,i,0),\ (2i,1,i)\}.$$

13.7 Find the eigenvalues and eigenvectors of the following matrices in $M_2(\mathbf{R})$:

a) $\begin{pmatrix} 1 & 0 \\ 0 & -1 \end{pmatrix}$ b) $\begin{pmatrix} 1 & 1 \\ 0 & 1 \end{pmatrix}$

c) $\begin{pmatrix} 1 & 0 \\ 1 & 1 \end{pmatrix}$ d) $\begin{pmatrix} 1 & 1 \\ 1 & 1 \end{pmatrix}$.

13.8 Find the eigenvalues and eigenvectors of the matrix

$$\begin{pmatrix} 1 & a \\ b & 1 \end{pmatrix} \in M_2(\mathbf{C}),$$

where a and b are real parameters.

13.9 Let $F : \mathbf{R}^3 \to \mathbf{R}^3$ be the linear operator

$$F(x,y,z) = (y - z,\ -x + 2y - z,\ x - y + 2z).$$

Prove that F is diagonalizable by finding a basis $\boldsymbol{b}$ of $\mathbf{R}^3$ consisting of eigenvectors of F. Find the matrix of F in this basis.

13.10 Calculate the eigenvalues and their algebraic and geometric multiplicities for each of the following matrices in $M_3(\mathbf{R})$, and deduce whether or not they are diagonalizable:

a) $\begin{pmatrix} -6 & 2 & -5 \\ -4 & 4 & -2 \\ 10 & -3 & 8 \end{pmatrix}$ b) $\begin{pmatrix} -8 & -13 & -14 \\ -6 & -5 & -8 \\ 14 & 17 & 21 \end{pmatrix}$

c) $\begin{pmatrix} 1 & 0 & 0 \\ 0 & -3 & -15 \\ 0 & 2 & 8 \end{pmatrix}$ d) $\begin{pmatrix} -5 & 4 & -1 \\ -4 & 1 & -2 \\ 8 & -4 & 3 \end{pmatrix}$

e) $\begin{pmatrix} 13 & 59 & 34 \\ 10 & 40 & 24 \\ -18 & -79 & -46 \end{pmatrix}$ f) $\begin{pmatrix} -4 & 2 & 5 \\ 0 & -44 & -120 \\ 0 & 16 & 44 \end{pmatrix}$.

13.11 Find the eigenvalues of the matrix

$$\begin{pmatrix} 0 & 1 & 0 & 0 \\ 0 & 0 & 1 & 0 \\ 0 & 0 & 0 & 1 \\ 1 & 0 & 0 & 0 \end{pmatrix} \in M_4(\mathbf{C}).$$

13.12 Find $a, b, c, d, e, f \in \mathbf{R}$ given that $(1,1,1)$, $(1,0,-1)$, $(1,-1,0)$ $\in \mathbf{R}^3$ are eigenvectors of the matrix

$$\begin{pmatrix} 1 & 1 & 1 \\ a & b & c \\ d & e & f \end{pmatrix}.$$

13.13 Let $A \in M_n(\mathbf{K})$. Prove that if A has an eigenvalue λ then $a\lambda + b$ is an eigenvalue of the matrix $aA + b\mathbf{I}_n$, where a and b are scalars.

13.14 Let $A, B \in M_n(\mathbf{K})$. Prove that if $M \in \mathrm{GL}_n(\mathbf{K})$ is such that $B = M^{-1}AM$ then, for each integer $k \geq 0$ one has $B^k = M^{-1}A^kM$. In particular, if A and B are similar then so are A^k and B^k for each integer $k \geq 0$.

13.15 Use the preceding exercise to calculate F^5, where $F : \mathbf{R}^3 \to \mathbf{R}^3$ is the linear operator defined in Exercise 13.9.

14

Transformation groups

The concept of a 'group', and in particular the idea of a 'transformation group', is of fundamental importance in geometry. We do not attempt a systematic treatment of group theory, which is better left to an algebra text, but instead limit ourselves to giving some essential definitions necessary for the most important geometric examples.

Definition 14.1

A *group* is a pair $(G, \cdot)$ consisting of a non-empty set G and a *binary operation* on G, that is, a map $\cdot : G \times G \to G$, that associates to every pair $(g, g') \in G \times G$ an element $g \cdot g' \in G$, called the *product* of g and g', such that the following axioms are satisfied:

[G1] (*Associativity*) $(g \cdot g') \cdot g'' = g \cdot (g' \cdot g'')$, for every $g, g', g'' \in G$.

[G2] (*Existence of an identity element*) There is an element $e \in G$ satisfying $e \cdot g = g$ for every $g \in G$.

[G3] (*Existence of an inverse*) For every $g \in G$ there is an element $g^{-1} \in G$ satisfying $g \cdot g^{-1} = g^{-1} \cdot g = e$.

A group $(G, \cdot)$ is said to be *commutative* or *abelian* if it satisfies the additional axiom:

[G4] (*Commutativity*) $g \cdot g' = g' \cdot g$, for every $g, g' \in G$.

In an abelian group, the binary operation is usually denoted by $+$ and called 'addition'.

When it is clear from the context what operation is being used on G then the group $(G, \cdot)$ is usually denoted simply G.

An important example of a group is the set $\mathcal{T}(S)$ of all the bijections of a non-empty set S into itself, also called *transformations* of S. To any pair $(f, g) \in \mathcal{T}(S) \times \mathcal{T}(S)$ there is associated their composite $f \circ g \in \mathcal{T}(S)$, and this defines a binary operation on $\mathcal{T}(S)$ which satisfies axioms G1, G2 and G3, with $e = \mathbf{1}_S$ (the proof of this is left to the reader). The pair $(\mathcal{T}(S), \circ)$ is therefore a group, called the *group of transformations of* S.

A vector space $\mathbf{V}$ is an abelian group by defining the operation to be addition of vectors: the four axioms are satisfied because $(\mathbf{V}, +)$ satisfies axioms VS1 - VS4 — see Chapter 1. The group $(\mathbf{V}, +)$ is called the *additive group associated to the vector space* $\mathbf{V}$.

From the properties of matrices that we have studied, it follows that the set $\mathrm{GL}_n(\mathbf{K})$ of all invertible $n \times n$ matrices with entries in $\mathbf{K}$ is a group under the operation of matrix multiplication; it is called the *general linear group of order n over* $\mathbf{K}$. The identity element in $\mathrm{GL}_n(\mathbf{K})$ is $\mathbf{I}_n$.

Definition 14.2

A subset H of a group G is called a *subgroup* of G if the following conditions are satisfied:

[SG1] If $h, h' \in H$ then their product $h \cdot h' \in H$.
[SG2] The identity $e \in H$.
[SG3] If $h \in H$ then $h^{-1} \in H$.

It is easy to see that if $F \subset H \subset G$ and H is a subgroup of G, then F a subgroup of G if and only if it is a subgroup of H. Furthermore, if H and H' are two subgroups of G then their intersection $H \cap H'$ is also a subgroup of G, and so, as we have just pointed out, it is also a subgroup of both H and H'.

The subgroups of $\mathrm{GL}_n(\mathbf{K})$ are called the *linear groups of order n*. For example, the set $\mathrm{SL}_n(\mathbf{K})$ consisting of those matrices $A \in \mathrm{GL}_n(\mathbf{K})$ with $\det(A) = 1$ is a linear group, called the *special linear group of order n*.

Another example of a linear group is the *orthogonal group of order n*, denoted $\mathrm{O}(n)$, consisting of *orthogonal matrices* (those matrices $A \in M_n(\mathbf{R})$ satisfying $A^t A = \mathbf{I}_n$). By definition, an orthogonal matrix is invertible, and $A^{-1} = A^t$. Since $\det(A^t) = \det(A)$, an orthogonal matrix has the property that $\det(A)^2 = 1$, so $\det(A) = \pm 1$.

Let us check that $\mathrm{O}(n)$ is indeed a group. If $A, B \in \mathrm{O}(n)$, then

$$(AB)^t(AB) = B^t(A^tA)B = B^t\mathbf{I}_nB = B^tB = \mathbf{I}_n.$$

Thus $AB \in \mathrm{O}(n)$, and condition SG1 is satisfied. SG2 is obvious. To check SG3, let $A \in \mathrm{O}(n)$. Then

$$(A^{-1})^t(A^{-1}) = (A^t)^tA^t = AA^t = \mathbf{I}_n,$$

and so $A^{-1} \in \mathrm{O}(n)$.

The subset of $\mathrm{O}(n)$ of matrices with determinant equal to 1 is a subgroup, called the *special orthogonal group*, denoted $\mathrm{SO}(n)$. That it is a subgroup of $\mathrm{GL}_n(\mathbf{R})$ follows because it is the intersection of two subgroups:

$$\mathrm{SO}(n) = \mathrm{O}(n) \cap \mathrm{SL}_n(\mathbf{R}).$$

The orthogonal and special orthogonal groups are of great importance in Euclidean geometry (as well as in other fields, such as classical mechanics). We will return to them frequently in Part II.

The study of complex groups is particularly relevant in relativistic and quantum mechanics. An important example of a complex linear group is $\mathrm{U}(n)$, the *unitary group of order n*. It consists of all unitary $n \times n$ matrices $A = (a_{ij}) \in \mathrm{GL}_n(\mathbf{C})$, that is complex matrices A satisfying

$$A^*A = \mathbf{I}_n,$$

where $A^* = \overline{A}^t = (\overline{a}_{ji})$ — the transpose of the complex conjugate matrix. The conditions SG1, SG2 and SG3 are checked in a similar way as for the orthogonal case we have already seen.

Since $\det(A^*) = \overline{\det(A)}$ for any $A \in M_n(\mathbf{C})$, it follows from the definition that $\overline{\det(A)}\det(A) = 1$, and so any $A \in \mathrm{U}(n)$ satisfies

$$|\det(A)| = 1.$$

The matrices $A \in \mathrm{U}(n)$ for which $\det(A) = 1$ form in turn a subgroup of $\mathrm{U}(n)$ called the *special unitary group of order n*, and denoted $\mathrm{SU}(n)$. Clearly, one has

$$\mathrm{SO}(n) = \mathrm{SU}(n) \cap \mathrm{GL}_n(\mathbf{R}).$$

Other examples of linear groups will be introduced at a later stage.

A subgroup G of the group $\mathcal{T}(S)$ of transformations of a set S is called *a transformation group of* S. Clearly, $\mathcal{T}(S)$ is itself a transformation group of S. Another example is the subset $\{\mathbf{1}_S\}$ consisting of only the identity.

Let $s \in S$, and denote by $\mathcal{T}(S)_s$ the set of all transformations $f \in \mathcal{T}(S)$ for which $f(s) = s$. If $f, g \in \mathcal{T}(S)$ then $(f \circ g)(s) = f(g(s)) = f(s) = s$, and so $f \circ g \in \mathcal{T}(S)_s$. Clearly, $\mathbf{1}_S \in \mathcal{T}(S)_s$ and moreover, if $f \in \mathcal{T}(S)_s$ then $f^{-1}(s) = s$ and so $f^{-1} \in \mathcal{T}(S)_s$. Thus $\mathcal{T}(S)_s$ is a transformation group of S; it is called the *stabilizer of* s *in* S.

If H is a transformation group of S and $s \in S$, then the intersection $H \cap \mathcal{T}(S)_s$ is called the *stabilizer of* s *in* H, denoted H_s.

Definition 14.3
Let G and G' be two groups. A map $\omega : G \to G'$ is a *homomorphism* if $\omega(f \cdot g) = \omega(f) \cdot \omega(g)$ for every $f, g \in G$. A bijective homomorphism is called an *isomorphism*; in this case ω^{-1} is also an isomorphism. If an isomorphism $\omega : G \to G'$ exists, then G and G' are said to be *isomorphic*.

An important class of examples is provided by the groups of linear transformations of a vector space.

The set $\mathrm{GL}(\mathbf{V})$ of all automorphisms of a vector space $\mathbf{V}$ is a transformation group of $\mathbf{V}$ (the proof is immediate) called the *general linear group of* $\mathbf{V}$. Every transformation group of the vector space $\mathbf{V}$ consisting of linear transformations is a subgroup of $\mathrm{GL}(\mathbf{V})$.

Let $\dim(\mathbf{V}) = n \geq 1$, and let $\{\mathbf{e}_1, \ldots, \mathbf{e}_n\}$ be a basis of $\mathbf{V}$. Associating to any automorphism of $\mathbf{V}$ the matrix with respect to this basis gives a homomorphism of groups $\mathrm{GL}(\mathbf{V}) \to \mathrm{GL}_n(\mathbf{K})$. Proposition 12.3 tells us that this is actually an isomorphism.

The geometrically interesting class of transformations of an affine space are the 'affine transformations' which we now introduce.

Definition 14.4
Let $\mathbf{V}$ and $\mathbf{V}'$ be two $\mathbf{K}$-vector spaces, $\mathbf{A}$ an affine space over $\mathbf{V}$ and $\mathbf{A}'$ an affine space over $\mathbf{V}'$. An *isomorphism of* $\mathbf{A}$ *on to* $\mathbf{A}'$ is a bijection $f : \mathbf{A} \to \mathbf{A}'$ such that there is an isomorphism $\phi : \mathbf{V} \to \mathbf{V}'$ satisfying

$$\overrightarrow{f(P)f(Q)} = \phi(\overrightarrow{PQ})$$

for every $P, Q \in \mathbf{A}$.

An *affine transformation of* $\mathbf{A}$ is an isomorphism of $\mathbf{A}$ with itself.

It follows immediately from the definition that the isomorphism $\phi : \mathbf{V} \to \mathbf{V}'$ is uniquely determined by f; it is called the *isomorphism associated to* f. If f is an affine transformation, then $\phi \in \mathrm{GL}(\mathbf{V})$, and it is called the *automorphism associated to* f.

If there is an isomorphism $f : \mathbf{A} \to \mathbf{A}'$ then the two spaces $\mathbf{A}$ and $\mathbf{A}'$ are said to be *isomorphic.* The identity map $\mathbf{1_A} : \mathbf{A} \to \mathbf{A}$, the inverse map $f^{-1} : \mathbf{A}' \to \mathbf{A}$ of an isomorphism f and the composite $g \circ f : \mathbf{A} \to \mathbf{A}''$ of two isomorphisms $f : \mathbf{A} \to \mathbf{A}'$ and $g : \mathbf{A}' \to \mathbf{A}''$, are also isomorphisms (these statements follow directly from Definition 14.4 and are left to the reader). Thus, *isomorphism is an equivalence relation between affine spaces.*

An important example of an isomorphism is found by considering an affine frame $O\mathbf{e}_1 \dots \mathbf{e}_n$ in an affine space $\mathbf{A}$ over a vector space $\mathbf{V}$. It is the map $f : \mathbf{A} \to \mathbf{A}^n(\mathbf{K})$ defined by

$$f(P(x_1, \dots, x_n)) = (x_1, \dots, x_n),$$

the map sending each point to its n-tuple of coordinates. For each pair of points $P(x_1, \dots, x_n), Q(y_1, \dots, y_n) \in \mathbf{A}$ one has

$$\overrightarrow{f(P)f(Q)} = (y_1 - x_1, \dots, y_n - x_n) = \phi_e(\overrightarrow{PQ}),$$

where $\phi_{\boldsymbol{e}} : \mathbf{V} \to \mathbf{A}^n(\mathbf{K})$ is the isomorphism which assoiates to each vector the n-tuple of its coordinates in the basis $\boldsymbol{e} = \{\mathbf{e}_1, \dots, \mathbf{e}_n\}$. Since it is bijective, f is an affine isomorphism, with associated isomorphism $\phi_{\boldsymbol{e}}$.

From this example it follows that *every affine space of dimension n over* $\mathbf{K}$ *is isomorphic to* $\mathbf{A}^n(\mathbf{K})$. Since isomorphism is an equivalence relation, and in particular transitive, *two affine spaces over* $\mathbf{K}$ *of the same dimension are isomorphic.*

Let $\mathbf{V}$ be a $\mathbf{K}$-vector space and let $\mathbf{A}$ be an affine space over $\mathbf{V}$. An affine transformation of $\mathbf{A}$ can be thought of intuitively as a transformation which is compatible with the affine structure and so leaves untouched those geometric properties deriving from vectors.

Lemma 14.5
Given any points $O, O' \in \mathbf{A}$ then for every $\phi \in \mathrm{GL}(\mathbf{V})$ there is one and only one affine transformation $f : \mathbf{A} \to \mathbf{A}$ such that $f(O) = O'$ and the automorphism associated to f is ϕ.

In particular, an affine transformation is uniquely determined by the automorphism of $\mathbf{V}$ and the image $f(O)$ of any one point $O \in \mathbf{A}$.

Proof
For every $P \in \mathbf{A}$ the identity $\overrightarrow{O'f(P)} = \phi(\overrightarrow{OP})$ determines uniquely the point $f(P) \in \mathbf{A}$. We thus obtain a map $f : \mathbf{A} \to \mathbf{A}$ which, as is easy to see, is an affine transformation of $\mathbf{A}$ with the requisite properties.

If $g : \mathbf{A} \to \mathbf{A}$ is another affine transformation with the same properties, then for every $P \in \mathbf{A}$ one has

$$\overrightarrow{f(O)f(P)} = \phi(\overrightarrow{OP}) = \overrightarrow{g(O)g(P)} = \overrightarrow{O'g(P)} = \overrightarrow{f(O)g(P)},$$

and so $g(P) = f(P)$.

The last assertion is an easy consequence of the first one. □

As we have already noticed, the identity $\mathbf{1_A} : \mathbf{A} \to \mathbf{A}$ is an affine transformation, with associated automorphism $\mathbf{1_V} \in \mathrm{GL}(\mathbf{V})$. For any affine transformation of $\mathbf{A}$ with associated automorphism ϕ, the inverse transformation f^{-1} is also affine with associated automorphism ϕ^{-1}. Finally, if f and g are affine transformations with associated automorphisms ϕ and ψ respectively, then $f \circ g$ is an affine transformation with associated automorphism $\phi \circ \psi$.

We see therefore, that the set of all affine transformations of $\mathbf{A}$ is a transformation group; it is called the *affine group of* $\mathbf{A}$, denoted $\mathrm{Aff}(\mathbf{A})$. In the particular case that $\mathbf{A} = \mathbf{A}^n(\mathbf{K})$, the group $\mathrm{Aff}(\mathbf{A}^n(\mathbf{K}))$ is called the *affine group of order n over* $\mathbf{K}$, and denoted $\mathrm{Aff}_n(\mathbf{K})$.

The subgroups of $\mathrm{Aff}(\mathbf{A})$ are called *affine transformation groups of* $\mathbf{A}$.

Examples 14.6

1. It is easy to check that for $n > 1$ the group $\mathrm{GL}_n(\mathbf{K})$ is not abelian. For example, for $n = 2$,

$$\begin{pmatrix} 0 & 1 \\ 1 & 0 \end{pmatrix} \begin{pmatrix} a & b \\ c & d \end{pmatrix} = \begin{pmatrix} c & d \\ a & b \end{pmatrix},$$

while

$$\begin{pmatrix} a & b \\ c & d \end{pmatrix}\begin{pmatrix} 0 & 1 \\ 1 & 0 \end{pmatrix} = \begin{pmatrix} b & a \\ d & c \end{pmatrix},$$

for every $a, b, c, d \in \mathbf{K}$. A similar proof can be given for $n > 2$ by showing that for the elementary matrices R^n_{ij} (see 3.3(6)) one has $R^n_{ij}A \neq AR^n_{ij}$.

Similarly, the groups $\mathrm{O}(n)$, $n \geq 2$, and $\mathrm{SO}(n)$, $n \geq 3$ are not abelian. On the other hand, it is easy to check that $\mathrm{SO}(2)$ is abelian, by using the fact that any $A \in \mathrm{SO}(2)$ can be written in the form (2.6).

To prove that $\mathrm{O}(2)$ is not abelian it is sufficient to notice that

$$\begin{pmatrix} 0 & 1 \\ 1 & 0 \end{pmatrix}\begin{pmatrix} 1 & 0 \\ 0 & -1 \end{pmatrix} = \begin{pmatrix} 0 & -1 \\ 1 & 0 \end{pmatrix},$$

while

$$\begin{pmatrix} 1 & 0 \\ 0 & -1 \end{pmatrix}\begin{pmatrix} 0 & 1 \\ 1 & 0 \end{pmatrix} = \begin{pmatrix} 0 & 1 \\ -1 & 0 \end{pmatrix}.$$

Each of these factors belong to $\mathrm{O}(2)$ but not to $\mathrm{SO}(2)$.

The subgroup $D_n(\mathbf{K})$ of $\mathrm{GL}_n(\mathbf{K})$ consisting of invertible diagonal matrices is an abelian subgroup of $\mathrm{GL}_n(\mathbf{K})$. This follows immediately from the fact that, for every $a_1, \dots, a_n, b_1, \dots, b_n \in \mathbf{K}$, one has

$$\mathrm{diag}(a_1, \dots, a_n)\,\mathrm{diag}(b_1, \dots, b_n) = \mathrm{diag}(a_1b_1, \dots, a_nb_n),$$

where $\mathrm{diag}(c_1, \dots, c_n)$ denotes the diagonal matrix with entries $c_1, \dots, c_n$ — see Exercise 2.11.

2. An important class of affine transformations of an affine space $\mathbf{A}$ are the 'translations'.

   Let $\mathbf{v} \in \mathbf{V}$. The *translation defined by* $\mathbf{v}$ is the affine transformation that associates to each $P \in \mathbf{A}$ the point $t_\mathbf{v}(P)$ which satisfies $\overrightarrow{Pt_\mathbf{v}(P)} = \mathbf{v}$.

   Let us check that $t_\mathbf{v}$ is indeed an affine transformation. For each $Q \in \mathbf{A}$ put $P = t_{-\mathbf{v}}(Q)$; one has $Q = t_\mathbf{v}(P)$ since $\overrightarrow{QP} = -\mathbf{v}$, and so $\overrightarrow{PQ} = \mathbf{v}$. Thus, $t_\mathbf{v}$ is a bijection whose inverse is $t_{-\mathbf{v}}$. Moreover,

$$\overrightarrow{t_\mathbf{v}(P)t_\mathbf{v}(Q)} = \overrightarrow{t_\mathbf{v}(P)P} + \overrightarrow{PQ} + \overrightarrow{Qt_\mathbf{v}(Q)} = -\mathbf{v} + \overrightarrow{PQ} + \mathbf{v} = \overrightarrow{PQ}$$

and so $t_{\mathbf{v}}$ is an affine transformation with associated automorphism $\mathbf{1_V}$.

Conversely, if $f : \mathbf{A} \to \mathbf{A}$ is an affine transformation with associated automorphism $\mathbf{1_V} \in \mathrm{GL}(\mathbf{V})$, then for every $P, Q \in \mathbf{A}$ one has

$$\overrightarrow{f(P)f(Q)} = \overrightarrow{PQ}, \tag{14.1}$$

and so

$$\overrightarrow{Pf(P)} = \overrightarrow{Qf(Q)} = \mathbf{v}$$

is independent of P, and so $f = t_{\mathbf{v}}$. Thus the translations are precisely those affine transformations with associated automorphism $\mathbf{1_V} \in \mathrm{GL}(\mathbf{V})$.

Equation (14.1) expresses the fact that translations send one ordered pair of points to another defining the same vector. Intuitively, the translations can be thought of as 'rigid movements' of the space.

The identity map is the translation $t_{\mathbf{0}}$ defined by the zero vector $\mathbf{0}$. The product of two translations $t_{\mathbf{v}}$ and $t_{\mathbf{w}}$ is $t_{\mathbf{v}} \circ t_{\mathbf{w}} = t_{\mathbf{v}+\mathbf{w}}$ which is also a translation. Indeed, for every $P \in \mathbf{A}$, putting $Q = t_{\mathbf{v}}(P)$ one has

$$\overrightarrow{P(t_{\mathbf{v}} \circ t_{\mathbf{w}}(P))} = \overrightarrow{Pt_{\mathbf{v}}(Q)} = \overrightarrow{PQ} + \overrightarrow{Qt_{\mathbf{v}}(Q)} = \mathbf{w} + \mathbf{v}.$$

As we have already seen, the inverse of $t_{\mathbf{v}}$ is $t_{-\mathbf{v}}$. The translations of $\mathbf{A}$ therefore constitute a group, called the *group of translations of* $\mathbf{A}$, and denoted $T_{\mathbf{A}}$.

Associating to any translation $t_{\mathbf{v}} \in T_{\mathbf{A}}$ the corresponding vector $\mathbf{v} \in \mathbf{V}$ defines a bijection

$$T_{\mathbf{A}} \to \mathbf{V}. \tag{14.2}$$

This bijection associates to the product of two translations the sum of the corresponding vectors. (14.2) is thus an isomorphism of the group $T_{\mathbf{A}}$ onto the additive group of $\mathbf{V}$.

3. Let $\mathbf{A}$ be an affine space over $\mathbf{V}$. Suppose we are given a point $O \in \mathbf{A}$. Consider the group $\mathrm{Aff}(\mathbf{A})_O = \{f \in \mathrm{Aff}(\mathbf{A}) \mid f(O) = O\}$. For each $f \in \mathrm{Aff}(\mathbf{A})_O$ denote by $\Phi(f) \in \mathrm{GL}(\mathbf{V})$ the associated automorphism. We thus obtain a map

$$\Phi : \mathrm{Aff}(\mathbf{A})_O \to \mathrm{GL}(\mathbf{V}). \tag{14.3}$$

By Lemma 14.5 it follows that each $f \in \mathrm{Aff}(\mathbf{A})_O$ is completely determined by $\Phi(f)$, and conversely that each $\phi \in \mathrm{GL}(\mathbf{V})$ is the image of some $f \in \mathrm{Aff}(\mathbf{A})_O$. Thus Φ is bijective. Since the automorphism associated to the composite of two affine transformations is the composite of their corresponding automorphisms, it follows that the map Φ is a group isomorphism.

If $c \in \mathbf{K}$ is non-zero, the affine transformation $\Phi^{-1}(c\mathbf{1}_{\mathbf{V}}) \in \mathrm{Aff}(\mathbf{A})_O$ is called the *homothety with centre O and scalar c*, and is denoted $\omega_{O,c}$. Then

$$\overrightarrow{O\omega_{O,c}(P)} = c\overrightarrow{OP}.$$

Note that $\mathbf{1}_{\mathbf{A}} = \omega_{O,1}$. Moreover, $(\omega_{O,c})^{-1} = \omega_{O,c^{-1}}$, because for each $P \in \mathbf{A}$ the point $Q = \omega_{0,c^{-1}}(P)$ satisfies

$$\overrightarrow{OQ} = c^{-1}\overrightarrow{OP}$$

and so

$$\overrightarrow{O\omega_{0,c}(Q)} = c\overrightarrow{OQ} = \overrightarrow{OP}$$

that is, $\omega_{O,c}(Q) = P$. Thus $\omega_{O,c} \circ \omega_{O,c^{-1}} = \mathbf{1}_{\mathbf{A}}$. Furthermore,

$$\omega_{O,c} \circ \omega_{O,d} = \omega_{O,cd}$$

since for $P \in \mathbf{A}$ and with $Q = \omega_{O,c}(P)$, one has

$$\overrightarrow{O\omega_{0,d}(Q)} = d\overrightarrow{OQ} = d(c\overrightarrow{OP}) = (cd)\overrightarrow{OP}.$$

In conclusion, the homotheties with centre O form a subgroup of $\mathrm{Aff}(\mathbf{A})_O$.

4. Suppose $\mathbf{A}$ is an affine space over $\mathbf{V}$ and we are given a point $C \in \mathbf{A}$. For every $P \in \mathbf{A}$, the symmetric point of P with respect to C — see Example 7.5(4) — is the point $\sigma_C(P)$ satisfying the vector identity

$$\overrightarrow{C\sigma_C(P)} = -\overrightarrow{CP}.$$

From this identity it follows that

$$\overrightarrow{\sigma_C(P)\sigma_C(Q)} = \overrightarrow{\sigma_C(P)C} + \overrightarrow{C\sigma_C(Q)} = \overrightarrow{CP} - \overrightarrow{CQ} = -\overrightarrow{PQ}$$

and so $\sigma_C : \mathbf{A} \to \mathbf{A}$ is an affine transformation with associated isomorphism $\mathbf{1_V}$.

It is easy to check that $\sigma_C \circ \sigma_C = \mathbf{1_A}$. For $\mathbf{A} = \mathbf{A}^n$ and $C = (c_1, \ldots, c_n)$, one has

$$\sigma_C(P) = (2c_1 - x_1, \ldots, 2c_n - x_n)$$

for every $P = (x_1, \ldots, x_n)$.

A subset S of $\mathbf{A}$ is said to be *symmetric with respect to* $C \in \mathbf{A}$ if $\sigma_C(P) \in S$ whenever $P \in S$. In this case C is called a *centre of symmetry of* S.

A set can have more than one centre of symmetry or it can have none at all. For example, every affine space is symmetric with respect to every one of its points.

Suppose that a subspace S has associated vector subspace $\mathbf{W}$, and let $C \in S$. If $P \in S$ then $\overrightarrow{CP} \in \mathbf{W}$ and $\sigma_C(P)$ satisfies $\overrightarrow{C\sigma_C(P)} = -\overrightarrow{CP}$, that is $\overrightarrow{C\sigma_C(P)} \in \mathbf{W}$ and $\sigma_C(P) \in S$. Thus S is symmetric with respect to C.

Lemma 14.7

Let $O \in \mathbf{A}$ *and* $f \in \mathrm{Aff}(\mathbf{A})$. *There exist* $\mathbf{v}, \mathbf{v}' \in \mathbf{V}$ *and* $g \in \mathrm{Aff}(\mathbf{A})_O$ *determined uniquely by* f, *for which*

$$\begin{aligned} f &= g \circ t_{\mathbf{v}} \\ f &= t_{\mathbf{v}'} \circ g. \end{aligned} \tag{4.4}$$

Proof

Put $\mathbf{v} = -\overrightarrow{Of^{-1}(O)}$, $\mathbf{v}' = \overrightarrow{Of(O)}$, $g = f \circ t_{-\mathbf{v}}$, and $g' = t_{-\mathbf{v}'} \circ f$. It is clear that $g \circ t_{\mathbf{v}} = f = t_{\mathbf{v}'} \circ g'$. Moreover,

$$\begin{aligned} g(O) &= (f \circ t_{-\mathbf{v}})(O) = f(f^{-1}(O)) = O \\ g'(O) &= (t_{-\mathbf{v}'} \circ f)(O) = (t_{-\mathbf{v}'})(f(O)) = O, \end{aligned}$$

that is, $g, g' \in \mathrm{Aff}(\mathbf{A})_O$. Since g and g' have the same associated automorphism, it now follows from Lemma 14.5 that they are the same. The identities $t_{\mathbf{v}} = g^{-1} \circ f$ and $t_{\mathbf{v}'} = f \circ g^{-1}$ show that also $\mathbf{v}$ and $\mathbf{v}'$ are uniquely determined. □

We now show how to describe the elements of $\mathrm{Aff}_n(\mathbf{K})$ explicitly.

Let $f : \mathbf{A}^n \to \mathbf{A}^n$ be an affine transformation, and let $\phi : \mathbf{K}^n \to \mathbf{K}^n$ be the associated isomorphism. Suppose that $A \in \mathrm{GL}_n(\mathbf{K})$ is the matrix of ϕ in the canonical basis, and that $f(\mathbf{0}) = \mathbf{c}$. By the definition of affine transformation, for each $\mathbf{x} \in \mathbf{A}^n$ one has

$$f(\mathbf{x}) - \mathbf{c} = f(\mathbf{x}) - f(\mathbf{0}) = A(\mathbf{x} - \mathbf{0}) = A\mathbf{x}.$$

The affine transformation f is therefore given by the formula

$$f(\mathbf{x}) = A\mathbf{x} + \mathbf{c}. \tag{14.5}$$

Conversely, for every $A \in \mathrm{GL}_n(\mathbf{K})$ and $\mathbf{c} \in \mathbf{K}^n$, the map $f_{A,\mathbf{c}} : \mathbf{A}^n \to \mathbf{A}^n$ defined by (14.5) is an affine transformation. Indeed,

$$\begin{aligned} \overrightarrow{f_{A,\mathbf{c}}(\mathbf{x}_1) f_{A,\mathbf{c}}(\mathbf{x}_2)} &= f_{A,\mathbf{c}}(\mathbf{x}_2) - f_{A,}(\mathbf{x}_1) \\ &= (A\mathbf{x}_2 + \mathbf{c}) - (A\mathbf{x}_1 + \mathbf{c}) = A(\mathbf{x}_2 - \mathbf{x}_1) \end{aligned}$$

and so $f_{A,\mathbf{c}}$ is an affine transformation, with associated isomorphism given by the matrix A in the canonical basis of $\mathbf{K}^n$.

Thus, $\mathrm{Aff}_n(\mathbf{K})$ is equal to *the set of all transformations* $f_{A,\mathbf{c}}$.

In the case $n = 1$, it follows that the affine transformations of $\mathbf{A}^1$ are the transformations of the form $f(x) = ax + c$ for some $a, c \in \mathbf{K}$ with $a \neq 0$.

Given two affine transformations $f_{A,\mathbf{c}}, f_{B,\mathbf{d}} \in \mathrm{Aff}_n(\mathbf{K})$, their product is

$$f_{B,\mathbf{d}} \circ f_{A,\mathbf{c}} = f_{BA,\mathbf{d}+B\mathbf{c}}, \tag{14.6}$$

while the inverse of $f_{A,\mathbf{c}}$ is

$$(f_{A,\mathbf{c}})^{-1} = f_{A^{-1},-A^{-1}\mathbf{c}}.$$

The affine transformations $f_{\mathbf{1},\mathbf{c}}$ are the translations of $\mathbf{A}^n$, which are denoted $t_{\mathbf{c}}$; thus,

$$t_{\mathbf{c}}(\mathbf{x}) = \mathbf{x} + \mathbf{c}.$$

The affine transformations belonging to the stabilizer $\mathrm{Aff}_n(\mathbf{K})_O$ of the origin are precisely those of the form $f_{A,\mathbf{0}}$, and they correspond bijectively to the matrices $A \in \mathrm{GL}_n(\mathbf{K})$. In this way, one obtains an identification of $\mathrm{Aff}_n(\mathbf{K})_O$ and $\mathrm{GL}_n(\mathbf{K})$.

In the case of $\mathbf{A}^n$, Lemma 14.7 states that avery affine transformation $f_{A,\mathbf{c}} \in \mathrm{Aff}_n(\mathbf{K})$ can always be writen as

$$f_{A,\mathbf{c}} = t_{\mathbf{c}} \circ f_{A,\mathbf{0}} = f_{A,\mathbf{0}} \circ t_{A^{-1}\mathbf{c}}.$$

The affine transformations of any other affine space $\mathbf{A}$ can be described explicitly in a way very similar to $\mathbf{A}^n$, once a frame of reference $O\mathbf{e}_1 \dots \mathbf{e}_n$ has been fixed.

Theorem 14.8
Fix a reference frame $O\mathbf{e}_1 \dots \mathbf{e}_n$ in the affine space $\mathbf{A}$ over $\mathbf{V}$. Every affine transformation $f \in \mathrm{Aff}(\mathbf{A})$ with associated automorphism ϕ, can be written in the form

$$f(P(x_1, \dots, x_n)) = Q(y_1, \dots, y_n)$$

with

$$\mathbf{y} = A\mathbf{x} + \mathbf{c}, \tag{14.7}$$

where $\mathbf{c} = (c_1, \dots, c_n)^t \in \mathbf{K}^n$ is the vector with coordinates $f(O)$, and $A = M_{\mathbf{e}}(\phi) \in \mathrm{GL}_n(\mathbf{K})$.

Conversely, every transformation $f : \mathbf{A} \to \mathbf{A}$ of the form (14.7) *for some $A \in \mathrm{GL}_n(\mathbf{K})$ and $\mathbf{c} \in \mathbf{K}^n$ is an affine transformation.*

In the particular case that $\mathbf{A} = \mathbf{A}^n$ and $O\mathbf{e}_1 \dots \mathbf{e}_n$ is the standard reference frame, we regain the description of affine transformations given above. The proof of the theorem is essentially identical to the preceding one, and is left to the reader.

Notice the analogy between (14.7) and (12.3) which expresses the change of coordinates from one reference frame to another. However, the two formulae describe two operations of a totally different nature: (12.3) relates the coordinates of a single point in two different frames, while (14.7) relates the coordinates of two different points in a single frame.

Corollary 14.9
Fix a reference frame $O\mathbf{e}_1 \dots \mathbf{e}_n$ in the affine space $\mathbf{A}$ over $\mathbf{V}$. The map

$$\mathrm{Aff}(\mathbf{A}) \to \mathrm{Aff}_n(\mathbf{K}) \tag{14.8}$$

which associates to an affine transformation f of $\mathbf{A}$ the affine transformation $f_{A,\mathbf{c}}$ of $\mathbf{A}^n$ given by (14.7) is an isomorphism of groups.

Proof
It follows from Theorem 14.8 that (14.8) is bijective, and so it suffices to show that it is a homomorphism. If $f, g \in \mathrm{Aff}(\mathbf{A})$, with associated

automorphisms ϕ, ψ respectively, are given by (14.7) and by

$$\mathbf{y} = B\mathbf{x} + \mathbf{d}$$

respectively, their product $g \circ f$ has associated automorphism $\psi \circ \phi$, which is represented by the matrix BA in the basis $\boldsymbol{e}$. Moreover, the point $(g \circ f)(O)$ has coordinates $B\mathbf{c} + \mathbf{d}$. Therefore the affine transformation of $\mathbf{A}^n$ corresponding to $g \circ f$ in (14.8) is $f_{BA,B\mathbf{c}+\mathbf{d}}$. That (14.8) is a homomorphism now follows from (14.6) and the fact that $f_{A,\mathbf{c}}$ and $f_{B,\mathbf{d}}$ correspond respectively to f and g. □

The subsets of an affine space $\mathbf{A}$ are often called *(affine) geometric figures of* $\mathbf{A}$.

Definition 14.10

Two geometric figures $\mathbf{F}, \mathbf{F}' \subset \mathbf{A}$ are said to be *affinely equivalent* if there is an affine transformation taking $\mathbf{F}$ to $\mathbf{F}'$, that is there exists $f \in \mathrm{Aff}(\mathbf{A})$ with $f(\mathbf{F}) = \mathbf{F}'$.

An *affine property* of a figure $\mathbf{F}$ is a property which is common to all figures that are affinely equivalent to $\mathbf{F}$.

If, for example, $\mathbf{F}$ is a finite set of points, then the number of points in $\mathbf{F}$ is an affine property as every affine transformation is a bijection. If $\mathbf{F}$ is an affine subspace then its dimension is an affine property, as is verified by the following result.

Proposition 14.11

Let $\mathbf{F}$ *be an affine subspace of* $\mathbf{A}$ *and* $f \in \mathrm{Aff}(\mathbf{A})$. *The image* $f(\mathbf{F})$ *is also an affine subspace, with* $\dim(f(\mathbf{F})) = \dim(\mathbf{F})$.

Proof

Let $\mathbf{F}$ be the subspace passing through $Q \in \mathbf{A}$ with associated vector subspace $\mathbf{W}$, and the automorphism associated to f be $\phi \in \mathrm{GL}(\mathbf{V})$. Then $\phi(\mathbf{W})$ is a vector subspace of $\mathbf{V}$ and $\dim(\phi(\mathbf{W})) = \dim(\mathbf{W})$. Moreover, for every $P \in \mathbf{F}$ we have $\overrightarrow{f(Q)f(P)} = \phi(\overrightarrow{QP}) \in \phi(\mathbf{W})$. Therefore $f(P)$ belongs to the subspace S passing through $f(Q)$ and having associated vector subspace $\phi(\mathbf{W})$. Conversely, for every $R \in S$ one has $\overrightarrow{Qf^{-1}(R)} = \phi^{-1}(\overrightarrow{f(Q)R}) \in \mathbf{W}$, that is $f^{-1}(R) \in \mathbf{F}$, and so $R \in f(\mathbf{F})$. Thus, $f(\mathbf{F}) = S$, and the assertion is proved. □

Proposition 14.12
Let $\mathbf{A}$ *be an affine space over* $\mathbf{V}$ *of dimension* n, *and* $\{P_0, P_1, \ldots, P_n\}$ *and* $\{Q_0, Q_1, \ldots, Q_n\}$ *be two* $(n+1)$*-tuples of independent points. Then there exists a unique affine transformation* $f : \mathbf{A} \to \mathbf{A}$ *satisfying* $f(P_i) = Q_i$ *for* $i = 0, 1, \ldots, n$.

Proof
By the assumption of independence, the two sets of vectors

$$\{\overrightarrow{P_0P_1}, \overrightarrow{P_0P_2}, \ldots, \overrightarrow{P_0P_n}\} \quad \text{and} \quad \{\overrightarrow{Q_0Q_1}, \overrightarrow{Q_0Q_2}, \ldots, \overrightarrow{Q_0Q_n}\}$$

form bases of $\mathbf{V}$. Therefore the unique linear operator $\phi : \mathbf{V} \to \mathbf{V}$ for which $\phi(\overrightarrow{P_0P_i}) = \overrightarrow{Q_0Q_i}$ $(i = 1, 2, \ldots, n)$ is an isomorphism. Define $f : \mathbf{A} \to \mathbf{A}$ by

$$\overrightarrow{Q_0f(P)} = \phi(\overrightarrow{P_0P})$$

for each $P \in \mathbf{A}$. Clearly f is a bijection, and satisfies the identity

$$\begin{aligned} \overrightarrow{f(P)f(P')} &= \overrightarrow{Q_0f(P')} - \overrightarrow{Q_0f(P)} = \phi(\overrightarrow{P_0P'}) - \phi(\overrightarrow{P_0P}) \\ &= \phi(\overrightarrow{P_0P'} - \overrightarrow{P_0P}) = \phi(\overrightarrow{PP'}), \end{aligned}$$

and so f is an affine transformation. Moreover,

$$\overrightarrow{Q_0f(P_0)} = \phi(\overrightarrow{P_0P_0}) = \mathbf{0},$$

and so $f(P_0) = Q_0$. Finally, for each $i = 1, \ldots, n$ one has

$$\overrightarrow{Q_0f(P_i)} = \phi(\overrightarrow{P_0P_i}) = \overrightarrow{Q_0Q_i}$$

by definition of ϕ; therefore $f(P_i) = Q_i$.

The uniqueness of f follows from the uniqueness of ϕ, together with the condition $f(P_0) = Q_0$ and Lemma 14.5. □

Corollary 14.13
Let $\mathbf{A}$ *be an* n*-dimensional affine space over* $\mathbf{V}$. *Then:*

1) *For each* $1 \leq k \leq n+1$, *any two* k*-tuples of independent points of* $\mathbf{A}$ *are affinely equivalent.*
2) *Two affine subspaces of* $\mathbf{A}$ *are affinely equivalent if and only if they have the same dimension.*

Proof

1) Let $\{P_0, P_1, \ldots, P_{k-1}\}$ and $\{Q_0, Q_1, \ldots, Q_{k-1}\}$ be two k-tuples of independent points. Then there are points $P_k, \ldots, P_n$ and $Q_k, \ldots, Q_n$ such that $\{P_0, P_1, \ldots, P_n\}$ and $\{Q_0, Q_1, \ldots, Q_n\}$ are two $(n+1)$-tuples of independent points. By Proposition 14.12 there is an affine transformation which sends $\{P_0, P_1, \ldots, P_{k-1}\}$ to $\{Q_0, Q_1, \ldots, Q_{k-1}\}$.

2) If the affine subspaces S and S' are affinely equivalent then by Proposition 14.11 they have the same dimension.

Conversely, suppose that $\dim(S) = \dim(S') = s$. It is possible to find $s+1$ independent points $P_0, P_1, \ldots, P_s \in S$ such that $S = \overline{P_0 P_1 \ldots P_s}$. Similarly there are independent points $Q_0, Q_1, \ldots, Q_s \in S'$ such that $S' = \overline{Q_0 Q_1 \ldots Q_s}$. Let f be an affine transformation for which $f(P_i) = Q_i$ for $i = 0, \ldots, s$. Then $f(S) = S'$. Indeed, $f(S)$ contains $Q_0, \ldots, Q_s$ and is an affine subspace one has $S' \subset f(S)$. However, they have the same dimension, so they must be equal. □

EXERCISES

14.1 Prove that the groups SO(2) and $\mathbf{U}(1)$ are isomorphic.

14.2 For each of the following matrices A find $\overline{A}$, A^t and A^*:

a) $\begin{pmatrix} 1 & 0 \\ 0 & -1 \end{pmatrix}$ b) $\begin{pmatrix} i & 0 \\ 0 & i \end{pmatrix}$

c) $\begin{pmatrix} 1 & i \\ i & 1 \end{pmatrix}$ d) $\begin{pmatrix} \sqrt{2}/2 & \sqrt{2}i/2 \\ \sqrt{2}i/2 & \sqrt{2}/2 \end{pmatrix}$

e) $\begin{pmatrix} 3/5 & 4i/5 \\ 4i/5 & 3/5 \end{pmatrix}$ f) $\begin{pmatrix} 1 & 1+i \\ 1-i & -1 \end{pmatrix}$.

14.3 Which of the matrices in the preceding exercise are unitary?

14.4 Let $A \in \mathbf{U}(n)$. Prove that each of the matrices A^t, $\overline{A}$ and A^* is unitary.

14.5 Suppose that $\mathcal{P}$, $\mathcal{Q}$ and $\mathcal{R}$ are lines in an affine plane $\mathbf{A}$ which do not all three belong to the same pencil and no two of which are parallel, and similarly $\mathcal{P}'$, $\mathcal{Q}'$ and $\mathcal{R}'$ are another three lines in $\mathbf{A}$ with the same properties. Prove that there is an affine transformation $f : \mathbf{A} \to \mathbf{A}$ such that $f(\mathcal{P}) = \mathcal{P}'$, $f(\mathcal{Q}) = \mathcal{Q}'$ and $f(\mathcal{R}) = \mathcal{R}'$.

14.6 Suppose that $\mathcal{P}_1, \mathcal{P}_2, \mathcal{P}_3$ and $\mathcal{Q}_1, \mathcal{Q}_2, \mathcal{Q}_3$ are triples of planes in a 3-dimensional affine space $\mathbf{A}$ such that both $\mathcal{P}_1 \cap \mathcal{P}_2 \cap \mathcal{P}_3$ and $\mathcal{Q}_1 \cap \mathcal{Q}_2 \cap \mathcal{Q}_3$ consist of just one point. Prove that there exists $f \in \mathrm{Aff}(\mathbf{A})$ with $f(\mathcal{P}_i) = \mathcal{Q}_i$ for $i = 1, 2, 3$.

14.7 In each of the following cases find the affine transformation $f : \mathbf{A}^2 \to \mathbf{A}^2$ which satisfies the given conditions.
a) $f(0,0) = (1,-1)$, $f(1,0) = (3,-1)$, $f(0,1) = (2,2)$
b) $f(2,1) = (1,2)$, $f(-1,-1) = (1,1)$, $f(0,1) = (2,-1)$
c) $f(\mathcal{P}) = \mathcal{P}'$, $f(\mathcal{Q}) = \mathcal{Q}'$, $f(\mathcal{R}) = (\mathcal{R}')$ with $\mathcal{P} : X = 1$, $\mathcal{Q} : Y = X$, $\mathcal{R} : Y = -2$, $\mathcal{P}' : 2X - Y = 0$, $\mathcal{Q}' : X + Y = 0$, $\mathcal{R}' : 2X + Y = 1$.

14.8 Let $\mathbf{A}$ be an affine plane. Prove:
a) any two half-lines in $\mathbf{A}$ are affinely equivalent. (*Hint*: show that a point and a non-zero vector can be transformed into any other point and non-zero vector.)
b) Any two segements in $\mathbf{A}$ are affinely equivalent.
c) Any two half-planes in $\mathbf{A}$ are affinely equivalent.
d) Any two triangles in $\mathbf{A}$ are affinely equivalent.

14.9 Let $\mathbf{A}$ be a real affine space. Prove the following.
a) Any two half-spaces in $\mathbf{A}$ are affinely equivalent.
b) If $\mathbf{U} \subset \mathbf{A}$ is convex, then any subset of $\mathbf{A}$ affinely equivalent to $\mathbf{U}$ is also convex. In other words, convexity is an affine property.
c) Every affine transformation maps the mid-point of a segment to the mid-point of the image segment.

14.10 Let $\mathbf{b} \in \mathbf{A}^n$ and $c \in \mathbf{K}^*$. Prove that $\omega_{\mathbf{b},c} = T_{c\mathbf{1},\mathbf{b}(1-c)}$.

14.11 Let $n \geq 1$ be an integer. Two integers $a, b \in \mathbf{Z}$ are said to be *congruent modulo* n if $b - a$ is divisible by n. Prove that congruence modulo n is an equivalence relation on $\mathbf{Z}$.

The set of equivalence classes (also called *residue classes*) modulo n is denoted $\mathbf{Z}/n\mathbf{Z}$, and consists of the n elements $\overline{0}, \overline{1}, \ldots, \overline{n-1}$, which are the classes of $0, 1, \ldots, n-1$. Prove that addition in $\mathbf{Z}$ induces an operation in $\mathbf{Z}/n\mathbf{Z}$ with respect to which $\mathbf{Z}/n\mathbf{Z}$ is an abelian group.

14.12 Let G be a group. A subset $\tilde{G}$ of G is said to be a *system of generators* of G if every element of G can be expressed as a product of a finite number of elements of $\tilde{G}$ and their inverses. If $\tilde{G}$ is a finite set then G is said to be *finitely generated*. If $\tilde{G} = \{g\}$ consists of only one element then G is said to be a *cyclic group* and g a *generator*. Prove that

a) Every cyclic group is abelian.

b) $\mathbf{Z}$ (with operation $+$) and $\mathbf{Z}/n\mathbf{Z}$ are cyclic groups.

c) Every cyclic group is isomorphic to $\mathbf{Z}$ or to $\mathbf{Z}/n\mathbf{Z}$ for some $n \geq 1$.

d) The set of n-th roots of 1 is a cyclic subgroup of $\mathbf{C}$ (any generator is called a *primitive* n-th root).

14.13 Let $n \geq 2$ be an integer. Define on the Cartesian product

$$\frac{\mathbf{Z}}{n\mathbf{Z}} \times \frac{\mathbf{Z}}{2\mathbf{Z}}$$

an operation as follows:

$$\begin{aligned}(\overline{a}_1, \overline{0})(\overline{a}_2, \overline{0}) &= (\overline{a}_1 + \overline{a}_2, \overline{0})\\ (\overline{a}_1, \overline{0})(\overline{a}_2, \overline{1}) &= (\overline{a}_1 + \overline{a}_2, \overline{1}) = (\overline{a}_1, \overline{1})(-\overline{a}_2, \overline{0})\\ (\overline{a}_1, \overline{1})(\overline{a}_2, \overline{1}) &= (\overline{a}_1 - \overline{a}_2, \overline{0}).\end{aligned}$$

Prove that

a) with this operation, $\frac{\mathbf{Z}}{n\mathbf{Z}} \times \frac{\mathbf{Z}}{2\mathbf{Z}}$ is a group whose identity element is $e = (\overline{0}, \overline{0})$. This group, denoted D_{2n}, is called the *dihedral group of order* $2n$;

b) putting $x = (\overline{1}, \overline{0})$, and $y = (\overline{0}, \overline{1})$, one has

$$D_{2n} = \{x^0 = e, x, \ldots, x^{n-1}, y, xy, \ldots, x^{n-1}y\};$$

c) x generates a subgroup of D_{2n} isomorphic to $\mathbf{Z}/n\mathbf{Z}$;

d) y generates a subgroup of D_{2n} isomorphic to $\mathbf{Z}/2\mathbf{Z}$;

e) D_{2n} is not abelian.

14.14 Let $n \geq 2$ be an integer. The set S_n of all permutations of a set of n elements is a group of transformations called the *symmetric group on n elements*. Prove that S_n is not abelian if $n \geq 3$.

Part II
Euclidean Geometry

15

Bilinear forms and quadratic forms

In affine geometry the only geometric properties that make sense are those which do not make use of metric notions. In Part II we will see how notions such as distance, angle and perpendicular can only be defined if one introduces into affine space a finer structure, that of Euclidean space. To do this we will need to introduce some further linear algebra, the theory of bilinear and quadratic forms, which we begin studying in this chapter. We will also deal with some topics which are beyond those needed here, but are of fundamental importance in mathematics.

Definition 15.1
Let $\mathbf{V}$ be a vector space over $\mathbf{K}$. A map

$$b : \mathbf{V} \times \mathbf{V} \to \mathbf{K}$$

is a *bilinear form on* $\mathbf{V}$ if it is linear in each of its arguments. That is if it satisfies:

[BF1] $b(\mathbf{v} + \mathbf{v}', \mathbf{w}) = b(\mathbf{v}, \mathbf{w}) + b(\mathbf{v}', \mathbf{w})$
[BF2] $b(\mathbf{v}, \mathbf{w} + \mathbf{w}') = b(\mathbf{v}, \mathbf{w}) + b(\mathbf{v}, \mathbf{w}')$
[BF3] $b(k\mathbf{v}, \mathbf{w}) = b(\mathbf{v}, k\mathbf{w}) = kb(\mathbf{v}, \mathbf{w})$

for every $\mathbf{v}, \mathbf{v}', \mathbf{w}, \mathbf{w}' \in \mathbf{V}$ and $k \in \mathbf{K}$. The bilinear form is *symmetric* if

$$b(\mathbf{v}, \mathbf{w}) = b(\mathbf{w}, \mathbf{v}) \quad \text{for every} \quad \mathbf{v}, \mathbf{w} \in \mathbf{V};$$

it is *skew-symmetric* if

$$b(\mathbf{v}, \mathbf{w}) = -b(\mathbf{w}, \mathbf{v}) \quad \text{for every} \quad \mathbf{v}, \mathbf{w} \in \mathbf{V}.$$

If b is skew-symmetric, then $b(\mathbf{v},\mathbf{v}) = -b(\mathbf{v},\mathbf{v}) = 0$ for every $\mathbf{v} \in \mathbf{V}$. Conversely, a bilinear form satisfying $b(\mathbf{v},\mathbf{v}) = 0$ for every $\mathbf{v} \in \mathbf{V}$ is skew-symmetric. This can be seen as follows:

$$\begin{aligned} 0 &= b(\mathbf{v}+\mathbf{w},\mathbf{v}+\mathbf{w}) = b(\mathbf{v},\mathbf{v}) + b(\mathbf{v},\mathbf{w}) + b(\mathbf{w},\mathbf{v}) + b(\mathbf{w},\mathbf{w}) \\ &= b(\mathbf{v},\mathbf{w}) + b(\mathbf{w},\mathbf{v}). \end{aligned}$$

Examples 15.2

1. A trivial example of a bilinear form on any vector space is the *zero form* $0(\mathbf{v},\mathbf{w}) = 0$ for all $\mathbf{v},\mathbf{w} \in \mathbf{V}$. The zero form is both symmetric and skew-symmetric.

2. Let $A = (a_{ij}) \in M_n(\mathbf{K})$. Considering the vectors in $\mathbf{K}^n$ as column n-vectors one obtains a bilinear form on $\mathbf{K}^n$ by putting

$$b(\mathbf{x},\mathbf{y}) = \mathbf{x}^t A\mathbf{y} = \sum_{i,j} a_{ij}x_i y_j$$

for every $\mathbf{x} = (x_1,\ldots,x_n)^t$, $\mathbf{y} = (y_1,\ldots,y_n)^t$. It follows easily from the basic properties of matrix multiplication that this form is indeed bilinear — see Proposition 2.2(1). If $\{\mathbf{E}_1,\ldots,\mathbf{E}_n\}$ is the canonical basis of $\mathbf{K}^n$ then, for each i,j,

$$b(\mathbf{E}_i,\mathbf{E}_j) = a_{ij}.$$

For example, take $\mathbf{K} = \mathbf{R}$, $n = 3$ and

$$A = \begin{pmatrix} 1 & 0 & -2 \\ 0 & 0 & \frac{1}{2} \\ -1 & \pi & 0 \end{pmatrix}.$$

The corresponding bilinear form on $\mathbf{R}^3$ is

$$b(\mathbf{x},\mathbf{y}) = x_1y_1 - 2x_1y_3 + \tfrac{1}{2}x_2y_3 - x_3y_1 + \pi x_3y_2.$$

This bilinear form is not symmetric because, for example,

$$b(\mathbf{E}_1,\mathbf{E}_3) = -2 \neq -1 = b(\mathbf{E}_3,\mathbf{E}_1).$$

If $A = \mathbf{I}_n$ then one obtains the *standard symmetric form* on $\mathbf{K}^n$,

$$b(\mathbf{x},\mathbf{y}) = \mathbf{x}^t\mathbf{y} = x_1y_1 + x_2y_2 + \cdots + x_ny_n. \tag{15.1}$$

If $n = 2k$ is even, and we take $A = \mathbf{J}_k$, where

$$\mathbf{J}_k = \begin{pmatrix} \mathbf{0}_k & \mathbf{I}_k \\ -\mathbf{I}_k & \mathbf{0}_k \end{pmatrix},$$

we obtain the *standard alternating form on* $\mathbf{K}^n$,

$$\begin{aligned} b(\mathbf{x}, \mathbf{y}) &= x_1 y_{k+1} + x_2 y_{k+2} + \cdots + x_k y_{2k} \\ &\quad - x_{k+1} y_1 - x_{k+2} y_2 + \cdots - x_{2k} y_k. \end{aligned}$$

Let $b : \mathbf{V} \times \mathbf{V} \to \mathbf{K}$ be a bilinear form. The bilinearity of b gives rise to two linear maps from $\mathbf{V}$ to its dual $\mathbf{V}^*$ as follows.

For each $\mathbf{v} \in \mathbf{V}$ the map $b_{\mathbf{v}} : \mathbf{V} \to \mathbf{K}$ defined by

$$b_{\mathbf{v}}(\mathbf{w}) = b(\mathbf{v}, \mathbf{w}),$$

for each $\mathbf{w} \in \mathbf{V}$, is linear, since

$$\begin{aligned} b_{\mathbf{v}}(c_1 \mathbf{w}_1 + c_2 \mathbf{w}_2) &= b(\mathbf{v}, c_1 \mathbf{w}_1 + c_2 \mathbf{w}_2) = c_1 b(\mathbf{v}, \mathbf{w}_1) + c_2 b(\mathbf{v}, \mathbf{w}_2) \\ &= c_1 b_{\mathbf{v}}(\mathbf{w}_1) + c_2 b_{\mathbf{v}}(\mathbf{w}_2). \end{aligned}$$

Thus $b_{\mathbf{v}} \in \mathbf{V}^*$, so we have a map

$$\delta_b : \mathbf{V} \to \mathbf{V}^*$$

defined by $\delta_b(\mathbf{v}) = b_{\mathbf{v}}$.

This map δ_b is linear, also because of the bilinearity of b. Indeed, for each $\mathbf{v}_1, \mathbf{v}_2 \in \mathbf{V}$ and $c_1, c_2 \in \mathbf{K}$,

$$\begin{aligned} [\delta_b(c_1 \mathbf{v}_1 + c_2 \mathbf{v}_2)](\mathbf{w}) &= b_{c_1 \mathbf{v}_1 + c_2 \mathbf{v}_2}(\mathbf{w}) = b(c_1 \mathbf{v}_1 + c_2 \mathbf{v}_2, \mathbf{w}) \\ &= c_1 b(\mathbf{v}_1, \mathbf{w}) + c_2 b(\mathbf{v}_2, \mathbf{w}) \\ &= c_1 b_{\mathbf{v}_1}(\mathbf{w}) + c_2 b_{\mathbf{v}_2}(\mathbf{w}) \\ &= [c_1 \delta_b(\mathbf{v}_1) + c_2 \delta_b(\mathbf{v}_2)](\mathbf{w}) \end{aligned}$$

for every $\mathbf{w} \in \mathbf{V}$. That is, $\delta_b(c_1 \mathbf{v}_1 + c_2 \mathbf{v}_2) = c_1 \delta_b(\mathbf{v}_1) + c_2 \delta_b(\mathbf{v}_2)$.

In a similar way b gives rise to another map $\delta'_b : \mathbf{V} \to \mathbf{V}^*$ defined by putting $\delta'_b(\mathbf{w}) = b'_{\mathbf{w}}$ where for each $\mathbf{w} \in \mathbf{V}$, $b'_{\mathbf{w}} : \mathbf{V} \to \mathbf{K}$ is the linear functional

$$b'_w(\mathbf{v}) = b(\mathbf{v}, \mathbf{w})$$

for every $\mathbf{v} \in \mathbf{V}$.

The reader should check that $\delta_b = \delta'_b$ if and only if b is symmetric.

Definition 15.3

Let $\mathbf{V}$ be a $\mathbf{K}$-vector space of dimension n, $\{\mathbf{e}_1, \ldots, \mathbf{e}_n\}$ a basis and let $b : \mathbf{V} \times \mathbf{V} \to \mathbf{K}$ be a bilinear form on $\mathbf{V}$. The *matrix of b* (or *representing b*) with respect to the basis $\{\mathbf{e}_1, \ldots, \mathbf{e}_n\}$ is the matrix $A = (a_{ij}) \in M_n(\mathbf{K})$ defined by

$$a_{ij} = b(\mathbf{e}_i, \mathbf{e}_j), \quad 1 \leq i, j \leq n.$$

The matrix A determines the bilinear form b. To see this write

$$\mathbf{v} = x_1\mathbf{e}_1 + \cdots + x_n\mathbf{e}_n, \quad \mathbf{w} = y_1\mathbf{e}_1 + \cdots + y_n\mathbf{e}_n,$$

then

$$\begin{aligned}
b(\mathbf{v}, \mathbf{w}) &= b(x_1\mathbf{e}_1 + \cdots + x_n\mathbf{e}_n, y_1\mathbf{e}_1 + \cdots + y_n\mathbf{e}_n) \\
&= b(x_1\mathbf{e}_1, y_1\mathbf{e}_1 + \cdots + y_n\mathbf{e}_n) + \cdots \\
&\quad + b(x_n\mathbf{e}_n, y_1\mathbf{e}_1 + \cdots + y_n\mathbf{e}_n) \\
&= x_1 b(\mathbf{e}_1, y_1\mathbf{e}_1 + \cdots + y_n\mathbf{e}_n) + \cdots \\
&\quad + x_n b(\mathbf{e}_n, y_1\mathbf{e}_1 + \cdots + y_n\mathbf{e}_n) \\
&= x_1[b(\mathbf{e}_1, y_1\mathbf{e}_1) + \cdots + b(\mathbf{e}_1, y_n\mathbf{e}_n)] + \cdots \\
&\quad + x_n[b(\mathbf{e}_n, y_1\mathbf{e}_1) + \cdots + b(\mathbf{e}_n, y_n\mathbf{e}_n)] \\
&= x_1[y_1 b(\mathbf{e}_1, \mathbf{e}_1) + \cdots + y_n b(\mathbf{e}_1, \mathbf{e}_n)] + \cdots \\
&\quad + x_n[y_1 b(\mathbf{e}_n, \mathbf{e}_1) + \cdots + y_n b(\mathbf{e}_n, \mathbf{e}_n)] \\
&= \sum_{i,j} x_i y_j b(\mathbf{e}_i, \mathbf{e}_j) = \mathbf{x}^t A \mathbf{y},
\end{aligned}$$

where $\mathbf{x}$ and $\mathbf{y}$ denote the column vectors with the coordinates of $\mathbf{v}$ and $\mathbf{w}$ respectively.

Conversely, for a fixed basis $\{\mathbf{e}_1, \ldots, \mathbf{e}_n\}$ of $\mathbf{V}$, any square matrix $A \in M_n(\mathbf{K})$ determines a bilinear form b on $\mathbf{V}$ by putting

$$b(\mathbf{v}, \mathbf{w}) = \sum_{i,j} a_{ij} x_i y_j = \mathbf{x}^t A \mathbf{y},$$

for every $\mathbf{v}(x_1, \ldots, x_n)$ and $\mathbf{w}(y_1, \ldots, y_n)$. This is clearly bilinear, because

$$\begin{aligned}
&b(\mathbf{v} + \mathbf{v}', \mathbf{w}) = (\mathbf{x} + \mathbf{x}')^t A \mathbf{y} = \mathbf{x} A \mathbf{y} + \mathbf{x}'^t A \mathbf{y} = b(\mathbf{v}, \mathbf{w}) + b(\mathbf{v}', \mathbf{w}), \\
&b(\mathbf{v}, \mathbf{w} + \mathbf{w}') = \mathbf{x}^t A \mathbf{y} + \mathbf{y}' = \mathbf{x}^t A \mathbf{y} + \mathbf{x}^t A \mathbf{y}' = b(\mathbf{v}, \mathbf{w}) + b(\mathbf{v}, \mathbf{w}'), \\
&b(k\mathbf{v}, \mathbf{w}) = (k\mathbf{x})^t A \mathbf{y} = k\mathbf{x}^t A \mathbf{y} = kb(\mathbf{v}, \mathbf{w}),
\end{aligned}$$

$$b(\mathbf{v}, k\mathbf{w}) = \mathbf{x}^t A(k\mathbf{y}) = k\mathbf{x}^t A\mathbf{y} = kb(\mathbf{v}, \mathbf{w}).$$

It is clear that the bilinear form just defined has A again as its matrix with respect to the basis $\{\mathbf{e}_1, \ldots, \mathbf{e}_n\}$.

Notice that

$$b(\mathbf{w}, \mathbf{v}) = \mathbf{y}^t A\mathbf{x} = \mathbf{x}^t A^t \mathbf{y},$$

and so $b(\mathbf{v}, \mathbf{w}) = b(\mathbf{w}, \mathbf{v})$ if and only if $A = A^t$, that is A is a symmetric matrix. Similarly, b is skew-symmetric if and only if A is skew-symmetric.

Suming up, we have the following proposition.

Proposition 15.4

Let $\mathbf{V}$ be a finite dimensional vector space and $\mathbf{e} = \{\mathbf{e}_1, \ldots, \mathbf{e}_n\}$ a basis. Associating to each bilinear form its matrix with respect to $\mathbf{e}$ gives rise to a bijection between the set Bil($\mathbf{V}$) *of bilinear forms on $\mathbf{V}$ and the set $M_n(\mathbf{K})$. This bijection induces a bijection of the sets of symmetric and skew-symmetric bilinear forms with the sets of symmetric and skew-symmetric matrices, respectively.*

The bijection described in Proposition 15.4 depends on the choice of basis; that is, the matrices representing a given bilinear form in two different bases are in general different. Let us see exactly how they differ.

Let $b : \mathbf{V} \times \mathbf{V} \to \mathbf{K}$ be a bilinear form and let $\boldsymbol{e} = \{\mathbf{e}_1, \ldots, \mathbf{e}_n\}$ and $\boldsymbol{f} = \{\mathbf{f}_1, \ldots, \mathbf{f}_n\}$ be two bases of $\mathbf{V}$. Let

$$\begin{aligned} A &= (a_{ij}) = (b(\mathbf{e}_i, \mathbf{e}_j)) \\ B &= (b_{ij}) = (b(\mathbf{f}_i, \mathbf{f}_j)) \end{aligned}$$

be the matrices representing b in these two bases. If $\mathbf{v}, \mathbf{w} \in \mathbf{V}$ are two arbitrary vectors, with coordinates

$$\begin{aligned} \mathbf{v} &= x_1\mathbf{v}_1 + \cdots + x_n\mathbf{v}_n = x'_1\mathbf{f}_1 + \cdots + x'_n\mathbf{f}_n, \\ \mathbf{w} &= y_1\mathbf{v}_1 + \cdots + y_n\mathbf{v}_n = y'_1\mathbf{f}_1 + \cdots + y'_n\mathbf{f}_n, \end{aligned}$$

in each of the bases, respectively, then

$$b(\mathbf{v}, \mathbf{w}) = \mathbf{x}^t A\mathbf{y} = \mathbf{x}'^t B\mathbf{y}'. \tag{15.2}$$

Putting $M = M_{\boldsymbol{e},\boldsymbol{f}}(\mathbf{1_V})$ gives $\mathbf{x} = M\mathbf{x}'$ and $\mathbf{y} = M\mathbf{y}'$. Substituting into (15.2) gives

$$\mathbf{x}'^t B\mathbf{y}'(M\mathbf{x}')^t A(M\mathbf{y}') = \mathbf{x}'^t M^t AM\mathbf{y}'. \tag{15.3}$$

Since (15.3) holds for every $\mathbf{x}', \mathbf{y}' \in \mathbf{K}^n$ it follows that

$$B = M^t AM. \tag{15.4}$$

Equation (15.4) expresses the relationship between the matrices A and B which represent the bilinear form b with respect to two different bases $\boldsymbol{e}$ and $\boldsymbol{f}$.

Conversely, if A is a matrix representing the bilinear form b with respect to the basis $\boldsymbol{e}$, and if $M \in \mathrm{GL}_n(\mathbf{K})$ is any invertible square matrix, then there is a basis $\boldsymbol{f}$ such that $M = M_{\boldsymbol{e},\boldsymbol{f}}(\mathbf{1_V})$. Thus, $B = M^t AM$ is the matrix representing b with respect to the basis $\boldsymbol{f}$.

One says that two matrices $A, B \in M_n(\mathbf{K})$ are *congruent* if there exists $M \in \mathrm{GL}_n(\mathbf{K})$ satisfying

$$B = M^t AM.$$

We leave to the reader the task of showing that congruency of matrices is an equivalence relation.

We have shown the following result.

Proposition 15.5

Let $\mathbf{V}$ be a $\mathbf{K}$-vector space of dimension n. Two matrices represent the same bilinear form b on $\mathbf{V}$ with respect to two bases if and only if they are congruent.

It follows from Proposition 5.3(2) that two congruent matrices have the same rank. Thus, by Proposition 15.5, the rank r of a matrix which represents a bilinear form b with respect to some basis does not depend on the basis, but only on b. We call r the *rank of the bilinear form b.*

If b has rank $r = \dim(\mathbf{V})$, it is said to be a *non-degenerate* bilinear form. Otherwise it is a *degenerate* bilinear form. The following proposition gives several characterizations of non-degenerate bilinear forms.

Proposition 15.6

Let $\mathbf{V}$ *be a finite dimensional* $\mathbf{K}$*-vector space and let* $b : \mathbf{V} \times \mathbf{V} \to \mathbf{K}$ *be a bilinear form. The following properties are equivalent:*

1) b *is non-degenerate.*
2) *For every* $\mathbf{v} \neq \mathbf{0}$ *in* $\mathbf{V}$ *there is a* $\mathbf{w} \in \mathbf{V}$ *for which* $b(\mathbf{v}, \mathbf{w}) \neq 0$.
3) *For every* $\mathbf{w} \neq \mathbf{0}$ *in* $\mathbf{V}$ *there is a* $\mathbf{v} \in \mathbf{V}$ *for which* $b(\mathbf{v}, \mathbf{w}) \neq 0$.
4) *The map* $\delta_b : \mathbf{V} \to \mathbf{V}^*$ *is an isomorphism.*
5) *The map* $\delta'_b : \mathbf{V} \to \mathbf{V}^*$ *is an isomorphism.*

Proof

Fix a basis $\boldsymbol{e} = \{\mathbf{e}_1, \ldots, \mathbf{e}_n\}$ of $\mathbf{V}$ and let $A \in M_n(\mathbf{K})$ be the matrix of b with resect to $\boldsymbol{e}$.

(1) $\Rightarrow$ (2) If $\operatorname{rk} A = n$ and $\mathbf{x} \neq \mathbf{0}$ is the vector of coordinates of $\mathbf{v}$, then $\mathbf{x}^t A \neq (0, \ldots, 0)$, and so there is a $\mathbf{y} \in \mathbf{K}^n$ such that $\mathbf{x}^t A \mathbf{y} \neq 0$. The vector $\mathbf{w}$ with coordinates $\mathbf{y}$ is such that $b(\mathbf{v}, \mathbf{w}) \neq 0$.

(2) $\Rightarrow$ (1) By hypothesis, for every $\mathbf{x} \neq \mathbf{0}$ there is a $\mathbf{y}$ for which $\mathbf{x}^t A \mathbf{y} \neq 0$. This implies that $\mathbf{x}^t A \neq (0, \ldots, 0)$ for every $\mathbf{x} \neq \mathbf{0}$, and this in turn implies that $\operatorname{rk} A = n$.

(1) $\Leftrightarrow$ (3) This is proved similarly.

(2) $\Rightarrow$ (4) Since $\dim(\mathbf{V}) = \dim(\mathbf{V}^*)$ it suffices to show that $\ker(\delta_b) = \langle 0 \rangle$. So, let $\mathbf{v} \in \mathbf{V}$ be such that $b_{\mathbf{v}}$ is the zero functional. Then

$$0 = b_{\mathbf{v}}(\mathbf{w}) = b(\mathbf{v}, \mathbf{w})$$

for every $\mathbf{w} \in \mathbf{V}$. This contradicts (2) unless $\mathbf{v} = \mathbf{0}$.

(4) $\Rightarrow$ (2) For every $\mathbf{v} \in \mathbf{V}$ with $\mathbf{v} \neq \mathbf{0}$, one has $\delta_b(\mathbf{v}) = b_{\mathbf{v}} \neq 0$, and so there is a $\mathbf{w} \in \mathbf{V}$ for which $0 \neq b_{\mathbf{v}}(\mathbf{w}) = b(\mathbf{v}, \mathbf{w})$.

The equivalence of (3) and (5) is proved analogously. □

From now on we consider only *symmetric* bilinear forms, which are of particular importance for the geometrical notions we will be developing.

Definition 15.7

Let b be a symmetric bilinear form on a vector space $\mathbf{V}$ and let $\mathbf{v} \in \mathbf{V}$. A vector $\mathbf{w}$ is said to be *orthogonal to* $\mathbf{v}$ if $b(\mathbf{v}, \mathbf{w}) = 0$. In this case, one also says that the *two vectors are orthogonal.*

Suppose we are given a symmetric bilinear form b on $\mathbf{V}$, and let S

be any non-empty subset of $\mathbf{V}$. The set of vectors orthogonal to every vector in S is denoted $S^\perp$:

$$S^\perp = \{\mathbf{w} \in \mathbf{V} \mid b(\mathbf{v}, \mathbf{w}) = 0 \quad \text{for every } \mathbf{v} \in S\}.$$

It is easy to check that $S^\perp$ is a vector subspace of $\mathbf{V}$. Indeed, if $\mathbf{w}, \mathbf{w}' \in \mathbf{V}$ and $k \in \mathbf{K}$ then

$$\begin{aligned} b(\mathbf{v}, \mathbf{w} + \mathbf{w}') &= b(\mathbf{v}, \mathbf{w}) + b(\mathbf{v}, \mathbf{w}') = 0 + 0 = 0, \\ b(\mathbf{v}, k\mathbf{w}) &= kb(\mathbf{v}, \mathbf{w}) = k0 = 0, \end{aligned}$$

for every $\mathbf{v} \in S$. The space $S^\perp$ is called the *subspace orthogonal to* S. If $S = \{\mathbf{v}\}$ then we usually write $\mathbf{v}^\perp$ rather than $\{\mathbf{v}\}^\perp$.

Two subspaces $\mathbf{U}$ and $\mathbf{W}$ of $\mathbf{V}$ are said to be *orthogonal* if $\mathbf{U} \subset \mathbf{W}^\perp$; it follows immediately from the symmetry of b that this is equivalent to $\mathbf{W} \subset \mathbf{U}^\perp$. The subspace $\mathbf{V}^\perp$ is called the *radical* of $\mathbf{V}$. It follows from Proposition 15.6 that b is non-degenerate if and only if $\mathbf{V}^\perp = \langle \mathbf{0} \rangle$.

A vector $\mathbf{v} \in \mathbf{V}$ is said to be *isotropic* if $\mathbf{v} \in \mathbf{v}^\perp$, that is if $b(\mathbf{v}, \mathbf{v}) = 0$. Clearly $\mathbf{0}$ is isotropic. If $\mathbf{v}$ is isotropic then for any $k \in \mathbf{K}$,

$$b(k\mathbf{v}, k\mathbf{v}) = k^2 b(\mathbf{v}, \mathbf{v}) = k^2 0 = 0$$

and so the subspace $\langle \mathbf{v} \rangle$ consists entirely of isotropic vectors.

Suppose $\mathbf{v}$ is not isotropic. For any $\mathbf{w} \in \mathbf{V}$ define

$$a_{\mathbf{v}}(\mathbf{w}) = b(\mathbf{v}, \mathbf{w}) / b(\mathbf{v}, \mathbf{v}). \tag{15.6}$$

Then

$$b(\mathbf{v}, \mathbf{w} - a_{\mathbf{v}}(\mathbf{w})\mathbf{v}) = 0$$

that is $\mathbf{w} - a_{\mathbf{v}}(\mathbf{w})\mathbf{v} \in \mathbf{v}^\perp$. Since,

$$\mathbf{w} = a_{\mathbf{v}}(\mathbf{w})\mathbf{v} + (\mathbf{w} - a_{\mathbf{v}}(\mathbf{w})\mathbf{v})$$

it follows that $\mathbf{V} = \langle \mathbf{v} \rangle + \mathbf{v}^\perp$. On the other hand, $\langle \mathbf{v} \rangle \cap \mathbf{v}^\perp = \langle \mathbf{0} \rangle$ as $\mathbf{v}$ is not orthogonal to itself. Thus for any non-isotropic vector $\mathbf{v} \in \mathbf{V}$ we have

$$\mathbf{V} = \langle \mathbf{v} \rangle \oplus \mathbf{v}^\perp. \tag{15.7}$$

The scalar $a_{\mathbf{v}}(\mathbf{w})$ defined in (15.6) is sometimes called the *Fourier coefficient of* $\mathbf{w}$ *with respect to* $\mathbf{v}$. Note that $a_{\mathbf{v}}(\mathbf{w})$ is only defined if $\mathbf{v}$ is not isotropic.

If $\mathbf{V}$ is finite dimensional and $\boldsymbol{e} = \{\mathbf{e}_1, \ldots, \mathbf{e}_n\}$ is a basis whose vectors are pairwise orthogonal, that is, if $b(\mathbf{e}_i, \mathbf{e}_j) = 0$ whenever $i \neq j$, then $\boldsymbol{e}$ is called a *diagonalizing* or *orthogonal basis* for b.

If $\boldsymbol{e}$ is an orthogonal basis for b, then the matrix $A = (a_{ij})$ of b with respect to $\boldsymbol{e}$ is a diagonal matrix. This is because if $i \neq j$ then $a_{ij} = b(\mathbf{e}_i, \mathbf{e}_j) = 0$. In such a basis, the bilinear form can therefore be written in the following simple manner:

$$b(\mathbf{v}, \mathbf{w}) = a_{11}x_1y_1 + a_{22}x_2y_2 + \cdots + a_{nn}x_ny_n. \tag{15.8}$$

Note that if a diagonalizing basis exists, it is not unique. For example any basis of the form $\{\lambda_1\mathbf{e}_1, \ldots, \lambda_n\mathbf{e}_n\}$, where $\lambda_1, \ldots, \lambda_n \in \mathbf{K}^*$, is also diagonalizing.

The *quadratic form determined by* (or *associated to*) b is the map $q : \mathbf{V} \to \mathbf{K}$, defined by

$$q(\mathbf{v}) = b(\mathbf{v}, \mathbf{v}), \quad \text{for every } \mathbf{v} \in \mathbf{V}.$$

If, for example, b is the standard (symmetric) bilinear form on $\mathbf{K}^n$, the associated quadratic form is

$$q(\mathbf{x}) = x_1^2 + x_2^2 + \cdots + x_n^2,$$

which is called the *standard quadratic form on* $\mathbf{K}^n$.

Proposition 15.8
Let $\mathbf{V}$ be a vector space over $\mathbf{K}$ on which there is given a symmetric bilinear form $b : \mathbf{V} \times \mathbf{V} \to \mathbf{K}$. The quadratic form q determined by b satisfies the following two conditions:

$$\begin{aligned} q(k\mathbf{v}) &= k^2 q(\mathbf{v}) \\ 2b(\mathbf{v}, \mathbf{w}) &= q(\mathbf{v} + \mathbf{w}) - q(\mathbf{v}) - q(\mathbf{w}) \end{aligned}$$

for every $k \in \mathbf{K}$ and every $\mathbf{v}, \mathbf{w} \in \mathbf{V}$.

Proof
The first property is an immediate consequence of BF3. Furthermore,

$$\begin{aligned} q(\mathbf{v} + \mathbf{w}) - q(\mathbf{v}) - q(\mathbf{w}) &= b(\mathbf{v} + \mathbf{w}, \mathbf{v} + \mathbf{w}) - b(\mathbf{v}, \mathbf{v}) - b(\mathbf{w}, \mathbf{w}) \\ &= b(\mathbf{v}, \mathbf{w}) + b(\mathbf{w}, \mathbf{v}) = 2b(\mathbf{v}, \mathbf{w}). \end{aligned}$$

□

From Proposition 15.8 it follows in particular that a quadratic form q determines uniquely the symmetric bilinear form b to which it is associated, since b can be expressed in terms of q. It is called the *polar bilinear form* of the quadratic form q. Consequently, assigning a symmetric bilinear form to a vector space $\mathbf{V}$ is equivalent to assigning the associated quadratic form.

If $\mathbf{V}$ is finite dimensional one says that a quadratic form q has *rank* r if its polar bilinear form b has rank r.

If $\{\mathbf{e}_1, \dots, \mathbf{e}_n\}$ is a basis for $\mathbf{V}$ and if $A = (a_{ij})$ is the symmetric matrix which represents the bilinear form b, then for the associated quadratic form, one has for every $\mathbf{v}(x_1, \dots, x_n) \in \mathbf{V}$,

$$q(\mathbf{v}) = \mathbf{x}^t A\mathbf{x} = \sum_{i,j} a_{ij}x_i x_j.$$

Thus $q(\mathbf{v}) = Q(\mathbf{x})$, where

$$Q(\mathbf{X}) = \mathbf{X}^t A\mathbf{X} = \sum_{i,j} a_{ij}X_i X_j \tag{15.9}$$

is a homogeneous polynomial of degree 2 in the unknowns $X_1, \dots, X_n$, which are represented by a column vector $\mathbf{X} = (X_1 \ \dots \ X_n)^t$. One says that $Q(\mathbf{X})$ *represents the quadratic form* q *in the basis* $\boldsymbol{e}$.

If, for example, $\dim(\mathbf{V}) = 3$ and the matrix of b with repect to the basis $\{\mathbf{e}_1, \mathbf{e}_2, \mathbf{e}_3\}$ is

$$\begin{pmatrix} 2 & -1 & 5 \\ -1 & 0 & 1/3 \\ 5 & 1/3 & -3 \end{pmatrix},$$

then

$$Q(\mathbf{X}) = 2X_1^2 - 2X_1X_2 + 10X_1X_3 + \tfrac{2}{3}X_2X_3 - 3X_3^2.$$

Note that any homogeneous polynomial of degree 2 in n unknowns

$$Q(\mathbf{X}) = \sum_{1 \le i \le j \le n} q_{ij}X_iX_j$$

can be represented in the form (15.9) for some symmetric matrix A. Thus, for a given basis $\boldsymbol{e}$ of $\mathbf{V}$, Q represents the quadratic form whose polar bilinear form is the one represented by A. The matrix A is given by the formula

$$\begin{aligned} a_{ii} &= q_{ii}, \quad i = 1, \dots, n \\ a_{ij} = a_{ji} &= \tfrac{1}{2}q_{ij}, \quad i < j. \end{aligned}$$

A homogeneous polynomial $Q(\mathbf{X})$ of degree 2 can of course be considered as a function $Q : \mathbf{K}^n \to \mathbf{K}$. Obviously, Q is the quadratic form whose associated polynomial with respect to the canonical basis is $Q(\mathbf{X})$ itself. Thus we often identify the polynomial $Q(\mathbf{X})$ with the quadratic form Q. Such a polynomial is often called an *n-ary quadratic form* (*binary* if $n = 2$, *ternary* if $n = 3$, etc.).

If A is a diagonal matrix, the polynomial (15.9) has no 'mixed terms' $q_{ij}X_iX_j$ with $i \neq j$, and so is of the form

$$Q(\mathbf{X}) = a_{11}X_1^2 + a_{22}X_2^2 + \cdots + a_{nn}X_n^2. \qquad (15.10)$$

Thus a basis $\boldsymbol{e}$ of $\mathbf{V}$ diagonalizes the bilinear form b if and only if the polynomial representing the associated quadratic form q is of the form (15.10). In this case, we say that $\boldsymbol{e}$ is a *diagonalizing basis* for q. In Chapter 16 we will prove that every quadratic form on a finite dimensional vector space has a diagonalizing basis.

Obviously, two homogeneous polynomials of degree 2 $Q(\mathbf{X}) = \mathbf{X}^tA\mathbf{X}$ and $R(\mathbf{X}) = \mathbf{X}^tB\mathbf{X}$ in the indeterminate $\mathbf{X} = (X_1 \;\ldots\; X_n)^t$ represent the same quadratic form on $\mathbf{V}$ in two different bases if and only if the symmetric matrices A and B are congruent. Indeed, by Proposition 15.5, being congruent is a necessary and sufficient condition for the matrices A and B to represent the same bilinear form in two different bases.

Suppose that a symmetric bilinear form $b : \mathbf{V} \times \mathbf{V} \to \mathbf{K}$ is given on a vector space $\mathbf{V}$, and the associated quadratic form is q. Let $\mathbf{W}$ be a vector subspace of $\mathbf{V}$. Then b induces a map

$$b' : \mathbf{W} \times \mathbf{W} \to \mathbf{K}$$

which obviously satisfies the conditions given in the definition, and is therefore a bilinear form on $\mathbf{W}$. Moreover b' is symmetric since b is. In the same manner one sees that the quadratic form q' associated to b' coincides with the restriction of q to $\mathbf{W}$, and thus this restriction of a quadratic form is also a quadratic form.

Complements 15.10

1. Let $\mathbf{U}$ and $\mathbf{W}$ be subspaces of a vector space $\mathbf{V}$ such that $\mathbf{V} = \mathbf{U} \oplus \mathbf{W}$. Suppose we are given two bilinear forms $h : \mathbf{U} \times \mathbf{U} \to \mathbf{K}$ and $k : \mathbf{W} \times \mathbf{W} \to \mathbf{K}$. Define $h \oplus k : \mathbf{V} \times \mathbf{V} \to \mathbf{K}$ by

$$(h \oplus k)((\mathbf{u}, \mathbf{w}), (\mathbf{u}', \mathbf{w}')) = h(\mathbf{u}, \mathbf{u}') + k(\mathbf{w}, \mathbf{w}').$$

The map $h \oplus k$ is a bilinear form, called the *direct sum of h and k*. If h and k are both symmetric or skew-symmetric, then $h \oplus k$ is symmetric or skew-symmetric, respectively. The proofs of these facts are left to the reader.

2. Let $b : \mathbf{V} \times \mathbf{V} \to \mathbf{K}$ be a bilinear form, $e = \{\mathbf{e}_1, \ldots, \mathbf{e}_n\}$ a basis of $\mathbf{V}$, and $A = (a_{ij})$ the matrix of b with respect to e. Let $e^* = \{\eta_1, \ldots, \eta_n\}$ be the basis of $\mathbf{V}^*$ dual to e. Then A^t is the matrix representing the linear map

$$\delta_b : \mathbf{V} \to \mathbf{V}^*$$

with respect to the bases e and e^*. Recall that δ_b is defined by $\delta_b(\mathbf{v}) = b_{\mathbf{v}}$, where $b_{\mathbf{v}} \in \mathbf{V}^*$ is the functional

$$b_{\mathbf{v}}(\mathbf{w}) = b(\mathbf{v}, \mathbf{w}).$$

To prove the preceding statement, one has to show that for each $i = 1, \ldots n$, one has

$$\delta_b(\mathbf{e}_i) = \sum_{t=1}^{n} a_{it}\eta_t. \tag{15.11}$$

However, for each i we have

$$[\delta_b(\mathbf{e}_i)](\mathbf{e}_j) = b(\mathbf{e}_i, \mathbf{e}_j) = a_{ij},$$

while

$$[\sum_{t=1}^{n} a_{it}\eta_t](\mathbf{e}_j) = \sum_{t=1}^{n} a_{it}\delta_{tj} = a_{ij}.$$

Since the linear functionals on the left and right hand sides of (15.11) take the same values on the basis e they are in fact equal.

In the same way, one can show that A is the matrix representing the linear map $\delta'_b : \mathbf{V} \to \mathbf{V}'$ with respect to the bases e and e^*.

3. Let $\mathbf{U}$ be a $\mathbf{K}$-vector space of dimension 2. Supppose we are given a non-degenerate symmetric bilinear form h on $\mathbf{V}$ for which there is a non-zero isotropic vector. Then h is said to be a *hyperbolic form* on $\mathbf{U}$ and the pair $(\mathbf{U}, h)$ is a *hyperbolic plane*.

If $(\mathbf{U}, h)$ is a hyperbolic plane then $\mathbf{U}$ possesses a basis $\{\mathbf{u}_1, \mathbf{u}_2\}$ consisting of two isotropic vectors for which $h(\mathbf{u}_1, \mathbf{u}_2) = 1$. Indeed, by definition, there exists a non-zero isotropic vector $\mathbf{u}_1$. Since h

is non-degenerate, there is a vector $\mathbf{v} \in \mathbf{U}$ with $h(\mathbf{u}_1, \mathbf{v}) = 1$. The vectors $\mathbf{u}_1$ and $\mathbf{v}$ are not proportional since if $\mathbf{v} = k\mathbf{u}_1$ for some $k \in \mathbf{K}$ then one would have $1 = h(\mathbf{u}_1, \mathbf{v}) = h(\mathbf{u}_1, k\mathbf{u}_1) = kh(\mathbf{u}_1, \mathbf{u}_1) = 0$ which is absurd. Putting $\mathbf{u}_2 = \mathbf{v} - h(\mathbf{v}, \mathbf{v})\mathbf{u}_1/2$ gives the required basis.

A basis $\{\mathbf{u}_1, \mathbf{u}_2\}$ with these properties is said to be *hyperbolic.* The matrix representing h with respect to a hyperbolic basis is

$$\begin{pmatrix} 0 & 1 \\ 1 & 0 \end{pmatrix}. \tag{15.12}$$

Conversely, it is clear that if the vector space $\mathbf{U}$ has a basis $\{\mathbf{u}_1, \mathbf{u}_2\}$ such that the matrix of the bilinear form is as given in (15.12) then $(\mathbf{U}, h)$ is a hyperbolic plane.

For example, the bilinear form

$$h(\mathbf{x}, \mathbf{y}) = x_1 y_2 + x_2 y_1$$

on $\mathbf{K}^2$ is hyperbolic, and the canonical basis is hyperbolic. Also the bilinear form

$$k(\mathbf{x}, \mathbf{y}) = \tfrac{1}{2} x_1 y_1 - \tfrac{1}{2} x_2 y_2$$

is hyperbolic, and a hyperbolic basis is given by $\{(1, 1), (1, -1)\}$. For the form k, the canonical basis is diagonalizing.

4. Let $\mathbf{V}$ be a vector space and $b : \mathbf{V} \times \mathbf{V} \to \mathbf{K}$ be a non-degenerate symmetric bilinear form for which $\mathbf{V}$ contains an isotropic vector $\mathbf{u} \neq \mathbf{0}$. Then there is a subspace $\mathbf{U}$ of $\mathbf{V}$ for which the pair $(\mathbf{U}, b_{\mathbf{U}})$ is a hyperbolic plane.

   Indeed, since b is non-degenerate there is a $\mathbf{v} \in \mathbf{V}$ for which $b(\mathbf{u}, \mathbf{v}) = 1$. The vector

   $$\mathbf{w} = \mathbf{v} - \frac{b(\mathbf{v}, \mathbf{v})}{2} \mathbf{u}$$

   is isotropic and such that $b(\mathbf{u}, \mathbf{w}) = 1$. Thus the subspace $\mathbf{U} = \langle \mathbf{u}, \mathbf{w} \rangle$ has the required properties.

5. Let $\mathbf{V}$ be a vector space and $q : \mathbf{V} \to \mathbf{K}$ a non-degenerate quadratic form. A scalar $\alpha \in \mathbf{K}$ is said to be *representable by q* if there exists $\mathbf{v} \in \mathbf{V}$ with $\mathbf{v} \neq \mathbf{0}$ for which $q(\mathbf{v}) = \alpha$.

If $\mathbf{V}$ is a complex vector space of dimension greater than 1, then every $\alpha \in \mathbf{C}$ is representable by q. Indeed, let $\mathbf{v} \in \mathbf{V}$ be such that $q(\mathbf{v}) = \alpha \neq 0$. If $\alpha \neq 0$ then let β be a square root of $\alpha q(\mathbf{v})^{-1}$, so that

$$q(\beta \mathbf{v}) = \beta^2 q(\mathbf{v}) = \alpha.$$

To represent 0, let $\{\mathbf{e}_1, \dots, \mathbf{e}_n\}$ be a basis of $\mathbf{V}$. Then, as in (15.9), $Q(\mathbf{X}) = \sum_{i,j} a_{ij} X_i X_j$ for some symmetric matrix (a_{ij}). In particular, if $\mathbf{x}$ has coordinates $(x_1, x_2, 0, \dots, 0)$ then $q(\mathbf{x}) = a_{11}x_1^2 + 2a_{12}x_1x_2 + a_{22}x_2^2$. Thus, whatever the values of a_{11}, a_{12}, a_{22}, there is a non-zero solution $(x_1, x_2, 0, \dots, 0)$ to the equation $q(\mathbf{x}) = 0$.

If $\mathbf{K} = \mathbf{R}$, $\mathbf{V} = \mathbf{R}^n, n \geq 1$ and $q(\mathbf{x}) = x_1^2 + \cdots + x_n^2$ is the standard quadratic form, then clearly no number $\alpha \leq 0$ can be represented by q. Instead, any real number $\alpha > 0$ can be represented by q: this can be proved in exactly the same way as for the complex case above, since $\alpha > 0$ has a real square root.

The notion of representability of a scalar by a quadratic form is important if $\mathbf{K} = \mathbf{Q}$. The set of scalars representable by a given quadratic form on $\mathbf{Q}^n$ is of considerable importance in arithmetic.

If $(\mathbf{U}, h)$ is a hyperbolic plane and $q : \mathbf{U} \to \mathbf{K}$ the quadratic form associated to h, then $q(\mathbf{U} \setminus \{\mathbf{0}\}) = \mathbf{K}$, and so every scalar is representable by q. Indeed, let $\{\mathbf{u}_1, \mathbf{u}_2\}$ be a hyperbolic basis of $\mathbf{U}$ and $\alpha \in \mathbf{K}$. Then

$$q(\mathbf{u}_1 + \frac{\alpha}{2}\mathbf{u}_2) = \alpha.$$

From this observation, together with 15.10(4) it follows that if q is a non-degenerate quadratic form on a vector space $\mathbf{V}$ such that there exists in $\mathbf{V}$ a non-zero isotropic vector, then every $\alpha \in \mathbf{K}$ is representable by q. In other words, if 0 is representable by q then so is every $\alpha \in \mathbf{K}$.

Taking for example, $\mathbf{K} = \mathbf{Q}$ and $\mathbf{V} = \mathbf{Q}^2$ we can deduce that every rational number can be expressed in the form

$$\alpha = \frac{1}{2}(x^2 - y^2), \quad \text{with } x, y \in \mathbf{Q}.$$

6. Let $b : \mathbf{V} \times \mathbf{V} \to \mathbf{K}$ be a non-degenerate symmetric bilinear form on $\mathbf{V}$, and let $\mathbf{U}$ be a subspace of $\mathbf{V}$. The orthogonal subspace $\mathbf{U}^\perp$ coincides with the orthogonal subspace of $\delta_b(\mathbf{U}) \subset \mathbf{V}^*$ as defined in 11.14(2). Since b is non-degenerate, δ_b is an isomorphism, and

so $\dim(\mathbf{U}) = \dim[\delta_b(\mathbf{U})]$. From (11.13) it follows that

$$\dim(\mathbf{U}^\perp) = n - \dim(\mathbf{U}). \tag{15.13}$$

Suupose that $\mathbf{U}$ does not contain any non-zero isotropic vectors, then $\mathbf{U} \cap \mathbf{U}^\perp = \langle \mathbf{0} \rangle$, and from (15.13) it follows that $\mathbf{V} = \mathbf{U} \oplus \mathbf{U}^\perp$. In particular, if $\mathbf{v}$ is a non-isotropic vector, putting $\mathbf{U} = \langle \mathbf{v} \rangle$ gives the identity (15.7) that we have already proved.

7. Let $b : \mathbf{V} \times \mathbf{V} \to \mathbf{K}$ be a symmetric bilinear form on $\mathbf{V}$, and denote by $\mathbf{I}_b(\mathbf{V}) \subset \mathbf{V}$ the set of vectors that are isotropic with respect to b. The set $\mathbf{I}_b(\mathbf{V})$ is called the *isotropic cone* of $\mathbf{V}$ (with respect to b). A subspace $\mathbf{U} \in \mathbf{V}$ is said to be *isotropic* if $\mathbf{U} \subset \mathbf{I}_b(\mathbf{V})$. Clearly $\langle \mathbf{0} \rangle$ is an isotropic subspace: the *trivial* isotropic subspace. If b is degenerate, then the radical $\mathbf{V}^\perp$ of $\mathbf{V}$ is a non-trivial isotropic subspace.

   If $\mathbf{U}$ is an isotropic subspace and $\mathbf{u}_1, \mathbf{u}_2 \in \mathbf{U}$ then, since $\mathbf{u}_1 + \mathbf{u}_2 \in \mathbf{U}$ one has

$$0 = b(\mathbf{u}_1 + \mathbf{u}_2, \mathbf{u}_1 + \mathbf{u}_2) = b(\mathbf{u}_1, \mathbf{u}_2),$$

   that is $\mathbf{u}_1, \mathbf{u}_2$ are orthogonal. Thus it follows that $\mathbf{U} \subset \mathbf{U}^\perp$. Conversely, if $\mathbf{U} \subset \mathbf{U}^\perp$ then it is obvious that $\mathbf{U}$ is isotropic.

   The form b is said to be *anisotropic* if $\mathbf{V}$ has no non-zero isotropic vectors, that is, if $\mathbf{I}_b(\mathbf{V}) = \{\mathbf{0}\}$. For example, the standard bilinear form on $\mathbf{R}^n$ is anistropic.

   Suppose that b is non-degenerate, and let $\mathbf{U}$ be an isotropic subspace. Then

$$\dim(\mathbf{U}) \leq \tfrac{1}{2}\dim(\mathbf{V}). \tag{15.14}$$

   Indeed, $\mathbf{U}$ isotropic implies that $\mathbf{U} \subset \mathbf{U}^\perp$, and so by (15.13) one has

$$\dim(\mathbf{U}) \leq \dim(\mathbf{U}^\perp) = \dim(\mathbf{V}) - \dim(\mathbf{U}),$$

   which implies (15.14).

EXERCISES

15.1 Which of the following are bilinear forms on $\mathbf{R}^n$:

a) $\langle \mathbf{x}, \mathbf{y} \rangle = \sum_{j=1}^n x_j |y_j|$

b) $\langle \mathbf{x}, \mathbf{y} \rangle = |\sum_{j=1}^{n} x_j y_j|$
c) $\langle \mathbf{x}, \mathbf{y} \rangle = (\sum_{i=1}^{n} x_i)(\sum_{j=1}^{n} y_j)$
d) $\langle \mathbf{x}, \mathbf{y} \rangle = \sqrt{\sum_{j=1}^{n} x_j^2 y_j^2}$
e) $\langle \mathbf{x}, \mathbf{y} \rangle = \sum_{j=1}^{n} (x_j + y_j)^2 - \sum_{j=1}^{n} x_j^2 - \sum_{j=1}^{n} y_j^2$.

15.2 In each of the following, determine the polar bilinear form associated to the given quadratic form $q : \mathbf{R}^2 \to \mathbf{R}$.
a) $q(x, y) = 3x^2 - 8xy - 3y^2$
b) $q(x, y) = 4x^2 - 9xy + 5y^2$
c) $q(x, y) = 4x^2 - 4xy + 7y^2$
d) $q(x, y) = x^2 - 2xy + y^2$
e) $q(x, y) = 3x^2 + 10xy + 3y^2$
f) $q(x, y) = 6xy$.

15.3 Determine the matrix and the rank of each of the quadratic forms given in Exercise 2.

15.4 In each of the following, determine the polar bilinear form associated to the given quadratic form $q : \mathbf{R}^3 \to \mathbf{R}$.
a) $q(x, y, z) = xz + xy + yz$
b) $q(x, y, z) = 2xy + y^2 - 2xz$
c) $q(x, y, z) = x^2 - 2xz - y^2 - z^2$
d) $q(x, y, z) = 5x^2 + 3y^2 + xz$
e) $q(x, y, z) = -x^2 - 4xy + 3y^2 + 2z^2$.

15.5 Determine the matrix and the rank of each of the quadratic forms given in Exercise 4.

16

Diagonalizing quadratic forms

In this chapter we consider the question of whether a diagonalizing basis exists for a symmetric bilinear form on a vector space. The principal result is the following theorem.

Theorem 16.1
Let $\mathbf{V}$ be a $\mathbf{K}$-vector space of dimension $n \geq 1$, and let b be a symmetric bilinear form on $\mathbf{V}$. Then there exists a diagonalizing basis for b. Equivalently, any symmetric matrix is congruent to a diagonal matrix.

Given the importance of the theorem, we give two proofs.

First proof
We proceed by induction on n. If $n = 1$ there is nothing to prove. Suppose then that $n \geq 2$, and that every quadratic form on a vector space of dimension less than n has a diagonalizing basis.

If b is the zero form there is nothing to prove, since in this case with respect to any basis the matrix of b is the zero matrix which is diagonal. Suppose therefore, that b is not the zero form, and so neither is the associated quadratic form. Thus there is a vector $\mathbf{e}_1 \in \mathbf{V}$ with $b(\mathbf{e}_1, \mathbf{e}_1) \neq 0$. From (15.7) it follows that $\mathbf{V} = \langle \mathbf{e}_1 \rangle \oplus \mathbf{e}_1^\perp$; in particular, $\dim(\mathbf{e}^\perp) = n - 1$.

By the inductive hypothesis the form b' induced by b on $\mathbf{e}_1^\perp$ has a diagonalizing basis, say $\{\mathbf{e}_2, \ldots, \mathbf{e}_n\}$. Then $e = \{\mathbf{e}_1, \mathbf{e}_2, \ldots, \mathbf{e}_n\}$ is a basis of $\mathbf{V}$. Indeed, $\mathbf{e}_2, \ldots, \mathbf{e}_n$ are linearly independent and $\mathbf{e}_1 \notin \langle \mathbf{e}_2, \ldots, \mathbf{e}_n \rangle = \mathbf{e}_1^\perp$, and so $\mathbf{e}_1, \mathbf{e}_2, \ldots, \mathbf{e}_n$ are linearly independent.

Furthermore, $b(\mathbf{e}_1, \mathbf{e}_j) = 0$ for each $j = 2, 3, \ldots, n$, since $\mathbf{e}_j \in \mathbf{e}_1^\perp$. Finally, $b(\mathbf{e}_i, \mathbf{e}_j) = b'(\mathbf{e}_i, \mathbf{e}_j) = 0$ for each $i \neq j$, $2 \leq i, j \leq n$ since $\{\mathbf{e}_2, \ldots, \mathbf{e}_n\}$ is a diagonalizing basis for b'. Thus $\boldsymbol{e}$ is a diagonalizing basis for b. □

Second proof (Lagrange)
This proof is by 'completing the square' and also involves induction on n. If $n = 1$ there is nothing to prove. Suppose therefore that $n \geq 2$, and that every symmetric bilinear form on a space of dimension less than n has a diagonalizing basis. Choose a basis $\boldsymbol{b} = \{\mathbf{b}_1, \ldots, \mathbf{b}_n\}$ of $\mathbf{V}$. If b is the zero form then $\boldsymbol{b}$ is diagonalizing and there is nothing to prove. Otherwise we can obtain from $\boldsymbol{b}$ a second basis $\boldsymbol{c} = \{\mathbf{c}_1, \ldots, \mathbf{c}_n\}$ for which $b(\mathbf{c}_1, \mathbf{c}_1) \neq 0$. Indeed, if there is some i for which $b(\mathbf{b}_i, \mathbf{b}_i) \neq 0$ then it suffices to exchange $\mathbf{b}_1$ and $\mathbf{b}_i$. If, on the other hand, $b(\mathbf{b}_i, \mathbf{b}_i) = 0$ for all i, then there are $i \neq j$ for which $b(\mathbf{b}_i, \mathbf{b}_j) \neq 0$ (otherwise b is the zero form), and again we can exchange these with $\mathbf{b}_1$ and $\mathbf{b}_2$ so that $b(\mathbf{b}_1, \mathbf{b}_2) \neq 0$. The new basis

$$\boldsymbol{c} = \{\mathbf{b}_1 + \mathbf{b}_2, \mathbf{b}_2, \ldots, \mathbf{b}_n\}$$

has the required property. In the basis $\mathbf{c}$, the quadratic form q associated to b has the form

$$q(\mathbf{v}(y_1, \ldots, y_n)) = h_{11}y_1^2 + 2\sum_{i=2}^{n} h_{1i}y_1y_i + \sum_{i,j=2}^{n} h_{ij}y_iy_j, \tag{16.1}$$

where $h_{ij} = b(\mathbf{c}_i, \mathbf{c}_j)$. Since $h_{11} = b(\mathbf{c}_1, \mathbf{c}_1) \neq 0$ we can rewrite (16.1) as

$$q(\mathbf{v}(y_1, \ldots, y_n)) = h_{11}(y_1 + \sum_{i=2}^{n} h_{11}^{-1}h_{1i}y_i)^2 + (\text{terms not involving } y_1).$$

We now change coordinates as follows

$$z_1 = y_1 + \sum_{i=2}^{n} h_{11}^{-1}h_{1i}y_i,\ z_2 = y_2, \ldots, z_n = y_n,$$

which corresponds to a change of basis from $\mathbf{c}$ to $\boldsymbol{d} = \{\mathbf{d}_1, \ldots, \mathbf{d}_n\}$ given by $\mathbf{d}_1 = \mathbf{c}_1$, and for $i > 1$, $\mathbf{d}_i = \mathbf{c}_i - h_{11}^{-1}h_{1i}\mathbf{c}_1$. In these coordinates, q has the form

$$q(\mathbf{v}(z_1, \ldots, z_n)) = h_{11}z_1^2 + q'(z_2, \ldots, z_n),$$

where q' is a homogeneous polynomial of degree 2 in $z_1, \dots, z_n$, and so defines a quadratic form on the space $\langle \mathbf{d}_2, \dots, \mathbf{d}_n \rangle$. By the inductive hypothesis, $\langle \mathbf{d}_2, \dots, \mathbf{d}_n \rangle$ has a basis $\{\mathbf{e}_2, \dots, \mathbf{e}_n\}$ which diagonalizes q'. Thus the basis $\{\mathbf{d}_1, \mathbf{e}_2, \dots, \mathbf{e}_n\}$ is diagonalizing for q. □

Theorem 16.1 asserts the existence of a basis $\boldsymbol{e}$ for which b takes the form (15.8), with $a_{11}, \dots, a_{nn} \in \mathbf{K}$. We will show that in the cases $\mathbf{K} = \mathbf{C}$ or $\mathbf{R}$ more precise results can be obtained. We begin with the case $\mathbf{K} = \mathbf{C}$ (in fact we consider more generally the case that $\mathbf{K}$ is algebraically closed).

Theorem 16.2

Suppose that $\mathbf{K}$ *is algebraically closed, and let* $\mathbf{V}$ *be a* $\mathbf{K}$*-vector space of dimension* $n \geq 1$. *Let* $b : \mathbf{V} \times \mathbf{V} \to \mathbf{K}$ *be a symmetric bilinear form. There exists a basis* $\{\mathbf{e}_1, \dots, \mathbf{e}_n\}$ *which diagonalizes* b*, and moreover is such that the matrix of* b *is of the form*

$$D = \begin{pmatrix} \mathbf{I}_r & \mathbf{0}_1 \\ \mathbf{0}_2 & \mathbf{0}_3 \end{pmatrix} \tag{16.2}$$

where r *is the rank of* b*, and* $\mathbf{0}_1 \in M_{r,n-r}(\mathbf{K})$, $\mathbf{0}_2 \in M_{n-r,r}(\mathbf{K})$ *and* $\mathbf{0}_3 \in M_{n-r,n-r}(\mathbf{K})$ *are zero matrices.*

Equivalently, every symmetric matrix of rank r *is congruent to the matrix* (16.2).

Proof

The equivalence of the two statements is obvious. We prove the first. By Theorem 16.1 there is a basis $\boldsymbol{f} = \{\mathbf{f}_1, \dots, \mathbf{f}_n\}$ for which the matrix of b is of the form

$$\begin{pmatrix} a_{11} & 0 & \dots & 0 \\ 0 & a_{22} & \dots & 0 \\ \vdots & \vdots & & \vdots \\ 0 & 0 & \dots & a_{nn} \end{pmatrix}.$$

After possibly changing the order of $\{\mathbf{f}_1, \dots, \mathbf{f}_n\}$ we can suppose that $a_{11}, \dots, a_{rr}$ are non-zero, and $a_{r+1\,r+1} = \cdots = a_{nn} = 0$. Since $\mathbf{K}$ is algebraically closed, there exist $\alpha_1, \dots, \alpha_r \in \mathbf{K}$ such that $\alpha_i^2 = a_{ii}$ for $i = 1, \dots, r$ Consider the vectors

$$\mathbf{e}_1 = \alpha_1^{-1}\mathbf{f}_1, \dots, \mathbf{e}_r = \alpha_r^{-1}\mathbf{f}_r, \mathbf{e}_{r+1} = \mathbf{f}_{r+1}, \dots, \mathbf{e}_n = \mathbf{f}_n.$$

Clearly $\boldsymbol{e} = \{\mathbf{e}_1, \ldots, \mathbf{e}_n\}$ is an orthogonal basis. Moreover,

$$\begin{aligned} b(\mathbf{e}_i, \mathbf{e}_i) &= \alpha_i^{-2} b(\mathbf{f}_i, \mathbf{f}_i) = \alpha_i^{-2} a_{ii} = 1 \quad \text{for } i = 1, \ldots, r \\ b(\mathbf{e}_i, \mathbf{e}_i) &= b(\mathbf{f}_i, \mathbf{f}_i) = 0 \quad \text{for } i = r+1, \ldots, n. \end{aligned}$$

Thus $\boldsymbol{e}$ has the required properties. □

In the preceding proof the fact that $\mathbf{K}$ is algebraically closed was used to prove the existence of the scalars α_i. In the case $\mathbf{K} = \mathbf{R}$ there is a rather weaker result.

Theorem 16.2 (Sylvester)
Let $\mathbf{V}$ be a real vector space of dimension $n \geq 1$, and let $b : \mathbf{V} \times \mathbf{V} \to \mathbf{R}$ be a symmetric bilinear form on $\mathbf{V}$ of rank $r \leq n$. Then there is an integer p with $0 \leq p \leq r$ depending only on b, and a basis $\boldsymbol{e} = \{\mathbf{e}_1, \ldots, \mathbf{e}_n\}$ of $\mathbf{V}$ with respect to which the matrix associated to b has the form

$$\begin{pmatrix} \mathbf{I}_p & \mathbf{0} & \mathbf{0} \\ \mathbf{0} & -\mathbf{I}_{r-p} & \mathbf{0} \\ \mathbf{0} & \mathbf{0} & \mathbf{0} \end{pmatrix}, \tag{16.3}$$

where the symbol $\mathbf{0}$ denotes zero matrices of appropriate sizes.

Equivalently, every symmetric matrix $A \in M_n(\mathbf{R})$ is congruent to a diagonal matrix of the form (16.3) in which $r = \operatorname{rk}(A)$ and p depends only on A.

Proof
The equivalence of the two statements is obvious. We prove the first.

By Theorem 16.1 there is a basis $\boldsymbol{f} = \{\mathbf{f}_1, \ldots, \mathbf{f}_n\}$ for which

$$q(\mathbf{v}) = a_{11} y_1^2 + \cdots + a_{nn} y_n^2$$

for each $\mathbf{v} = y_1 \mathbf{f}_1 + \cdots + y_n \mathbf{f}_n \in \mathbf{V}$. The number of non-zero coefficients a_{ii} is equal to the rank of the quadratic form q, and so depends only on q. After possibly reordering the basis, we can suppose that the first p coeffiecients are positive, the next $r - p$ are negative, and the remaining $n - r$ are zero. Then one has

$$a_{11} = \alpha_1^2, \ldots, a_{pp} = \alpha_p^2, \; a_{p+1\,p+1} = -\alpha_{p+1}^2, \ldots, a_{rr} = -\alpha_r^2$$

$$a_{r+1\,r+1} = \cdots = a_{nn} = 0,$$

for appropriate $\alpha_1, \ldots, \alpha_r \in \mathbf{R}$, which can be taken to be postive.

Exactly as in the proof of the preceding theorem one can show that with respect to the basis

$$\mathbf{e}_1 = \mathbf{f}_1/\alpha_1, \ldots, \mathbf{e}_r = \mathbf{f}_r/\alpha_r,$$

$$\mathbf{e}_{r+1} = \mathbf{f}_{r+1}, \ldots, \mathbf{e}_n = \mathbf{f}_n,$$

the matrix of b is that given in (16.3), and so the quadratic form q associated to b is

$$q(\mathbf{v}) = x_1^2 + \cdots + x_p^2 - x_{p+1}^2 - \cdots - x_r^2 \tag{16.4}$$

for each $\mathbf{v} = x_1\mathbf{e}_1 + \cdots + x_n\mathbf{e}_n \in \mathbf{V}$.

There remains only to prove that p depends only on b and not on the particular basis $\{\mathbf{e}_1, \ldots, \mathbf{e}_n\}$. Suppose that with respect to another basis $\boldsymbol{b} = \{\mathbf{b}_1, \ldots, \mathbf{b}_n\}$, q is

$$q(\mathbf{v}) = z_1^2 + \cdots + z_t^2 - z_{t+1}^2 - \cdots - z_r^2, \tag{16.5}$$

for each $\mathbf{v} = z_1\mathbf{b}_1 + \cdots + z_n\mathbf{b}_n$, for some integer $t \leq r$. We need to show that $t = p$. If $p \neq t$ we can suppose that $t < p$. Consider the subspaces of $\mathbf{V}$:

$$\begin{aligned} S &= \langle \mathbf{e}_1, \ldots, \mathbf{e}_p \rangle \\ T &= \langle \mathbf{b}_{t+1}, \ldots, \mathbf{b}_n \rangle. \end{aligned}$$

Since $\dim(S) + \dim(T) > n$ it follows from Grassman's formula that $S \cap T \neq \langle \mathbf{0} \rangle$, and so there is a $\mathbf{v} \in S \cap T$ with $\mathbf{v} \neq \mathbf{0}$. Then,

$$\mathbf{v} = x_1\mathbf{e}_1 + \cdots + x_p\mathbf{e}_p = z_{t+1}\mathbf{b}_{t+1} + \cdots + z_n\mathbf{b}_n.$$

Since $\mathbf{v} \neq \mathbf{0}$ it follows from (16.4) that

$$q(\mathbf{v}) = x_1^2 + \cdots + x_p^2 > 0,$$

and from (16.5) that

$$q(\mathbf{v}) = -z_{t+1}^2 - \cdots - z_r^2 < 0.$$

This is clearly a contradiction, so $t = p$. □

Expression (16.4) is called the *normal form* for the quadratic form q. The pair $(p, r-p)$ is called the *signature* of b, and of q.

A quadratic form on a real vector space $\mathbf{V}$ is said to be

positive definite	if	$q(\mathbf{v}) > 0$	for all $\mathbf{v} \neq \mathbf{0}$
positive semi-definite	if	$q(\mathbf{v}) \geq 0$	for all $\mathbf{v} \in \mathbf{V}$
negative definite	if	$q(\mathbf{v}) < 0$	for all $\mathbf{v} \neq \mathbf{0}$
negative semi-definite	if	$q(\mathbf{v}) \leq 0$	for all $\mathbf{v} \in \mathbf{V}$.

Clearly, if q is positive or negative definite, then it is also positive or negative semi-definite, respectively. If it is neither positive nor negative semi-definite, q is said to be *indefinite.*

The same terms are used to describe the polar bilinear form of q.

The normal forms corresponding to the different cases are as follows, with $0 \leq r \leq n$ for the semi-definite cases, and $0 < p < r \leq n$ for the indefinite case:

Type	*Normal form*	*Signature*
positive definite	$x_1^2 + \cdots + x_n^2$	$(n, 0)$
positive semi-definite	$x_1^2 + \cdots + x_r^2$	$(r, 0)$
negative definite	$-x_1^2 - \cdots - x_n^2$	$(0, n)$
negative semi-definite	$-x_1^2 - \cdots - x_r^2$	$(0, r)$
indefinite	$x_1^2 + \cdots + x_p^2 - x_{p+1}^2 - \cdots - x_r^2$	$(p, r-p)$

A symmetric matrix is said to be *positive definite, positive semi-definite, negative definite, negative semi-definite* or *indefinite* accordingly as its corresponding quadratic form.

From Sylvester's theorem it also follows that every equivalence class of real symmetric matrices of order n under congruency contains precisely one diagonal matrix of the form (16.3). That is, the matrices of the form (16.3) form a complete set of representatives for the equivalence classes of real symmetric matrices of order n under congruency.

In particular, a symmetric matrix of order n is positive definite if and only if it is congruent to the identity matrix $\mathbf{I}_n$, that is, if and only if there is a matrix $M \in \mathrm{GL}_n(\mathbf{R})$ such that $A = M^t \mathbf{I}_n M = M^t M$. Thus we have the following corollary:

Corollary 16.4

A symmetric matrix $A \in M_n(\mathbf{R})$ *is positive definite if and only if there is a matrix* $M \in \mathrm{GL}_n(\mathbf{R})$ *for which* $A = M^t M$.

The positive definite quadratic forms on real vector spaces are of central importance in Euclidean geometry, because, as we shall see in the following chapters, they allow the introduction of metric notions.

Other examples of quadratic forms on real vector spaces are important in geometry and physics. An example of this is the vector space $\mathbf{R}^4$ with the quadratic form

$$\lambda(x_1, x_2, x_3, x_4) = x_1^2 + x_2^2 + x_3^2 - x_4^2,$$

called the *Minkowski form.* λ is non-degenerate and indefinite of signature (3,1). The pair $(\mathbf{R}^4, \lambda)$ is called *Minkowski space*, and is of fundamental importance in special relativity.

EXERCISES

16.1 In each of the following cases find a basis with respect to which the given quadratic form on $\mathbf{C}^2$ has the form (16.2), and give the associated changes of coordinates:

a) $-x^2 + y^2$, b) $ix^2 - 2y^2$,
c) $4x^2 + 9y^2$, d) $-x^2 - 25y^2$.

16.2 In each of the following cases find a basis with respect to which the given quadratic form on $\mathbf{R}^3$ is in normal form, and calculate the signatures:

a) $4x^2 - 5y^2 + 12z^2$, b) $-x^2 + 9z^2$,
c) $-x^2 - y^2 + z^2$, d) $y^2 + 16z^2$.

16.3 Diagonalize each of the quadratic form of Exercise 15.2, determining the change of coordinates required, and the signatures of the forms.

16.4 For each of the forms of the preceding exercise, express the matrix B of the diagonalized form as $B = M^t AM$, where A is the matrix of the given form.

16.5 Diagonalize each of the quadratic forms of Exercise 15.4, and find their signature and the transformation required.

16.6 For each of the forms of the preceding exercise, express the matrix B of the diagonalized form as $B = M^t AM$, where A is the matrix of the given form.

17

Scalar product

Let $\mathbf{V}$ be a real vector space. A positive definite symmetric bilinear form on $\mathbf{V}$ is called a *scalar product*, or *inner product.* If $\mathbf{V}$ is endowed with a scalar product it is said to be a *Euclidean space.*

Note that the definition does not require that $\mathbf{V}$ be finite dimensional, and indeed there are infinite dimensional Euclidean vector spaces — see Example 17.8(2). However, we will be limiting ourselves to the finite dimensional case.

The standard bilinear form (15.3) on $\mathbf{R}^n$, $n \geq 1$, is a scalar product which from now on will be called the *standard scalar product* on $\mathbf{R}^n$; if $\mathbf{x}, \mathbf{y} \in \mathbf{R}^n$ we will denote their standard scalar product by $\mathbf{x} \cdot \mathbf{y}$. Thus

$$\mathbf{x} \cdot \mathbf{y} = x_1 y_1 + \cdots + x_n y_n = \mathbf{x}^t \mathbf{y}.$$

Given the standard inner product, $\mathbf{R}^n$ becomes a Euclidean vector space, called *Euclidean n-space.*

Let $\mathbf{V}$ be any Euclidean vector space. Denote the scalar product of two vectors $\mathbf{v}, \mathbf{w} \in \mathbf{V}$ by $\langle \mathbf{v}, \mathbf{w} \rangle$.[1]

Theorem 17.1 (Schwartz Inequality)
If $\mathbf{v}, \mathbf{w} \in \mathbf{V}$ *then*

$$\langle \mathbf{v}, \mathbf{w} \rangle^2 \leq \langle \mathbf{v}, \mathbf{v} \rangle \langle \mathbf{w}, \mathbf{w} \rangle \tag{17.1}$$

with equality if and only if $\mathbf{v}$ *and* $\mathbf{w}$ *are parallel.*

[1] Note that the brackets for the scalar product are set in bold type, while the similar brackets used to denote the vector space generated by a set of vectors are set in ordinary type.

Proof
If $\mathbf{w} = \mathbf{0}$ then (17.1) is obvious, both sides being zero. We can therefore suppose that $\mathbf{w} \neq \mathbf{0}$. For each $a, b \in \mathbf{R}$ one has

$$0 \leq \langle a\mathbf{v} + b\mathbf{w},\, a\mathbf{v} + b\mathbf{w}\rangle = a^2\langle \mathbf{v}, \mathbf{v}\rangle + 2ab\langle \mathbf{v}, \mathbf{w}\rangle + b^2\langle \mathbf{w}, \mathbf{w}\rangle$$

with equality if and only if $a\mathbf{v} + b\mathbf{w} = \mathbf{0}$, which occurs precisely when $\mathbf{v}$ and $\mathbf{w}$ are parallel, or $a = b = 0$. Putting $a = \langle \mathbf{w}, \mathbf{w}\rangle \neq 0$ and $b = -\langle \mathbf{v}, \mathbf{w}\rangle$ gives

$$0 \leq \langle \mathbf{w}, \mathbf{w}\rangle^2\langle \mathbf{v}, \mathbf{v}\rangle - 2\langle \mathbf{w}, \mathbf{w}\rangle\langle \mathbf{v}, \mathbf{w}\rangle^2 + \langle \mathbf{v}, \mathbf{w}\rangle^2\langle \mathbf{w}, \mathbf{w}\rangle.$$

Since $\mathbf{w} \neq \mathbf{0}$ it follows that $\langle \mathbf{w}, \mathbf{w}\rangle > 0$, so dividing by $\langle \mathbf{w}, \mathbf{w}\rangle$ gives

$$0 \leq \langle \mathbf{w}, \mathbf{w}\rangle\langle \mathbf{v}, \mathbf{v}\rangle - \langle \mathbf{v}, \mathbf{w}\rangle^2,$$

that is, the Schwartz inequality. □

For $\mathbf{v} \in \mathbf{V}$ we define the *length* or *norm* of $\mathbf{v}$ to be

$$\|v\| = \sqrt{\langle \mathbf{v}, \mathbf{v}\rangle}.$$

Using the norm, the Schwartz inequality can be expressed as

$$|\langle \mathbf{v}, \mathbf{w}\rangle| \leq \|\mathbf{v}\|\|\mathbf{w}\|, \tag{17.2}$$

which is obtained by taking the square root of of (17.1).

The norm of a vector enjoys the following three properties:

[N1] $\|v\| \geq 0$ for all $\mathbf{v} \in \mathbf{V}$, with equality if and only if $\mathbf{v} = \mathbf{0}$.
[N2] $\|r\mathbf{v}\| = |r|\|\mathbf{v}\|$ for every $\mathbf{v} \in \mathbf{V}$ and $r \in \mathbf{R}$.
[N3] $\|\mathbf{v} + \mathbf{w}\| \leq \|\mathbf{v}\| + \|\mathbf{w}\|$, for every $\mathbf{v}, \mathbf{w} \in \mathbf{V}$, with equality if and only if $\mathbf{v}$ and $\mathbf{w}$ are parallel.

Axiom N1 is a reformulation of the fact that the scalar product is positive definite. N2 follows by taking square roots of the following identity:

$$\langle r\mathbf{v}, r\mathbf{v}\rangle = r^2\langle \mathbf{v}, \mathbf{v}\rangle.$$

N3 is called the *triangle inequality*; it is proved as follows. By (17.2) one has

$$\begin{aligned} \|\mathbf{v} + \mathbf{w}\|^2 &= \langle \mathbf{v} + \mathbf{w},\, \mathbf{v} + \mathbf{w}\rangle = \|\mathbf{v}\|^2 + 2\langle \mathbf{v}, \mathbf{w}\rangle + \|\mathbf{w}\|^2 \\ &\leq \|\mathbf{v}\|^2 + 2\|\mathbf{v}\|\|\mathbf{w}\| + \|\mathbf{w}\|^2 = (\|\mathbf{v}\| + \|\mathbf{w}\|)^2, \end{aligned}$$

which is equivalent to N3.

By N1, a vector $\mathbf{v}$ has length 0 if and only if $\mathbf{v} = \mathbf{0}$.

A vector of length 1 is called a *unit vector*. By N2, if $\mathbf{v} \neq \mathbf{0}$ then $\mathbf{v}/\|\mathbf{v}\|$ is a unit vector parallel to $\mathbf{v}$; one says that it is obtained by *normalizing* $\mathbf{v}$.

Recall that 2 vectors $\mathbf{v}$ and $\mathbf{w}$ are said to be *orthogonal*, or *perpendicular*, if $\langle \mathbf{v}, \mathbf{w} \rangle = 0$. It follows then that the vector $\mathbf{0}$ is orthogonal to every vetor in $\mathbf{V}$. Since the scalar product is positive definite, $\mathbf{0}$ is the only vector orthogonal to itself.

A finite set of non-zero vectors $\{\mathbf{v}_1, \ldots, \mathbf{v}_t\}$ is said to be an *orthogonal set* of vectors if $\langle \mathbf{v}_i, \mathbf{v}_j \rangle = 0$ for all $i \neq j$, $1 \leq i, j \leq t$.

$\{\mathbf{v}_1, \ldots, \mathbf{v}_t\}$ is said to be an *orthonormal set* of vectors if it is an orthogonal set, and moreover the vectors are all unit vectors. Clearly, an orthogonal set $\{\mathbf{v}_1, \ldots, \mathbf{v}_t\}$ can easily be made into an orthonormal set: $\{\mathbf{v}_1/\|\mathbf{v}_1\|, \ldots, \mathbf{v}_t/\|\mathbf{v}_t\|\}$.

If $\mathbf{V}$ is finite dimensional, then an *orthogonal basis*, or an *orthonormal basis*, is a basis $\{\mathbf{e}_1, \ldots, \mathbf{e}_n\}$ which is also an orthogonal set, or orthonormal set, respectively.

Proposition 17.2

Any set of orthogonal vectors is linearly independent. In particular, if $\dim(\mathbf{V}) = n$ *is finite, any orthogonal set of* n *vectors is a basis.*

Proof

Let $\{\mathbf{v}_1, \ldots, \mathbf{v}_t\}$ be an orthogonal set of vectors. If $a_1\mathbf{v}_1 + \cdots + a_t\mathbf{v}_t = \mathbf{0}$ then for each $i = 1, \ldots, t$

$$\begin{aligned} 0 &= \langle \mathbf{v}_i, a_1\mathbf{v}_1 + \cdots + a_t\mathbf{v}_t \rangle \\ &= a_1\langle \mathbf{v}_i, \mathbf{v}_1 \rangle + \cdots + a_t\langle \mathbf{v}_i, \mathbf{v}_t \rangle \\ &= a_i\langle \mathbf{v}_i, \mathbf{v}_i \rangle. \end{aligned}$$

Since $\langle \mathbf{v}_i, \mathbf{v}_i \rangle > 0$ it follows that $a_i = 0$. Consequently, $\mathbf{v}_1, \ldots, \mathbf{v}_t$ are linearly independent. □

It follows from Theorem 16.1 that if $\mathbf{V}$ is finite dimensional (of strictly positive dimension) then it has an orthogonal basis. Normalizing the vectors then leads to an orthonormal basis. The existence of orthonormal bases also follows from Sylvester's theorem, since the signature of the scalar product is $(n, 0)$.

If $\boldsymbol{e} = \{\mathbf{e}_1, \ldots, \mathbf{e}_n\}$ is an orthonormal basis of $\mathbf{V}$ then the matrix that represents the scalar product with respect to $\boldsymbol{e}$ is $\mathbf{I}_n$. Thus, for each $\mathbf{v} = \mathbf{x}_1\mathbf{e}_1 + \cdots + x_n\mathbf{e}_n$ and $\mathbf{w} = y_1\mathbf{e}_1 + \cdots + y_n\mathbf{e}_n$, one has,

$$\langle \mathbf{v}, \mathbf{w} \rangle = x_1 y_1 + \cdots + x_n y_n = \mathbf{x} \cdot \mathbf{y}.$$

In other words, with respect to an orthonormal basis, the scalar product of two vectors has the same expression as the standard scalar product of the column vectors with the same coordinates. In practice, orthonormal bases are convenient for just this reason, as calculations in terms of coordinates are greatly simplified.

The next result describes the matrix of a change of orthonormal basis.

Proposition 17.3
Let $\boldsymbol{e} = \{\mathbf{e}_1, \ldots, \mathbf{e}_n\}$ and $\boldsymbol{f} = \{\mathbf{f}_1, \ldots, \mathbf{f}_n\}$ be two bases of the Euclidean vector space $\mathbf{V}$, and suppose that $\mathbf{e}$ is orthonormal. The basis $\boldsymbol{f}$ is orthonormal if and only if the matrix $M_{\boldsymbol{e}, \boldsymbol{f}}(\mathbf{1_V})$ for the change of coordinates from $\mathbf{e}$ to $\boldsymbol{f}$, is orthogonal.

Proof
The columns of the matrix $M = M_{\boldsymbol{e}, \boldsymbol{f}}(\mathbf{1_V})$ are the coordinates of the vectors $\mathbf{f}_1, \ldots, \mathbf{f}_n$ with respect to $\boldsymbol{e}$. Thus $\boldsymbol{f}$ is orthonormal if and only if

$$M_{(i)} \cdot M_{(j)} = \langle \mathbf{f}_i, \mathbf{f}_j \rangle = \delta_{ij}. \tag{17.3}$$

These conditions translate precisely to $M^t M = \mathbf{I}_n$, that is $M \in \mathrm{O}(n)$. □

We now describe a simple method, called the *Gram-Schmidt process*, for constructing orthogonal and orthonormal sets and bases.

This method is based on the notion of the 'Fourier coefficient', which we introduced in Chapter 15. Recall that if $\mathbf{v} \in \mathbf{V}$ is non-zero then for any $\mathbf{w} \in \mathbf{V}$ the Fourier coefficient of $\mathbf{w}$ with respect to $\mathbf{v}$ is the scalar,

$$a_{\mathbf{v}}(\mathbf{w}) = \frac{\langle \mathbf{v}, \mathbf{w} \rangle}{\langle \mathbf{v}, \mathbf{v} \rangle}.$$

We define the *orthogonal projection* of $\mathbf{w}$ along $\mathbf{v}$ to be the vector $a_{\mathbf{v}}(\mathbf{w})\mathbf{v}$. This terminology is justified by the fact that

$$\langle \mathbf{w} - a_{\mathbf{v}}(\mathbf{w})\mathbf{v}, \mathbf{v} \rangle = \langle \mathbf{w}, \mathbf{v} \rangle - a_{\mathbf{v}}(\mathbf{w})\langle \mathbf{v}, \mathbf{v} \rangle = 0,$$

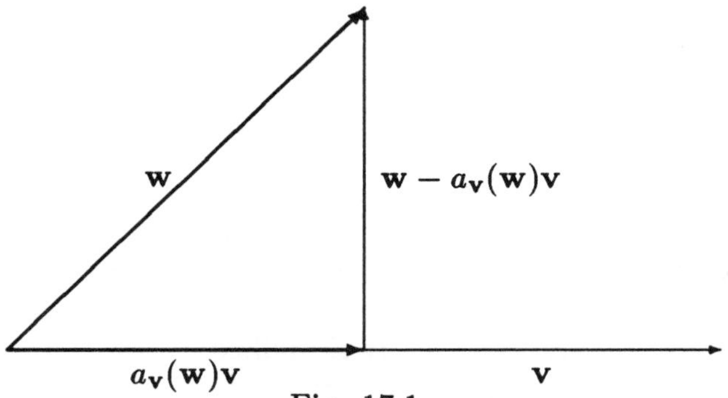

Fig. 17.1

that is, $\mathbf{w} - a_{\mathbf{v}}(\mathbf{w})\mathbf{v}$ is orthogonal to $\mathbf{v}$ — see Fig. 17.1.

In the case that $\mathbf{v}$ is a unit vector, the Fourier coefficient of any $\mathbf{w} \in \mathbf{V}$ with respect to $\mathbf{v}$ is given by

$$a_{\mathbf{v}}(\mathbf{w}) = \langle \mathbf{v}, \mathbf{w} \rangle;$$

the orthogonal projection of $\mathbf{w}$ along $\mathbf{v}$ is $\langle \mathbf{v}, \mathbf{w} \rangle \mathbf{v}$, and $\mathbf{w} - \langle \mathbf{v}, \mathbf{w} \rangle \mathbf{v}$ is orthogonal to $\mathbf{w}$.

Theorem 17.4 (Gram-Schmidt process)
Let $\mathbf{v}_1, \mathbf{v}_2, \mathbf{v}_3, \ldots$ be a sequence (finite or infinite) of vectors in the Euclidean vector space $\mathbf{V}$. Then

1) *There is a sequence (correspondingly finite with the same number of elements, or infinite) of vectors $\mathbf{w}_1, \mathbf{w}_2, \mathbf{w}_3, \ldots$ of $\mathbf{V}$ such that for each $k \geq 1$ one has*
   a) *$\langle \mathbf{v}_1, \mathbf{v}_2, \ldots, \mathbf{v}_k \rangle = \langle \mathbf{w}_1, \mathbf{w}_2, \ldots, \mathbf{w}_k \rangle$;*
   b) *the vectors $\mathbf{w}_1, \mathbf{w}_2, \ldots, \mathbf{w}_k$ are pairwise orthogonal.*

2) *If $\mathbf{u}_1, \mathbf{u}_2, \mathbf{u}_3, \ldots$ is another sequence satisfying conditions a) and b) for every $k \geq 1$, then there are non-zero scalars $c_1, c_2, \ldots$ for which $\mathbf{u}_k = c_k \mathbf{w}_k$, for every $k = 1, 2, 3, \ldots$.*

Proof
1) We construct the vectors $\mathbf{w}_1, \mathbf{w}_2, \ldots$ by induction on k. Take $\mathbf{w}_1 = \mathbf{v}_1$, which obviously satisfies (a) and (b) for $k = 1$. Let $k = t$, and suppose we have constructed $\mathbf{w}_1, \ldots, \mathbf{w}_t$ satisfying conditions (a) and

(b) for $k = t$. Define

$$\mathbf{w}_{t+1} = \mathbf{v}_{t+1} - \sum_{i=1}^{t}{}' a_{\mathbf{w}_i}(\mathbf{v}_{t+1})\mathbf{w}_i,$$

where $\sum'$ means that any i for which $\mathbf{w}_i = \mathbf{0}$ is excluded from the sum.

It follows from this definition that $\mathbf{v}_{t+1}$ is a linear combination of $\mathbf{w}_1, \ldots, \mathbf{w}_{t+1}$, and by the inductive hypothesis $\mathbf{v}_i \in \langle \mathbf{w}_1, \ldots, \mathbf{w}_t \rangle$. Consequently, $\langle \mathbf{v}_1, \ldots, \mathbf{v}_{t+1} \rangle \subset \langle \mathbf{w}_1, \ldots, \mathbf{w}_{t+1} \rangle$. On the other hand,

$$\mathbf{w}_{t+1} \in \langle \mathbf{w}_1, \ldots, \mathbf{w}_t, \mathbf{v}_{t+1} \rangle = \langle \mathbf{v}_1, \ldots, \mathbf{v}_{t+1} \rangle,$$

firstly by the definition of $\mathbf{w}_{t+1}$, and secondly by the inductive hypothesis. Thus, we also have $\langle \mathbf{w}_1, \ldots, \mathbf{w}_{t+1} \rangle \subset \langle \mathbf{v}_1, \ldots, \mathbf{v}_{t+1} \rangle$, and (a) is proved for $k = t + 1$.

For each $i = 1, \ldots, t$ for which $\mathbf{w}_i \neq \mathbf{0}$, we have

$$\begin{aligned}
\langle \mathbf{w}_{t+1}, \mathbf{w}_i \rangle &= \langle \mathbf{v}_{t+1} - \sum_{j=1}^{t}{}' a_{\mathbf{w}_j}(\mathbf{v}_{t+1})\mathbf{w}_j, \mathbf{w}_i \rangle \\
&= \langle \mathbf{v}_{t+1}, \mathbf{w}_i \rangle - a_{\mathbf{w}_i}(\mathbf{v}_{t+1})\langle \mathbf{w}_i, \mathbf{w}_i \rangle \\
&= \langle \mathbf{v}_{t+1}, \mathbf{w}_i \rangle - \langle \mathbf{v}_{t+1}, \mathbf{w}_i \rangle = 0.
\end{aligned}$$

Since, by the inductive hypothesis, $\langle \mathbf{w}_i, \mathbf{w}_j \rangle = 0$ for all $1 \leq i, j \leq t$, we have proved (b) for $k = t + 1$. Thus the proof of (1) is complete.
2) Again we proceed by induction on k. For $k = 1$ the statement is obvious. Suppose that $t \geq 1$, and that there are scalars $c_1, \ldots, c_t$ for which $\mathbf{u}_k = c_k \mathbf{w}_k$ for every $k \leq t$. By (a) we have that

$$\mathbf{u}_{t+1} = \mathbf{z} + c_{t+1}\mathbf{w}_{t+1},$$

for some $\mathbf{z} \in \langle \mathbf{w}_1, \ldots, \mathbf{w}_t \rangle$ and $c_{t+1} \in \mathbf{R}$. Since, by (b), both $\mathbf{u}_{t+1}$ and $\mathbf{w}_{t+1}$ are orthogonal to $\mathbf{z}$, so is $\mathbf{u}_{t+1} - c_{t+1}\mathbf{w}_{t+1} = \mathbf{z}$, whence $\mathbf{z} = \mathbf{0}$. □

Note that (b) of Theorem 17.4 does not assert that $\{\mathbf{w}_1, \ldots, \mathbf{w}_k\}$ is an orthogonal set of vectors, for it may happen that some of the $\mathbf{w}_i$ are zero. However, if all the $\mathbf{w}_i \neq \mathbf{0}$ then indeed $\{\mathbf{w}_1, \ldots, \mathbf{w}_k\}$ is an orthogonal set of vectors.

If $\mathbf{V}$ has finite dimension and $\{\mathbf{v}_1, \ldots, \mathbf{v}_n\}$ is a basis of $\mathbf{V}$, then the Gram-Schmidt process applied to $\mathbf{v}_1, \ldots, \mathbf{v}_n$ gives rise to vectors $\mathbf{w}_1, \ldots, \mathbf{w}_n$ which are pairwise orthogonal and generate $\mathbf{V}$, so form an orthogonal basis of $\mathbf{V}$.

Examples 17.5

1. Let $\{\mathbf{e}_1, \mathbf{e}_2, \mathbf{e}_3, \mathbf{e}_4\}$ be an orthonormal basis of a Euclidean vector space $\mathbf{V}$ of dimension 4. Consider the vectors,

$$\mathbf{v}_1(0,1,0,1), \quad \mathbf{v}_2(2,1,0,1), \quad \mathbf{v}_3(-1,0,0,1), \quad \mathbf{v}_4(0,0,1,0).$$

The Gram-Schmidt process applied to $\mathbf{v}_1, \mathbf{v}_2, \mathbf{v}_3, \mathbf{v}_4$ gives,

$$\begin{aligned}
\mathbf{w}_1 &= \mathbf{v}_1, \\
\mathbf{w}_2 &= \mathbf{v}_2 - \frac{\langle \mathbf{w}_1, \mathbf{v}_2 \rangle}{\langle \mathbf{w}_1, \mathbf{w}_1 \rangle}\mathbf{w}_1 = \mathbf{v}_2 - \tfrac{2}{2}\mathbf{w}_1 = 2\mathbf{e}_1, \\
\mathbf{w}_3 &= \mathbf{v}_3 - \frac{\langle \mathbf{w}_1, \mathbf{v}_3 \rangle}{\langle \mathbf{w}_1, \mathbf{w}_1 \rangle}\mathbf{w}_1 - \frac{\langle \mathbf{w}_2, \mathbf{v}_3 \rangle}{\langle \mathbf{w}_2, \mathbf{w}_2 \rangle}\mathbf{w}_2 \\
&= \mathbf{v}_3 - \tfrac{1}{2}\mathbf{w}_1 - \tfrac{-2}{4}\mathbf{w}_2 = -\tfrac{1}{2}\mathbf{e}_2 + \tfrac{1}{2}\mathbf{e}_4, \\
\mathbf{w}_4 &= \mathbf{v}_4 - \frac{\langle \mathbf{w}_1, \mathbf{v}_4 \rangle}{\langle \mathbf{w}_1, \mathbf{w}_1 \rangle}\mathbf{w}_1 - \frac{\langle \mathbf{w}_2, \mathbf{v}_4 \rangle}{\langle \mathbf{w}_2, \mathbf{w}_2 \rangle}\mathbf{w}_2 - \frac{\langle \mathbf{w}_3, \mathbf{v}_4 \rangle}{\langle \mathbf{w}_3, \mathbf{w}_3 \rangle}\mathbf{w}_3 \\
&= \mathbf{v}_4 - \tfrac{0}{2}\mathbf{w}_1 - \tfrac{0}{4}\mathbf{w}_2 - \tfrac{0}{1/2}\mathbf{w}_3 = \mathbf{v}_4.
\end{aligned}$$

The vectors $\mathbf{w}_1, \mathbf{w}_2, \mathbf{w}_3, \mathbf{w}_4$ form an orthogonal basis of $\mathbf{V}$. Normalizing these vectors gives the orthonormal basis

$$\left\{ \frac{\mathbf{e}_2}{\sqrt{2}} + \frac{\mathbf{e}_4}{\sqrt{2}}, \mathbf{e}_1, -\frac{\mathbf{e}_2}{\sqrt{2}} + \frac{\mathbf{e}_4}{\sqrt{2}}, \mathbf{e}_3 \right\}.$$

2. Consider $\mathbf{R}^4$ with the scalar product

$$\langle \mathbf{x}, \mathbf{y} \rangle = \tfrac{1}{2}x_1y_1 + \tfrac{1}{2}x_2y_2 + x_3y_3 + x_4y_4.$$

Applying the Gram-Schmidt process to the vectors

$$\mathbf{v}_1 = (1,1,-1,-1), \quad \mathbf{v}_2 = (1,1,1,1),$$

$$\mathbf{v}_3 = (-1,-1,-1,1), \quad \mathbf{v}_4 = (1,0,0,1)$$

gives

$$\begin{aligned}
\mathbf{w}_1 &= \mathbf{v}_1, \\
\mathbf{w}_2 &= \mathbf{v}_2 - \frac{\langle \mathbf{w}_1, \mathbf{v}_2 \rangle}{\langle \mathbf{w}_1, \mathbf{w}_1 \rangle}\mathbf{w}_1 = \mathbf{v}_2 - \tfrac{-1}{3}\mathbf{w}_1 = (\tfrac{4}{3}, \tfrac{4}{3}, \tfrac{2}{3}, \tfrac{2}{3}), \\
\mathbf{w}_3 &= \mathbf{v}_3 - \frac{\langle \mathbf{w}_1, \mathbf{v}_3 \rangle}{\langle \mathbf{w}_1, \mathbf{w}_1 \rangle}\mathbf{w}_1 - \frac{\langle \mathbf{w}_2, \mathbf{v}_3 \rangle}{\langle \mathbf{w}_2, \mathbf{w}_2 \rangle}\mathbf{w}_2
\end{aligned}$$

$$
\begin{aligned}
&= \mathbf{v}_3 - \tfrac{-1}{3}\mathbf{w}_1 - \tfrac{-4/3}{8/3}\mathbf{w}_2 = \mathbf{v}_3 + \tfrac{1}{3}\mathbf{w}_1 + \tfrac{1}{2}\mathbf{w}_2 = (00, -1, 1), \\
\mathbf{w}_4 &= \mathbf{v}_4 - \frac{\langle \mathbf{w}_1, \mathbf{v}_4\rangle}{\langle \mathbf{w}_1, \mathbf{w}_1\rangle}\mathbf{w}_1 - \frac{\langle \mathbf{w}_2, \mathbf{v}_4\rangle}{\langle \mathbf{w}_2, \mathbf{w}_2\rangle}\mathbf{w}_2 - \frac{\langle \mathbf{w}_3, \mathbf{v}_4\rangle}{\langle \mathbf{w}_3, \mathbf{w}_3\rangle}\mathbf{w}_3 \\
&= \mathbf{v}_4 - \tfrac{1/2}{3}\mathbf{w}_1 - \tfrac{-4/3}{8/3}\mathbf{w}_2 - \tfrac{1}{2}\mathbf{w}_3 \\
&= \mathbf{v}_4 + \tfrac{1}{6}\mathbf{w}_1 - \tfrac{1}{2}\mathbf{w}_2 - \tfrac{1}{2}\mathbf{w}_3 = (\tfrac{1}{2}, -\tfrac{1}{2}, 0, 0).
\end{aligned}
$$

Thus $\left\{(1,1,-1,-1), (\tfrac{4}{3}, \tfrac{4}{3}, \tfrac{2}{3}, \tfrac{2}{3}), (0,0,-1,1), (\tfrac{1}{2}, -\tfrac{1}{2}, 0, 0)\right\}$ is an orthonormal basis of $\mathbf{R}^4$ with respect to $\langle , \rangle$.

Proposition 17.6

Let $\mathbf{V}$ *be a Euclidean vector space, and* $\mathbf{W} \neq \langle \mathbf{0} \rangle$ *a finite dimensional subspace of* $\mathbf{V}$*. Then every element* $\mathbf{v} \in \mathbf{V}$ *can be expressed uniquely as a sum*

$$\mathbf{v} = \mathbf{w} + \mathbf{w}', \tag{17.4}$$

with $\mathbf{w} \in \mathbf{W}$ *and* $\mathbf{w}' \in \mathbf{W}^{\perp}$*. Thus,*

$$\mathbf{V} = \mathbf{W} \oplus \mathbf{W}^{\perp}.$$

Proof

Let $\{\mathbf{e}_1, \ldots, \mathbf{e}_t\}$ be an orthonormal basis of $\mathbf{W}$. If $\mathbf{v} \in \mathbf{V}$ put

$$
\begin{aligned}
\mathbf{w} &= \langle \mathbf{v}, \mathbf{e}_1\rangle \mathbf{e}_1 + \cdots + \langle \mathbf{v}, \mathbf{e}_t\rangle \mathbf{e}_t, \\
\mathbf{w}' &= \mathbf{v} - \mathbf{w}.
\end{aligned}
$$

Clearly, $\mathbf{w} \in \mathbf{W}$. Moreover, for each $i = 1, \ldots, t$,

$$\langle \mathbf{w}', \mathbf{e}_i\rangle = \langle \mathbf{v}, \mathbf{e}_i\rangle - \langle \mathbf{w}, \mathbf{e}_i\rangle = \langle \mathbf{v}, \mathbf{e}_i\rangle - \langle \mathbf{v}, \mathbf{e}_i\rangle\langle \mathbf{e}_i, \mathbf{e}_i\rangle = 0.$$

Thus, $\mathbf{w}'$ is perpendicular to each of $\mathbf{e}_1, \ldots, \mathbf{e}_t$, and so to every linear combination of them. That is, $\mathbf{w}' \in \mathbf{W}^{\perp}$, and so $\mathbf{v}$ can indeed be expressed as required.

If $\mathbf{v} = \mathbf{u} + \mathbf{u}'$ for some $\mathbf{u} \in \mathbf{W}$ and $\mathbf{u}' \in \mathbf{W}^{\perp}$, then

$$\mathbf{w} - \mathbf{u} = \mathbf{u}' - \mathbf{w}' \in \mathbf{W} \cap \mathbf{W}^{\perp}.$$

Thus $\mathbf{w} - \mathbf{u}$, being perpendicular to every element of $\mathbf{W}$, is perpendicular to itself, and so $\mathbf{w} - \mathbf{u} = \mathbf{0} = \mathbf{u}' - \mathbf{w}'$. That is, $\mathbf{u} = \mathbf{w}$ and $\mathbf{u}' = \mathbf{w}'$. □

The element $\mathbf{w}$ appearing on the right hand side of (17.4) is called the *orthogonal projection of* $\mathbf{v}$ *on the subspace* $\mathbf{W}$.

Expanding the identity $\langle \mathbf{v}, \mathbf{v} \rangle = \langle \mathbf{w} + \mathbf{w}', \mathbf{w} + \mathbf{w}' \rangle$ one gets

$$\|\mathbf{v}\|^2 = \|\mathbf{w}\|^2 + \|\mathbf{w}'\|^2,$$

namely, 'Pythagoras' theorem'.

The Schwartz inequality (17.1) allows one to define the notion of 'convex angle between two non-zero vector', as follows.

Let $\mathbf{v}, \mathbf{w} \in \mathbf{V}$ be two non-zero vectors. Inequality (17.2) can be written in the following form:

$$-\|\mathbf{v}\|\|\mathbf{w}\| \leq \langle \mathbf{v}, \mathbf{w} \rangle \leq \|\mathbf{v}\|\|\mathbf{w}\|.$$

Dividing through by $\|\mathbf{v}\|\|\mathbf{w}\|$, which is non-zero, gives

$$-1 \leq \frac{\langle \mathbf{v}, \mathbf{w} \rangle}{\|\mathbf{v}\|\|\mathbf{w}\|} \leq 1. \tag{17.5}$$

It follows from (17.5) and the properties of the cosine function that there is a unique $\theta \in [0, \pi]$ satisfying

$$\cos\theta = \frac{\langle \mathbf{v}, \mathbf{w} \rangle}{\|\mathbf{v}\|\|\mathbf{w}\|}.$$

We can therefore write,

$$\langle \mathbf{v}, \mathbf{w} \rangle = \|\mathbf{v}\|\|\mathbf{w}\| \cos\theta. \tag{17.6}$$

We call

$$\theta = \arccos\left(\frac{\langle \mathbf{v}, \mathbf{w} \rangle}{\|\mathbf{v}\|\|\mathbf{w}\|}\right) \tag{17.7}$$

the *convex (or, non-oriented) angle between (or formed by) the vectors* $\mathbf{v}$ *and* $\mathbf{w}$. From its definition, we see that this angle does not depend on the order of $\mathbf{v}$ and $\mathbf{w}$. Note, also from the definition, that two non-zero vectors are perpendicular if and only if the convex angle between them is $\pi/2$. Moreover, $\cos\theta = \langle \mathbf{v}, \mathbf{w} \rangle$ if $\mathbf{v}$ and $\mathbf{w}$ are unit vectors.

It will not have escaped the notice of more careful readers that the definition (17.7) of the convex angle between two non-zero vectors, making use of the inverse function of the trigonometric function $\cos\theta$, apparently uses elementary Euclidean geometry, where the trigonometric functions are usually defined. This would destroy our claim of developing geometry completely independently of elementary Euclidean geometry. However, this contradiction is only apparent. Indeed, all the trigonometric functions and their inverses can be defined independently of elementary geometry, namely by power series. For details, the reader should consult courses or texts in mathematical analysis. For our purposes, it is enough know that this is possible, and so the definition of convex angle between vectors in any Euclidean vector space given in (17.7) is independent of any elementary geometry. The convex angle so defined is a real number between 0 and π and does not *a priori* have the geometrical significance that the notion of angle has in ordinary plane geometry. The relation between this definition and the notion of angle in ordinary plane geometry is explained in Example 17.7. Readers should note that we will make free use of the principal properties of trigonometric functions.

Example 17.7
Let $\mathbf{V}$ be the space of geometric vectors in the ordinary plane, in which we suppose there is given a unit for measurements of the length of segments. For any $\mathbf{v} \in \mathbf{V}$ we denote the *length of* $\mathbf{V}$ by $|\mathbf{v}|$, defined to be the length of any representative of $\mathbf{v}$. If $\mathbf{v}, \mathbf{w} \in \mathbf{V}$ are both non-zero, we define the *convex (or non-oriented) angle between* $\mathbf{v}$ *and* $\mathbf{w}$ to be the convex angle θ ($0 \le \theta \le \pi$ if it is measured in radians) formed by representatives of $\mathbf{v}$ and $\mathbf{w}$ based at the same point, where we are using the definition of angle given in elementary Euclidean geometry. It is possible to define a scalar product on $\mathbf{V}$ in such a way that the length of a vector and the convex angle between two non-zero vectors obtained using the scalar product coincide with those we have just described in terms of elementary Euclidean geometry.

For each $\mathbf{v}, \mathbf{w} \in \mathbf{V}$ put

$$\mathbf{v} \bullet \mathbf{w} = \begin{cases} |v||w| \cos\theta & \text{if } \mathbf{v} \neq \mathbf{0} \neq \mathbf{w} \\ 0 & \text{otherwise.} \end{cases} \tag{17.8}$$

The $\bullet$ defined in (17.8) is a scalar product on $\mathbf{V}$; we leave the details to the reader.

Note that if $\mathbf{v}$ and $\mathbf{w}$ are unit vectors then their scalar product coincides with the cosine of the convex angle between them. Moreover, the norm $\|\mathbf{v}\|$ of a vector $\mathbf{v}$ coincides with its length $|\mathbf{v}|$. If two non-zero vectors $\mathbf{v}, \mathbf{w}$ satisfy $\mathbf{v} \bullet \mathbf{w} = 0$, then $\cos\theta = 0$, that is, the angle between them is $\pi/2$. Thus the definition of perpendicularity of two vectors using the scalar product (17.8) coincides with the usual one.

Let $\mathbf{v} = \overrightarrow{OA}, \mathbf{w} = \overrightarrow{OB} \in \mathbf{V}$, $\mathbf{v} \neq \mathbf{0}$. Let P be the foot of the perpendicular from B to the line containing O and A. It is easy to see that the Fourier coefficient $a_{\mathbf{v}}(\mathbf{w})$ is equal to the ratio of the lengths of $\overrightarrow{OP}$ and $\overrightarrow{OA}$, taken with sign $+$ or $-$ arcordingly as $\overrightarrow{OP}$ and $\overrightarrow{OA}$ have the same or opposite orientation. The orthogonal projection of $\mathbf{w}$ in the direction $\mathbf{v}$ is the vector $\overrightarrow{OP} = a_{\mathbf{v}}(\mathbf{w})\mathbf{v}$, and $\mathbf{w} - a_{\mathbf{v}}(\mathbf{w})\mathbf{v} = \overrightarrow{PB}$, which is perpendicular to $\mathbf{v}$.

Consider an orthonormal basis $\{\mathbf{i}, \mathbf{j}\}$ of $\mathbf{V}$.

For $\mathbf{v} = x_1\mathbf{i} + x_2\mathbf{j}$, the identity

$$|\mathbf{v}|^2 = x_1^2 + x_2^2$$

is nothing but Pythagoras' theorem.

Note also that Schwartz' inequality

$$|\mathbf{v} \bullet \mathbf{w}| \leq |\mathbf{v}||\mathbf{w}|$$

follows immediately from the fact that $|\cos\theta| \leq 1$.

If we introduce into the vector space of geometric vectors in ordinary space a unit of length for segments, then (17.8) defines in this case too a scalar product. Exactly the same discussion as above is valid in this case.

The notion of non-oriented angle defined in (17.7) is insufficient for developing linear algebra and geometry. The notion of 'oriented angle' is defined using properties of the real numbers, as follows.

Define on $\mathbf{R}$ an equivalence relation as follows. We will say that $\theta, \phi \in \mathbf{R}$ are *congruent modulo* 2π, or that θ is congruent to ϕ modulo 2π if $\theta - \phi = 2k\pi$ for some integer k; in this case we write $\theta \equiv \phi \pmod{2\pi}$.

It is easily verified that congruence is indeed an equivalence relation. Congruence classes are called *oriented angles*, or simply *angles*. By definition then, angles are subsets of $\mathbf{R}$ of the form

$$\{\ldots, \theta - 4\pi, \theta - 2\pi, \theta, \theta + 2\pi, \theta + 4\pi, \ldots\} = \{\theta + 2k\pi \mid k \in \mathbf{Z}\} \quad (17.9)$$

for some $\theta \in \mathbf{R}$. Since congruence classes form a partition[2] of $\mathbf{R}$, each real number determines uniquely the angle to which it belongs. The

[2]A partition of $\mathbf{R}$ is a collection of disjoint subsets whose union is $\mathbf{R}$.

elements of an angle are called its *determinations*; by definition, two determinations of the same angle differ by an integer multiple of 2π. Any angle has a unique determination θ_0 for which $0 \leq \theta_0 < 2\pi$, the so called *principal determination.* Abusing the terminology rather, we will often identify an angle with one of its determinations, that is with a real number, leaving implicit the fact that two real numbers represent the same angle if and only if they differ by an integer multiple of 2π.

Since every angle has a unique principal determination, there is a bijective correspondence between the set $\mathcal{D}$ of all angles and the half-closed interval $[0, 2\pi)$.

We previously defined a convex angle as a number in $[0, \pi]$. To each angle $\{\theta + 2k\pi \mid k \in \mathbf{Z}\}$ we can associate the real number

$$\min\{|\theta + 2k\pi| \mid k \in \mathbf{Z}\},$$

which, as is easy to check, is an element of $[0, \pi]$, called the *associated convex angle.*

Angles can be added by adding their determinations: if $\{\theta + 2k\pi \mid k \in \mathbf{Z}\}$ and $\{\phi + 2k\pi \mid k \in \mathbf{Z}\}$ are two angles, their sum is the angle

$$\{\theta + \phi + 2k\pi \mid k \in \mathbf{Z}\}.$$

With this binary operation of addition, the set $\mathcal{D}$ of angles forms a group whose identity element is $\{2k\pi \mid k \in \mathbf{Z}\}$, and the inverse of the angle $\{\theta + 2k\pi \mid k \in \mathbf{Z}\}$ is the angle $\{-\theta + 2k\pi \mid k \in \mathbf{Z}\}$.

For each $\theta \in \mathbf{R}$ consider the matrix

$$R_\theta = \begin{pmatrix} \cos\theta & -\sin\theta \\ \sin\theta & \cos\theta \end{pmatrix}.$$

From the identity $\cos^2\theta + \sin^2\theta = 1$, it follows that $R_\theta \in \mathrm{SO}(2)$. It is not hard to show, using the elementary properties of trigonometric functions, that for each $(a, b) \in \mathbf{R}^2$ such that $a^2 + b^2 = 1$ there is a θ for which $a = \cos\theta$, $b = \sin\theta$. Since every matrix in SO(2) is of the form (2.6), by varying θ the matrix R_θ describes all of SO(2).

Furthermore, since cosine and sine are periodic of period 2π, we have that $R_\theta = R_\phi$ if and only if θ and ϕ are two determinations of the same angle. Thus, to any angle there is associated a unique matrix R_θ, where θ is any determination of that angle. We thus have a bijective map from $\mathcal{D}$ to SO(2).

It is easy to show that this bijection is an isomorphism of groups. To this end, it is enough to note that for $\theta, \phi \in \mathbf{R}$

$$R_\theta R_\phi = R_{\theta+\phi}.$$

That is,

$$\begin{pmatrix} \cos\theta & -\sin\theta \\ \sin\theta & \cos\theta \end{pmatrix} \begin{pmatrix} \cos\phi & -\sin\phi \\ \sin\phi & \cos\phi \end{pmatrix} = \begin{pmatrix} \cos(\theta+\phi) & -\sin(\theta+\phi) \\ \sin(\theta+\phi) & \cos(\theta+\phi) \end{pmatrix}, \tag{17.11}$$

which is proved by multiplying the two matrices on the left and then using the formulae expressing $\cos(\theta+\phi)$ and $\sin(\theta+\phi)$ in terms of $\cos\theta, \cos\phi, \sin\theta$ and $\sin\phi$.

Consider now a two-dimensional vector space $\mathbf{V}$ with a fixed orthonormal basis $\{\mathbf{i}, \mathbf{j}\}$. Let $\mathbf{u}(u_1, u_2)$ and $\mathbf{v}(v_1, v_2)$ be two unit vectors, and let ϕ, ψ be such that

$$\begin{aligned} (u_1, u_2) &= (\cos\phi, \sin\phi) \\ (v_1, v_2) &= (\cos\psi, \sin\psi). \end{aligned}$$

Define the *oriented angle formed by the unit vectors* $\mathbf{u}$ *and* $\mathbf{v}$ to be the angle $\widehat{\mathbf{uv}}$ one of whose determinations is $\psi - \phi$.

If $\mathbf{a}, \mathbf{b}$ are two non-zero vectors in $\mathbf{V}$, the oriented angle between them is defined to be

$$\widehat{\mathbf{ab}} = \widehat{\frac{\mathbf{a}}{\|\mathbf{a}\|}\frac{\mathbf{b}}{\|\mathbf{b}\|}}.$$

The oriented angle between to vectors enjoys the following properties:

a) $\widehat{\mathbf{aa}} = 0$,
b) $\widehat{\mathbf{ab}} = -\widehat{\mathbf{ba}}$,
c) $\widehat{\mathbf{ab}} + \widehat{\mathbf{bc}} = \widehat{\mathbf{ac}}$,
d) $\widehat{(\lambda\mathbf{a})(\mu\mathbf{b})} = \widehat{\mathbf{ab}}$,

for all non-zero vectors $\mathbf{a}, \mathbf{b}, \mathbf{c}$, and for every $\lambda, \mu > 0$. Proofs are left to the reader.

It should be noted that although our definition of angle between two vectors depends on fixing an orthonormal basis of $\mathbf{V}$, in fact this angle depends only on the orientation of $\mathbf{V}$ defined by the given basis.

Indeed, let $\{\mathbf{i}', \mathbf{j}'\}$ be a second orthonormal basis of $\mathbf{V}$, defining the same orientation as $\{\mathbf{i}, \mathbf{j}\}$, and let

$$A = \begin{pmatrix} \cos\alpha & -\sin\alpha \\ \sin\alpha & \cos\alpha \end{pmatrix}$$

be the matrix defining the change of coordinates from $\{\mathbf{i},\mathbf{j}\}$ to $\{\mathbf{i}',\mathbf{j}'\}$. If

$$\begin{aligned}\mathbf{u} &= \cos\phi\mathbf{i} + \sin\phi\mathbf{j} = \cos\phi'\mathbf{i}' + \sin\phi'\mathbf{j}'\\ \mathbf{v} &= \cos\psi\mathbf{i} + \sin\psi\mathbf{j} = \cos\psi'\mathbf{i}' + \sin\psi'\mathbf{j}',\end{aligned}$$

then

$$\begin{pmatrix}\cos\phi'\\ \sin\phi'\end{pmatrix} = \begin{pmatrix}\cos\alpha & -\sin\alpha\\ \sin\alpha & \cos\alpha\end{pmatrix}\begin{pmatrix}\cos\phi\\ \sin\phi\end{pmatrix} = \begin{pmatrix}\cos(\phi+\alpha)\\ \sin(\phi+\alpha)\end{pmatrix}$$

and similarly,

$$\begin{pmatrix}\cos\psi'\\ \sin\psi'\end{pmatrix} = \begin{pmatrix}\cos\alpha & -\sin\alpha\\ \sin\alpha & \cos\alpha\end{pmatrix}\begin{pmatrix}\cos\psi\\ \sin\psi\end{pmatrix} = \begin{pmatrix}\cos(\psi+\alpha)\\ \sin(\psi+\alpha)\end{pmatrix}.$$

Thus, $\phi' - \psi' \equiv (\psi+\alpha) - (\phi+\alpha) \equiv \psi - \phi \pmod{2\pi}$, and so the two bases define the same angle between $\mathbf{u}$ and $\mathbf{v}$.

Complements 17.8

1. The set $\mathbf{C}$ of complex numbers is a real vector space of dimension 2. The basis $\{1, i\}$ identifies $\mathbf{C}$ with $\mathbf{R}^2$, associating to any complex number $a+ib$ the pair $(a,b) \in \mathbf{R}^2$. For $z = a+ib \in \mathbf{C}$, the modulus of z, defined by

$$|z| = \sqrt{z\bar{z}} = \sqrt{a^2+b^2},$$

is equal to $\|(a,b)\|$. To each $z \in \mathbf{C}^*$ we can associate the complex number $\frac{z}{|z|} = \frac{a}{|z|} + i\frac{b}{|z|}$ which has modulus 1, and so is of the form

$$\frac{z}{|z|} = \cos\theta + i\sin\theta.$$

The angle determined by θ is called the *argument* of z, and is denoted $\arg(z)$; the principal determination of $\arg(z)$ is often called the *principal argument* of z.

Since $z = |z|\frac{z}{|z|}$ it follows from what has already been said, that every non-zero complex number can be written in the form

$$z = |z|(\cos\theta + i\sin\theta), \qquad (17.12)$$

where θ is a determination of $\arg(z)$. (17.12) is a *trigonometric representation* of z.

If $z, w \in \mathbf{C}^*$ then $|zw| = |z||w|$, and $\arg(zw) = \arg(z) + \arg(w)$. This can be proved using the trigonometric representations as follows. Let $\theta \in \arg(z), \phi \in \arg(w)$. Then

$$\begin{aligned} zw &= |z|(\cos\theta + i\sin\theta)|w|(\cos\phi + i\sin\phi) \\ &= |z||w|(\cos\theta\cos\phi - \sin\theta\sin\phi + i(\cos\theta\sin\phi + \sin\theta\cos\phi)) \\ &= |z||w|[\cos(\theta+\phi) + i\sin(\theta+\phi)]. \end{aligned}$$

2. Consider the vector space $\mathbf{R}[X]$ of polynomials with real coefficients in one indeterminate X. For every $f(X), g(X) \in \mathbf{R}[X]$ put

$$\langle f, g\rangle = \int_{-1}^{1} f(x)g(x)\,dx.$$

It follows immediately from the basic properties of integration that $\langle,\rangle$ is a scalar product. With this scalar product, $\mathbf{R}[X]$ is a Euclidean vector space which does not have finite dimension.

EXERCISES

17.1 Prove that in a Euclidean vector space $(\mathbf{V}, \langle,\rangle)$ the following identities hold, for any $\mathbf{v}, \mathbf{w} \in \mathbf{V}$:
a) $\|\mathbf{v}+\mathbf{w}\|^2 + \|\mathbf{v}-\mathbf{w}\|^2 = 2\|\mathbf{v}\|^2 + 2\|\mathbf{w}\|^2$
b) $\|\mathbf{v}+\mathbf{w}\|^2 - \|\mathbf{v}-\mathbf{w}\|^2 = 4\langle\mathbf{v}, \mathbf{w}\rangle$.

17.2 Use the Gram-Schmidt process to find an orthonormal (with respect to the standard scalar product) basis for each of the subspaces of $\mathbf{R}^3$ generated by the following vectors.
a) $(1,1,1)$, $(1,0,1)$, $(3,2,3)$
b) $(1,1,1)$, $(-1,1,-1)$, $(1,0,1)$.

17.3 Show that putting

$$\langle\mathbf{x}, \mathbf{y}\rangle = x_1y_1 + 2x_2y_2 - x_1y_2 - x_2y_1 + x_3y_3$$

defines a scalar product on $\mathbf{R}^3$.

17.4 Repeat Ex. 2, using the inner product defined in Ex. 3.

17.5 Show that putting

$$\langle\mathbf{x}, \mathbf{y}\rangle = x_1y_1 + 6x_2y_2 - 2x_1y_2 - 2x_2y_1 + x_3y_3 - x_2y_3 - x_3y_2 + x_4y_4$$

defines a scalar product on $\mathbf{R}^4$.

17.6 Let $\mathbf{V}$ be a 4-dimensional Euclidean vector space, in which there is given an orthonormal basis e, and let $\mathbf{W}$ be the subspace generated by the vectors

$$\mathbf{w}_1(1,1,0,1),\ \mathbf{w}_2(1,-1,0,-1),\ \mathbf{w}_3(3,1,0,1).$$

Find $\dim(\mathbf{W})$, and an orthonormal basis of $\mathbf{W}$. Extend this basis to an orthonormal basis of $\mathbf{V}$.

17.7 In each of the following, find an orthonormal basis of the subspace of $\mathbf{R}^4$ generated by the given vectors.
a) $(2,0,0,1)$, $(1,2,2,3)$, $(10,-1,-\frac{1}{2},0)$, $(5,2,2,5)$
b) $(-1,0,1,1)$, $(2,1,1,4)$, $(0,1,3,6)$
c) $(1,1,-1,1)$, $(-2,-2,2,-2)$, $(2,1,1,2)$, $(3,1,1,1)$
d) $(1,1,0,-1)$, $(-2,1,\sqrt{3},5)$, $(4,4,\sqrt{3},2)$, $(-6,-3,0,3)$.

17.8 Find an orthonormal basis of $\mathbf{R}^4$ by applying the Gram-Schmidt process to the following system of vectors:

$$(0,1,0,1),\ (2,1,0,1),\ (-1,0,0,1),\ (0,0,1,0).$$

17.9 Using Gram-Schmidt, orthogonalize the canonical basis of $\mathbf{R}^4$ with respect to the following inner product:

$$\langle \mathbf{x}, \mathbf{y} \rangle = 2x_1y_1 + x_1y_2 + x_2y_1 + 2x_2y_2 + 2x_3y_3 + x_3y_4 + x_4y_3 + 2x_4y_4.$$

17.10 Prove that the complex numbers of modulus 1 form a subgroup of the multiplicative group $\mathbf{C}^*$, isomorphic to SO(2).

17.11 Let $\mathbf{V}$ be a Euclidean vector space and let $\mathbf{v} \in \mathbf{V}, \mathbf{v} \neq \mathbf{0}$. Prove that the map $f_{\mathbf{v}} : \mathbf{V} \to \langle \mathbf{v} \rangle$, associating to each $\mathbf{w} \in \mathbf{V}$ its orthogonal projection in the direction of $\mathbf{v}$, is linear.

17.12 Let $\mathbf{V}$ be a Euclidean vector space and let $\mathbf{v} \in \mathbf{V}, \mathbf{v} \neq \mathbf{0}$. The map

$$p_{\mathbf{v}} : \mathbf{V} \to \mathbf{v}^{\perp},$$

defined by $p_{\mathbf{v}}(\mathbf{w}) = \mathbf{w} - a_{\mathbf{v}}(\mathbf{w})\mathbf{v}$ is called the *projection of* $\mathbf{w}$ *on* $\mathbf{v}^{\perp}$. Prove that $p_{\mathbf{v}}$ is linear.

17.13 Prove that if two orthonormal bases $\{\mathbf{i},\mathbf{j}\}$ and $\{\mathbf{i}',\mathbf{j}'\}$ of a Euclidean vector space $\mathbf{V}$ of dimension 2, define opposite orientations, then they define opposite oriented angles.

18

Vector product

In this chapter, $\mathbf{V}$ is any 3-dimensional Euclidean vector space, with inner product $\langle , \rangle$, and $\{\mathbf{i}, \mathbf{j}, \mathbf{k}\}$ is a fixed orthonormal basis of $\mathbf{V}$.

Definition 18.1
Let $\mathbf{v}_1(x_1, y_1, z_1)$ and $\mathbf{v}_2(x_2, y_2, z_2)$ be two vectors in $\mathbf{V}$. The *vector product of* $\mathbf{v}_1$ *and* $\mathbf{v}_2$ is the vector with coordinates

$$(y_1 z_2 - z_1 y_2,\ z_1 x_2 - x_1 z_2,\ x_1 y_2 - y_1 x_2),$$

denoted $\mathbf{v}_1 \times \mathbf{v}_2$ (read $\mathbf{v}_1$ *cross* $\mathbf{v}_2$). These coordinates can be seen as the minors, taken with alternating signs $+, -, +$ of the matrix

$$\begin{pmatrix} x_1 & y_1 & z_1 \\ x_2 & y_2 & z_2 \end{pmatrix}.$$

The vector product thus associates to an ordered pair of vectors a third vector. It is therefore a map

$$\mathbf{V} \times \mathbf{V} \to \mathbf{V}.$$

Since it was defined using coordinates in a specified orthonormal basis $\{\mathbf{i}, \mathbf{j}, \mathbf{k}\}$, we must expect that the vector product depends on this basis. After looking at some of its properties, we will see that in fact the vector product depends only on the orientation of $\mathbf{V}$ defined by the chosen orthonormal basis.

The following theorem describes the principal properties of the vector product.

Theorem 18.2

For any $\mathbf{v}_1, \mathbf{v}_2, \mathbf{v}_3 \in \mathbf{V}$, *and for each* $c \in \mathbf{R}$ *one has:*

1) $\mathbf{v}_1 \times \mathbf{v}_2 = -(\mathbf{v}_2 \times \mathbf{v}_1)$,
2) $\mathbf{v}_1 \times (\mathbf{v}_2 + \mathbf{v}_3) = \mathbf{v}_1 \times \mathbf{v}_2 + \mathbf{v}_1 \times \mathbf{v}_3$,
3) $c(\mathbf{v}_1 \times \mathbf{v}_2) = (c\mathbf{v}_1) \times \mathbf{v}_2 = \mathbf{v}_1 \times (c\mathbf{v}_2)$,
4) $\langle \mathbf{v}_1, \mathbf{v}_1 \times \mathbf{v}_2 \rangle = 0$,
5) $\langle \mathbf{v}_2, \mathbf{v}_1 \times \mathbf{v}_2 \rangle = 0$,
6) $\|\mathbf{v}_1 \times \mathbf{v}_2\|^2 = \|\mathbf{v}_1\|^2 \|\mathbf{v}_2\|^2 - \langle \mathbf{v}_1, \mathbf{v}_2 \rangle^2$,
7) $\mathbf{v}_1 \times \mathbf{v}_2 = \mathbf{0}$ *if and only if* $\mathbf{v}_1$ *and* $\mathbf{v}_2$ *are parallel.*

Proof

The proofs of (1), (2), (3) and (7) are easy and so are left to the reader.

(4) is proved by noting that $\langle \mathbf{v}_1, \mathbf{v}_1 \times \mathbf{v}_2 \rangle$ is equal to the determinant of the matrix

$$\begin{pmatrix} x_1 & y_1 & z_1 \\ x_1 & y_1 & z_1 \\ x_2 & y_2 & z_2 \end{pmatrix},$$

which, having 2 equal rows, is zero. (5) is proved in the same manner.

To prove (6), we calculate that

$$\|\mathbf{v}_1 \times \mathbf{v}_2\|^2 = (y_1 z_2 - z_1 y_2)^2 + (z_1 x_2 - x_1 z_2)^2 + (x_1 y_2 - y_1 x_2)^2$$

and

$$\|\mathbf{v}_1\|^2 \|\mathbf{v}_2\|^2 - \langle \mathbf{v}_1, \mathbf{v}_2 \rangle^2 = \\ (x_1^2 + y_1^2 + z_1^2)(x_2^2 + y_2^2 + z_2^2) - (x_1 x_2 + y_1 y_2 + z_1 z_2)^2.$$

Expanding the two right hand sides shows them to be equal. □

Note that (6) can also be written in the form

$$\|\mathbf{v}_1 \times \mathbf{v}_2\|^2 = \begin{vmatrix} \langle \mathbf{v}_1, \mathbf{v}_1 \rangle & \langle \mathbf{v}_1, \mathbf{v}_2 \rangle \\ \langle \mathbf{v}_2, \mathbf{v}_1 \rangle & \langle \mathbf{v}_2, \mathbf{v}_2 \rangle \end{vmatrix}.$$

Corollary 18.3

8) *If* $\mathbf{v}_1$ *and* $\mathbf{v}_2$ *are linearly independent, then any vector orthogonal to both* $\mathbf{v}_1$ *and* $\mathbf{v}_2$ *is a scalar multiple of* $\mathbf{v}_1 \times \mathbf{v}_2$.
9) *If* $\mathbf{v}_1$ *and* $\mathbf{v}_2$ *are linearly independent, then* $\{\mathbf{v}_1, \mathbf{v}_2, \mathbf{v}_1 \times \mathbf{v}_2\}$ *is a basis of* $\mathbf{V}$ *with the same orientation as* $\{\mathbf{i}, \mathbf{j}, \mathbf{k}\}$.

Proof

8) Since $\mathbf{v}_1$ and $\mathbf{v}_2$ are linearly independent and $\dim(\mathbf{V}) = 3$,

$$\dim(\{\mathbf{v}_1, \mathbf{v}_2\}^\perp) = \dim(\langle \mathbf{v}_1, \mathbf{v}_2 \rangle) = 1.$$

By (4) and (5) we have $\mathbf{v}_1 \times \mathbf{v}_2 \in \{\mathbf{v}_1, \mathbf{v}_2\}^\perp$, and by (7) $\mathbf{v}_1 \times \mathbf{v}_2 \neq \mathbf{0}$. Thus $\mathbf{v}_1 \times \mathbf{v}_2$ generates $\{\mathbf{v}_1, \mathbf{v}_2\}^\perp$.

9) The linear independence of $\mathbf{v}_1, \mathbf{v}_2, \mathbf{v}_1 \times \mathbf{v}_2$ follows from that of $\mathbf{v}_1, \mathbf{v}_2$ and the fact that $\mathbf{v}_1 \times \mathbf{v}_2 \in \{\mathbf{v}_1, \mathbf{v}_2\}^\perp$. Since $\dim(\mathbf{V}) = 3$ we can conclude that $\{\mathbf{v}_1, \mathbf{v}_2, \mathbf{v}_1 \times \mathbf{v}_2\}$ is a basis.

To show that $\{\mathbf{v}_1, \mathbf{v}_2, \mathbf{v}_1 \times \mathbf{v}_2\}$ has the same orientation as $\{\mathbf{i}, \mathbf{j}, \mathbf{k}\}$ it suffices to observe that the matrix for changing coordinates from $\{\mathbf{v}_1, \mathbf{v}_2, \mathbf{v}_1 \times \mathbf{v}_2\}$ to $\{\mathbf{i}, \mathbf{j}, \mathbf{k}\}$ has determinant $\|\mathbf{v}_1 \times \mathbf{v}_2\|^2$. □

It is now fairly easy to ascertain the dependence of $\mathbf{v}_1 \times \mathbf{v}_2$ on the chosen basis $\{\mathbf{i}, \mathbf{j}, \mathbf{k}\}$ of $\mathbf{V}$.

If $\mathbf{v}_1$ and $\mathbf{v}_2$ are linearly dependent, then $\mathbf{v}_1 \times \mathbf{v}_2 = \mathbf{0}$ independently of the chosen basis.

If, on the other hand, $\mathbf{v}_1, \mathbf{v}_2$ are linearly independent, then $\|\mathbf{v}_1 \times \mathbf{v}_2\|$ is uniquely determined by (6), and $\mathbf{v}_1 \times \mathbf{v}_2$ belongs to $\{\mathbf{v}_1, \mathbf{v}_2\}^\perp$, which is of dimension 1. Thus a different choice of basis can change only the sign of $\mathbf{v}_1 \times \mathbf{v}_2$, and by (9) this occurs if and only if the bases have different orientations. In sum, $\mathbf{v}_1 \times \mathbf{v}_2$ *depends only on the orientation of the Euclidean space* $\mathbf{V}$ *defined by the basis* $\{\mathbf{i}, \mathbf{j}, \mathbf{k}\}$.

Property (6) has the following interesting interpretation.

Proposition 18.4

Suppose that $\mathbf{v}, \mathbf{w}$ *are linearly independent, and write* $\mathbf{v} = \mathbf{a} + \mathbf{b}$, *where* $\mathbf{a}$ *is parallel to* $\mathbf{w}$ *and* $\mathbf{b}$ *is orthogonal to* $\mathbf{w}$. *Then*

$$\|\mathbf{v} \times \mathbf{w}\| = \|\mathbf{b}\|\|\mathbf{w}\|.$$

Proof

Using properties (2), (7) and (6) one has

$$\|\mathbf{v} \times \mathbf{w}\| = \|(\mathbf{a} + \mathbf{b}) \times \mathbf{w}\| =$$
$$\|(\mathbf{a} \times \mathbf{w}) + (\mathbf{b} \times \mathbf{w})\| = \|\mathbf{b} \times \mathbf{w}\| = \|\mathbf{b}\|\|\mathbf{w}\|.$$

□

Note that if $\mathbf{V}$ is the vector space of geometric vectors in ordinary space, Proposition 18.4 states that $\|\mathbf{v} \times \mathbf{w}\|$ is equal to the area of the parallelogram formed from the vectors $\mathbf{v}$ and $\mathbf{w}$ — see Fig. 18.1.

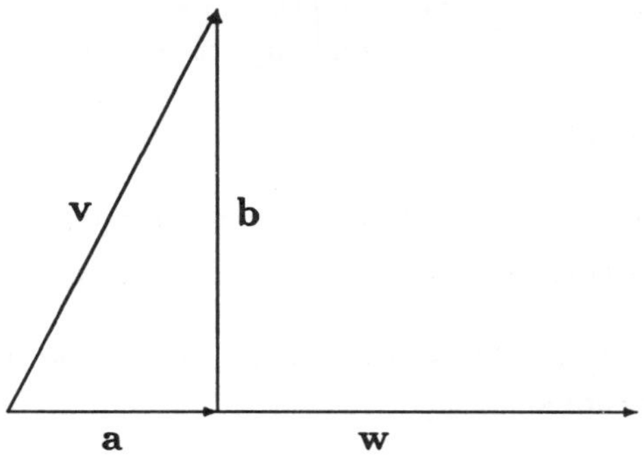

Fig. 18.1

Given three vectors $\mathbf{v}_1, \mathbf{v}_2, \mathbf{v}_3 \in \mathbf{V}$ the scalar $\langle \mathbf{v}_1, \mathbf{v}_2 \times \mathbf{v}_3 \rangle$ is called the *mixed product* of $\mathbf{v}_1, \mathbf{v}_2$, and $\mathbf{v}_3$.

If the coordinates of the three vectors $\mathbf{v}_1, \mathbf{v}_2, \mathbf{v}_3$ are, respectively, $(x_1, y_1, z_1), (x_2, y_2, z_2), (x_3, y_3, z_3)$ then

$$\langle \mathbf{v}_1, \mathbf{v}_2 \times \mathbf{v}_3 \rangle = \begin{vmatrix} x_1 & y_1 & z_1 \\ x_2 & y_2 & z_2 \\ x_3 & y_3 & z_3 \end{vmatrix}.$$

From this expression for $\langle \mathbf{v}_1, \mathbf{v}_2 \times \mathbf{v}_3 \rangle$ it follows immediately that exchanging two of the factors of the mixed product changes its sign, and that $\langle \mathbf{v}_1, \mathbf{v}_2 \times \mathbf{v}_3 \rangle = 0$ if and only if $\mathbf{v}_1, \mathbf{v}_2, \mathbf{v}_3$ are linearly dependent.

We conclude his brief chapter with an identity due to Lagrange.

Proposition 18.5 (Lagrange's identity)
For any $\mathbf{v}, \mathbf{w}, \mathbf{a}, \mathbf{b} \in \mathbf{V}$ *one has*

$$\langle \mathbf{v} \times \mathbf{w}, \mathbf{a} \times \mathbf{b} \rangle = \begin{vmatrix} \langle \mathbf{v}, \mathbf{a} \rangle & \langle \mathbf{v}, \mathbf{b} \rangle \\ \langle \mathbf{w}, \mathbf{a} \rangle & \langle \mathbf{w}, \mathbf{b} \rangle \end{vmatrix}. \tag{18.1}$$

Proof

If the coordinates of $\mathbf{v}, \mathbf{w}, \mathbf{a}, \mathbf{b}$ are (v_1, v_2, v_3), (w_1, w_2, w_3), (a_1, a_2, a_3), (b_1, b_2, b_3), then the left hand side of (18.1) is

$$\begin{vmatrix} v_2 & v_3 \\ w_2 & w_3 \end{vmatrix} \begin{vmatrix} a_2 & a_3 \\ b_2 & b_3 \end{vmatrix} + \begin{vmatrix} v_1 & v_3 \\ w_1 & w_3 \end{vmatrix} \begin{vmatrix} a_1 & a_3 \\ b_1 & b_3 \end{vmatrix} + \begin{vmatrix} v_1 & v_2 \\ w_1 & w_2 \end{vmatrix} \begin{vmatrix} a_1 & a_2 \\ b_1 & b_2 \end{vmatrix},$$

while the right hand side is

$$\begin{aligned} &(v_1a_1 + v_2a_2 + v_3a_3)(w_1b_1 + w_2b_2 + w_3b_3) \\ &\quad -(v_1b_1 + v_2b_2 + v_3b_3)(w_1a_1 + w_2a_2 + w_3a_3). \end{aligned}$$

Lagrange's identity (18.1) is now verified by expanding and comparing these two expressions. □

19

Euclidean spaces

Let $\mathbf{E}$ be a real affine space with associated vector space $\mathbf{V}$. We will say that $\mathbf{E}$ is a *Euclidean space* if $\mathbf{V}$ is endowed with a scalar product, that is if $\mathbf{V}$ is a Euclidean vector space. We will continue to use the notation $\langle \mathbf{v}, \mathbf{w} \rangle$ for the scalar product of two vectors $\mathbf{v}, \mathbf{w} \in \mathbf{V}$.

The affine numerical n-space $\mathbf{A}^n(\mathbf{R})$ becomes a Euclidean space if $\mathbf{R}^n$ is given the standard scalar product. This Euclidean space, denoted $\mathbf{E}^n$, is called the Euclidean numerical n-space. It's associated vector space is the Euclidean numerical n-space.

A system of coordinates $\mathbf{e}_1 \dots \mathbf{e}_n$ in the Euclidean space $\mathbf{E}$ for which $\{\mathbf{e}_1, \dots, \mathbf{e}_n\}$ is an orthonormal basis of $\mathbf{V}$ is called a *Cartesian coordinate system*, or *Cartesian frame*. It is natural to use Cartesian coordinate systems when studying Euclidean spaces, as they facilitate considerably the calculations.

It is worth noting at this point that the change of coordinates formula for Cartesian frames takes a particular form, because of the fact that the corresponding bases are orthonormal. Let $O\mathbf{e}_1 \dots \mathbf{e}_n$ and $O'\mathbf{e}'_1 \dots \mathbf{e}'_n$ be two Cartesian frames in $\mathbf{E}$, and let $P \in \mathbf{E}$ be a point with coordinates $\mathbf{x} = (x_1 \dots x_n)^t$ with respect to the first frame, and $\mathbf{x}' = (x'_1 \dots x'_n)^t$ with respect to the second. Then,

$$\mathbf{x}' = A\mathbf{x} + c,$$

where $A = M_{\mathbf{e},\mathbf{e}'}(\mathbf{1_V})$ is an orthogonal matrix, and $\mathbf{c} = (c_1 \dots c_n)^t$ is determined by the identity

$$\overrightarrow{O'O} = c_1\mathbf{e}'_1 + \cdots c_n\mathbf{e}'_n.$$

If $n = 2$, that is if $\mathbf{E}$ is a Euclidean plane, the matrix A is of the form (2.6) or (2.7) accordingly as the two frames have the same or opposite orientations.

Using the scalar product defined in $\mathbf{V}$ it is possible to introduce into $\mathbf{E}$ such metric notions as distance, angle and area.

Definition 19.1

Let P, Q be two points in the Euclidean space $\mathbf{E}$. The *distance* between P and Q, denoted by $d(P,Q)$, is

$$d(P,Q) = \|\overrightarrow{PQ}\|.$$

If $O\mathbf{e}_1 \dots \mathbf{e}_n$ is a Cartesian coordinate system for $\mathbf{E}$, and if the given points are $P(x_1, \dots, x_n)$ and $Q(y_1, \dots, y_n)$ respectively, then

$$d(P,Q) = \sqrt{(y_1 - x_1)^2 + \cdots + (y_n - x_n)^2}.$$

In the case that $\mathbf{E} = \mathbf{V}_a$, where $\mathbf{V}$ is a Euclidean vector space, then

$$d(\mathbf{v}, \mathbf{w}) = \|\mathbf{w} - \mathbf{v}\|.$$

Proposition 19.2

Distance has the following properties:

[MS1] $d(P,Q) \geq 0$, *for every* $P, Q \in \mathbf{E}$, *and* $d(P,Q) = 0$ *if and only if* $P = Q$;

[MS2] $d(P,Q) = d(Q,P)$ *for all* $P, Q \in \mathbf{E}$;

[MS3] $d(P,Q) + d(Q,R) \leq d(P,R)$, *for all* $P, Q, R \in \mathbf{E}$.

This proposition follows immediately form the definition of scalar product.

The three properties of Proposition 19.2 can be taken as the axioms for a more general class of space, called 'metric spaces'.

To be precise, a *metric space* is a set X on which there is defined a notion of distance — or metric — that is, a map $d : X \times X \rightarrow \mathbf{R}$ satisfying the three conditions MS1, MS2 and MS3.

Proposition 19.2 asserts that with distance defined in Definition 19.2, every Euclidean space is a metric space. The converse, however, is far from true: not every metric space is a Euclidean space. Indeed metric spaces are much more general and we will not be studying their geometry in this text. To see that they are more general, it is enough to be aware that any subset of a metric space is itself a metric space; in particular, *every* subset of a Euclidean space **E** is a metric space.

One can also define the convex angle between two lines using the notion of convex angle between two vectors, defined using the scalar product in **V**.

Given a line ℓ in **E**, a direction vector for ℓ of length 1 is called a *unit direction vector* of ℓ. There are exactly two unit direction vectors for any line, one the opposite of the other.

To define the convex angle between two lines ℓ and ℓ_1, we need to choose direction vectors $\mathbf{a}$ and $\mathbf{a}_1$ of ℓ and ℓ_1 respectively. The *convex angle* ϕ *between* ℓ and ℓ_1 is defined to be the convex angle between $\mathbf{a}$ and $\mathbf{a}_1$. Thus the convex angle between ℓ and ℓ_1 is defined by the two conditions

$$0 \leq \phi \leq \pi, \quad \cos\phi = \frac{\langle \mathbf{a}, \mathbf{a}_1 \rangle}{\|\mathbf{a}\|\|\mathbf{a}_1\|}.$$

This definition depends upon the choice of the direction vectors $\mathbf{a}$ and $\mathbf{a}_1$: the convex angle ϕ changes to $\pi - \phi$ if one of the direction vectors is multiplied by a negative scalar. This is nothing new: in elementary plane Euclidean geometry to define the angle between two lines it is necessary to specify the direction of each line, which is the smae as choosing direction vectors.

Two lines in **E** are said to be *orthogonal*, or *perpendicular*, if the convex angle between them is $\pi/2$, that is if one (and hence every) direction vector for one line is orthogonal to one (and hence to every) direction vector of the other line.

It is likewise possible to define distances and convex angles between subspaces of a Euclidean space **E** more generally. It is not, however, our intention to develop such a general theory, but we will limit ourselves to Euclidean spaces of dimensions 2 and 3. It should by now be clear to the reader that these possess essentially all the geometric properties of the ordinary plane and space.

Let **E** be a Euclidean plane in which there is given a Cartesian frame $O\mathbf{ij}$. Consider a line ℓ, with Cartesian equation

$$AX + BY + C = 0. \tag{19.1}$$

Since $\mathbf{a}(-B, A)$ is a direction vector of ℓ, it follows that the vectors

$$\pm\mathbf{u} = \pm\frac{\mathbf{a}}{\|\mathbf{a}\|}$$

with coordinates

$$\pm\left(\frac{-B}{\sqrt{A^2+B^2}}, \frac{A}{\sqrt{A^2+B^2}}\right)$$

are the unit direction vectors of ℓ.

A non-zero vector $\mathbf{m}$ is said to be *orthogonal* (or *perpendicular* or *normal*) to ℓ if it is orthogonal to one (and hence all) of its direction vectors. It is clear that since $\dim(\mathbf{V}) = 2$, if two vectors are both orthogonal to ℓ then they are parallel. Moreover there are exactly two unit vectors orthogonal to ℓ, they are called the *unit normals vectors to* ℓ.

Since $\mathbf{a}(-B, A)$ is a direction vector of ℓ, it follows that the vector $\mathbf{n}(A, B)$ is orthogonal to ℓ. Thus the unit normal vectors to ℓ are

$$\pm\mathbf{v} = \pm\frac{\mathbf{n}}{\|\mathbf{n}\|}$$

with coordinates

$$\pm\left(\frac{A}{\sqrt{A^2+B^2}}, \frac{B}{\sqrt{A^2+B^2}}\right).$$

Given a point $Q(a, b) \in \ell$, then

$$A(X-a) + B(Y-b) = 0 \tag{19.2}$$

is a Cartesian equation for ℓ .

Since equation (19.2) depends only on a, b, A and B, one sees that ℓ is determined by a point $Q(a, b)$ and a normal vector $\mathbf{n}(A, B)$: any line in $\mathbf{E}$ can thus be determined by giving a point on it and a direction perpendicular to it.

Let ℓ and ℓ_1 be two lines with Cartesian equations (19.1) and

$$A_1X + B_1Y + C_1 = 0. \tag{19.3}$$

respectively. The convex angle ϕ between them is defined by the conditions $0 \leq \phi \leq \pi$ and

$$\cos\phi = \frac{\langle \mathbf{a}, \mathbf{a}_1\rangle}{\|\mathbf{a}\|\|\mathbf{a}_1\|} = \frac{AA_1 + BB_1}{\sqrt{A^2+B^2}\sqrt{A_1^2+B_1^2}},$$

where $\mathbf{a}(-B, A)$ and $\mathbf{a}_1(-B_1, A_1)$ are the direction vectors of ℓ and ℓ_1 respectively.

In particular, ℓ and ℓ_1 are perpendicular if and only if

$$AA_1 + BB_1 = 0.$$

Suppose we are given a line ℓ with Cartesian equation (19.1), and a point $P_0(x_0, y_0) \in \mathbf{E}$. Consider the line ℓ_1 through P_0 and perpendicular to ℓ. Clearly such a line exists and is unique. Since ℓ_1 is not parallel to ℓ, they have precisely one point N in common: N is the *base of the perpendicular from* P_0 *to* ℓ.

The distance $d(P_0, N)$ is called the *distance from* P_0 *to* ℓ, which we denote $d(P_0, \ell)$.

It is easy to calculate $d(P_0, \ell)$ by first finding an equation for ℓ_1, then finding the coordinates of N, and finally calculating $d(P_0, N)$. However, there is a more efficient method given by the following.

Proposition 19.3

The distance $d(P_0, \ell)$ *of a point* $P_0(x_0, y_0) \in \mathbf{E}$ *from the line* ℓ *with equation (19.1), is given by the formula*

$$d(P_0, \ell) = \frac{|Ax_0 + By_0 + C|}{\sqrt{A^2 + B^2}}.$$

Proof

Note that if $\mathbf{v}$ is a unit normal to ℓ and if $Q(a, b)$ is any point on ℓ, then

$$d(P_0, \ell) = |\langle \mathbf{v}, \overrightarrow{QP_0} \rangle|,$$

see Fig. 19.1.

Choose the normal vector

$$\mathbf{v}\left(\frac{A}{\sqrt{A^2 + B^2}}, \frac{B}{\sqrt{A^2 + B^2}}\right).$$

Since $\overrightarrow{QP_0}$ has coordinates $(x_0 - a), (y_0 - b)$, and using the fact that $C = -(Aa + Bb)$ we obtain

$$d(P_0, \ell) = \frac{|A(x_0 - a) + B(y_0 - b)|}{\sqrt{A^2 + B^2}} = \frac{|Ax_0 + By_0 + C|}{\sqrt{A^2 + B^2}}.$$

□

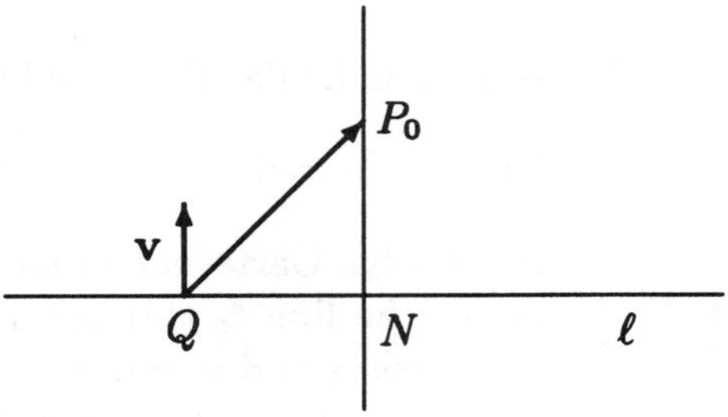

Fig. 19.1

If ℓ and ℓ' are two parallel lines in $\mathbf{E}$, their distance $d(\ell, \ell')$ is by definition $d(P, \ell')$ where P is any point of ℓ. To find $d(\ell, \ell')$ it is thus sufficient to find any point $P \in \ell$ and use the formula of the proposition.

Let $\mathbf{E}$ be a 3-dimensional Euclidean space with associated vector space $\mathbf{V}$. Let $O\mathbf{ijk}$ be a fixed Cartesian coordinate system.

Consider a plane $\mathcal{P}$ in $\mathbf{E}$ with equation

$$AX + BY + CZ + D = 0. \tag{19.4}$$

The vector $\mathbf{m}$ is said to be *orthogonal* (or *perpendicular*, or *normal*) to $\mathcal{P}$ if it is perpendicular to every vector in the vector subspace associated of $\mathcal{P}$. If, furthermore, $\mathbf{m}$ is a unit vector, it is called a *unit normal to* $\mathcal{P}$.

Since $\dim(\mathbf{E}) = 3$, any two vectors perpendicular to $\mathcal{P}$ are themselves parallel. Thus $\mathcal{P}$ has precisely two unit normals.

Fix a point $P_0(x_0, y_0, z_0) \in \mathcal{P}$ and consider the Cartesian equation for $\mathcal{P}$

$$A(X - x_0) + B(Y - y_0) + C(Z - z_0) = 0. \tag{19.5}$$

Every vector belonging to the vector subspace associated to $\mathcal{P}$ is of the form $\overrightarrow{P_0P}$ for $P \in \mathcal{P}$. By (19.5) we see that the vector $\mathbf{n}(A, B, C)$ is normal to $\mathcal{P}$. It follows that the two vectors $\pm\mathbf{v} = \pm\mathbf{n}/\|\mathbf{n}\|$ with coordinates

$$\pm\left(\frac{A}{\sqrt{A^2+B^2+C^2}}, \frac{B}{\sqrt{A^2+B^2+C^2}}, \frac{C}{\sqrt{A^2+B^2+C^2}}\right)$$

are the unit normals to $\mathcal{P}$.

Equation (19.5) also shows that a plane is determined by a point on it and a normal vector.

Convex angle between planes

The convex angle between two planes in $\mathbf{E}$ is defined using their normal vectors. Let $\mathcal{P}$ and $\mathcal{P}_1$ be two planes, with equations (19.4) and

$$A_1X + B_1Y + C_1Z + D_1 = 0. \tag{19.6}$$

The *convex angle between* $\mathcal{P}$ *and* $\mathcal{P}_1$ is the convex angle between the vectors $\mathbf{n}(A, B, C)$ and $\mathbf{n}_1(A_1, B_1, C_1)$. It is therefore the angle ϕ satisfying $0 \leq \phi \leq \pi$ and

$$\cos\phi = \frac{\langle \mathbf{n}, \mathbf{n}_1\rangle}{\|\mathbf{n}\|\|\mathbf{n}_1\|} = \frac{AA_1 + BB_1 + CC_1}{\sqrt{A^2 + B^2 + C^2}\sqrt{A_1^2 + B_1^2 + C_1^2}}. \tag{19.7}$$

Note that the definition of ϕ depends upon the choice of normals $\mathbf{n}$ and $\mathbf{n}_1$, and so on the choice of equations (19.4) and (19.6). Multiplying one or other of these two equations by a negative number changes ϕ into $\pi - \phi$.

If $\phi = \pi/2$ the planes are said to be *perpendicular* or *orthogonal.* This occurs if and only if

$$AA_1 + BB_1 + CC_1 = 0.$$

Angle between a line and a plane

Let $\mathcal{P}$ be a plane with equation (19.4), and let ℓ be a line with direction vector $\mathbf{a}(l, m, n)$. The *angle between* $\mathcal{P}$ *and* ℓ is the angle whose determination is

$$\phi = \psi - \frac{\pi}{2},$$

where ψ is the convex angle between the vectors $\mathbf{n}(A, B, C)$ and $\mathbf{a}$. Thus ϕ is defined by the conditions

$$-\tfrac{\pi}{2} \leq \phi \leq \tfrac{\pi}{2}$$

and

$$\sin\phi = \frac{Al + Bm + Cn}{\sqrt{A^2 + B^2 + C^2}\sqrt{l^2 + m^2 + n^2}}.$$

If $\phi = \pm\pi/2$, then $\mathcal{P}$ and ℓ are said to be *perpendicular* (or *orthogonal*).

Note that by putting $\sin\phi = 0$ one retrieves the condition for $\mathcal{P}$ and ℓ to be parallel

$$Al + Bm + Cn = 0$$

given already in Proposition 10.2.

Distance between a point and a plane

The distance of a point to a plane is defined similarly to the distance from a point to a line in the Euclidean plane.

Let $\mathcal{P}$ be the plane defined by (19.4), and let $P_0(x_0, y_0, z_0) \in \mathbf{E}$ be a point. Consider the line ℓ passing through P_0 and perpendicular to $\mathcal{P}$. Denote by $N = \ell \cap \mathcal{P}$ the foot of the perpendicular to $\mathcal{P}$ from P_0. Define the *distance from* P_0 *to* $\mathcal{P}$ by

$$d(P_0, \mathcal{P}) = d(P_0, N).$$

One can show that

$$d(P_0, \mathcal{P}) = \frac{|Ax_0 + By_0 + Cz_0 + D|}{\sqrt{A^2 + B^2 + C^2}}.$$

The proof, being similar to that of Proposition 19.4, is left to the reader.

Distance between a line and a plane parallel to it

Let ℓ be a line and $\mathcal{P}$ a plane parallel to it. The distance between them is defined to be the distance $d(P, \mathcal{P})$, where P is any point of the line ℓ. To find $d(\ell, \mathcal{P})$ it is therefore sucfficient to find the coordinates of any point P belonging to ℓ and then applying the formula for $d(P, \mathcal{P})$.

Distance between a point and a line

Let ℓ be a line and $P_0 \in \mathbf{E}$ a point. Consider the plane $\mathcal{P}$ passing through P_0 and perpendicular to ℓ, and let $N = \ell \cap \mathcal{P}$. The distance of P_0 from ℓ is defined to be

$$d(P_0, \ell) = d(P_0, N).$$

If ℓ passes throught the point $Q(a, b, c)$ and has direction vector $\mathbf{a}(l, m, n)$ and $P_0(x_0, y_0, z_0) \in \mathbf{E}$, then we have the following formula:

$$d(P_0, \ell) = \frac{\sqrt{\begin{vmatrix} y_0 - b & z_0 - c \\ m & n \end{vmatrix}^2 + \begin{vmatrix} x_0 - a & z_0 - c \\ l & n \end{vmatrix}^2 + \begin{vmatrix} x_0 - a & y_0 - b \\ l & m \end{vmatrix}^2}}{\sqrt{l^2 + m^2 + n^2}}.$$

Indeed,

$$\overrightarrow{QP_0} = \overrightarrow{QN} + \overrightarrow{NP_0},$$

and $\overrightarrow{QN}$ is parallel to $\mathbf{a}$, while $\overrightarrow{NP_0}$ is perpendicular to $\mathbf{a}$.

By Proposition 18.4 it follows that

$$d(P_0, \ell) = \|\overrightarrow{NP_0}\| = \frac{\|\mathbf{a} \times \overrightarrow{QP_0}\|}{\|\mathbf{a}\|}.$$

Expanding this equality gives the desired result.

Distance between two lines

Given two non-coplanar lines ℓ and ℓ_1 in $\mathbf{E}$ the distance between them, denoted $d(\ell, \ell_1)$, is defined as follows.

Suppose that ℓ passes through the point $Q(a, b, c)$ with direction vector $\mathbf{a}(l, m, n)$ and ℓ_1 passes through $Q_1(a_1, b_1, c_1)$ with direction vector $\mathbf{a}_1(l_1, m_1, n_1)$. Observe that there is a unique pair of points, $N \in \ell$ and $N_1 \in \ell_1$, such that the line that contains them is perpendicular both to ℓ and to ℓ_1. This condition of perpendicularity translates to

$$\begin{aligned} \langle \overrightarrow{NN_1}, \mathbf{a} \rangle &= 0 \\ \langle \overrightarrow{NN_1}, \mathbf{a}_1 \rangle &= 0. \end{aligned} \tag{19.8}$$

Writing

$$\begin{aligned} \overrightarrow{ON} &= \overrightarrow{OQ} + t\mathbf{a} \\ \overrightarrow{ON_1} &= \overrightarrow{OQ_1} + t_1\mathbf{a}_1, \end{aligned}$$

gives

$$\overrightarrow{NN_1} = \overrightarrow{QQ_1} + t_1\mathbf{a}_1 - t\mathbf{a}$$

and so (19.8) becomes

$$\begin{aligned} \langle \overrightarrow{QQ_1} + t_1\mathbf{a}_1 - t\mathbf{a}, \mathbf{a} \rangle &= 0 \\ \langle \overrightarrow{QQ_1} + t_1\mathbf{a}_1 - t\mathbf{a}, \mathbf{a}_1 \rangle &= 0, \end{aligned}$$

that is

$$\begin{aligned} \langle \overrightarrow{QQ_1}, \mathbf{a} \rangle + \langle t_1\mathbf{a}_1, \mathbf{a} \rangle - \langle t\mathbf{a}, \mathbf{a} \rangle &= 0 \\ \langle \overrightarrow{QQ_1}, \mathbf{a}_1 \rangle + \langle t_1\mathbf{a}_1, \mathbf{a}_1 \rangle - \langle t\mathbf{a}, \mathbf{a}_1 \rangle &= 0. \end{aligned} \tag{19.9}$$

Since $\mathbf{a}$ and $\mathbf{a}_1$ are linearly independent, then by Theorem 18.2(6),

$$\begin{vmatrix} \langle \mathbf{a}, \mathbf{a} \rangle & \langle \mathbf{a}_1, \mathbf{a} \rangle \\ \langle \mathbf{a}, \mathbf{a}_1 \rangle & \langle \mathbf{a}_1, \mathbf{a}_1 \rangle \end{vmatrix} = \|\mathbf{a} \times \mathbf{a}_1\|^2 \neq 0.$$

Thus system (19.9) has a unique solution (t, t_1), and so there is a unique pair (N, N_1) satisfying the required conditions.

The line ℓ' passing through N and N_1 is the *common perpendicular to ℓ and ℓ_1* — see Fig. 19.2.

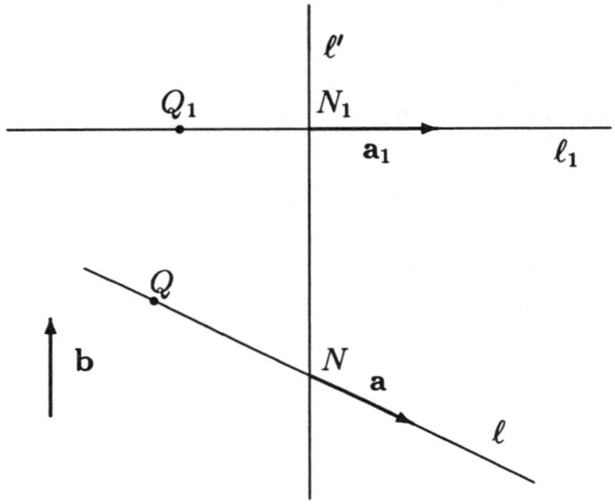

Fig. 19.2

The *distance between ℓ and ℓ_1* is defined to be

$$d(\ell, \ell_1) = d(N, N_1).$$

The following formula allows us to calculate easily the distance between two given lines

$$d(\ell, \ell_1) = \frac{\left\| \begin{matrix} a - a_1 & b - b_1 & c - c_1 \\ l & m & n \\ l_1 & m_1 & n_1 \end{matrix} \right\|}{\sqrt{\begin{vmatrix} m & n \\ m_1 & n_1 \end{vmatrix}^2 \begin{vmatrix} l & n \\ l_1 & n_1 \end{vmatrix}^2 \begin{vmatrix} l & m \\ l_1 & m_1 \end{vmatrix}^2}}. \tag{19.10}$$

(The numerator is the absolute value of the determinant.)

This formula is proved as follows.

Let $\mathbf{b}$ be a unit vector in the direction of the common perpendicular to ℓ and ℓ_1. Then

$$\langle \mathbf{b}, \overrightarrow{QQ_1} \rangle = \langle \mathbf{b}, \overrightarrow{QN} \rangle + \langle \mathbf{b}, \overrightarrow{NQ_1} \rangle = \langle \mathbf{b}, \overrightarrow{NQ_1} \rangle,$$

and so

$$d(\ell, \ell_1) = |\langle \mathbf{b}, \overrightarrow{NQ_1} \rangle| = |\langle \mathbf{b}, \overrightarrow{QQ_1} \rangle|.$$

On the other hand, since $\mathbf{b}$ is perpendicular to $\mathbf{a}$ and $\mathbf{a}_1$, then

$$\mathbf{b} = \pm \frac{\mathbf{a} \times \mathbf{a}_1}{\|\mathbf{a} \times \mathbf{a}_1\|}.$$

Thus,

$$d(\ell, \ell_1) = \frac{|\langle \mathbf{a} \times \mathbf{a}_1, \overrightarrow{QQ_1} \rangle|}{\|\mathbf{a} \times \mathbf{a}_1\|}.$$

Expanding this expression gives formula (19.10).

To find the Cartesian equations of the line ℓ' perpendicular to both ℓ and ℓ_1 one proceeds as follows.

We know $\mathbf{a} \times \mathbf{a}_1$, the direction vector of ℓ'. Denote its coordinates by $(\beta_1, \beta_2, \beta_3)$. The condition that the line passing through a point $P(X, Y, Z)$ with direction vector $\mathbf{a} \times \mathbf{a}_1$ be coplanar with ℓ and ℓ_1 are,

$$\begin{vmatrix} X - a & Y - b & Z - c \\ \beta_1 & \beta_2 & \beta_3 \\ l & m & n \end{vmatrix} = 0,$$

$$\begin{vmatrix} X - a_1 & Y - b_1 & Z - c_1 \\ \beta_1 & \beta_2 & \beta_3 \\ l_1 & m_1 & n_1 \end{vmatrix} = 0.$$

Since these equations represent two distinct planes and are satisfied by the points of ℓ', they form the Cartesian equations of ℓ'.

Complements 19.4

1. Suppose we are given a Cartesian frame $O\mathbf{e}_1 \ldots \mathbf{e}_n$ in the Euclidean space $\mathbf{E}$, and a point $C(c_1, \ldots, c_n)$. Let $r > 0$.

   The *sphere with centre C and radius r* is the subset $\mathbf{S}(C, r)$ of $\mathbf{E}$ consisting of those points $P(x_1, \ldots, x_n)$ such that $d(C, P) = r$; that is such that

$$(x_1 - c_1)^2 + \cdots (x_n - c_n^2) = r^2. \tag{19.11}$$

The *ball with centre C and radius r* is the subset $\mathbf{B}(C,r)$ of $\mathbf{E}$ consisting of those points $P(x_1,\ldots,x_n)$ such that $d(C,P) \leq r$; that is such that

$$(x_1 - c_1)^2 + \cdots (x_n - c_n^2) \leq r^2.$$

In the case that $\mathbf{E}$ is a plane, then $\mathbf{S}(C,r)$ and $\mathbf{B}(C,r)$ are the circle and disk, respectively, with centre C and radius r.

If $\dim(\mathbf{E}) = 1$, then $\mathbf{B}(C,r)$ is the segement of length $2r$ and centre C, while $\mathbf{S}(C,r)$ is the set consisting of its two endpoints.

When $\mathbf{E} = \mathbf{E}^n$, then one uses the symbols $\mathbf{S}^{n-1}$ and $\mathbf{B}^n$ to denote $\mathbf{S}(\mathbf{0},1)$ and $\mathbf{B}(\mathbf{0},1)$ respectively.

From (19.11) it follows that $\mathbf{S}(C,r)$ is the set of all points in $\mathbf{E}$ whose coordinates are the solutions of the equation

$$(X_1 - c_1)^2 + \cdots + (X_n - c_n)^2 = r^2 \tag{19.12}$$

in the unknowns $X_1,\ldots,X_n$. Equation (19.12) is called the *Cartesian equation for the sphere* $\mathbf{S}(C,r)$. Taking r^2 to the left-hand side and expanding the brackets, gives

$$X_1^2 + \cdots + X_n^2 + d_1X_1 + \cdots + d_nX_n + d = 0, \tag{19.13}$$

where $d_i = -2c_i$, $i = 1,\ldots,n$, and $d = c_1^2 + \cdots + c_n^2 - r^2$. By their definition, and the fact that $r^2 > 0$, the coeffiecients $d_1,\ldots,d_n,d$ in (19.13) satisfy the relation

$$\frac{d_1^2}{4} + \frac{d_2^2}{4} + \cdots + \frac{d_n^2}{4} - d > 0. \tag{19.14}$$

Conversely, any equation of the form (19.13) whose coefficients satisfy (19.14), is the Cartesian equation of a sphere $\mathbf{S}$ whose centre $C(c_1,\ldots,c_n)$ and radius r are

$$c_i = -\frac{d_i}{2}, \quad r = \sqrt{\frac{d_1^2}{4} + \frac{d_2^2}{4} + \cdots + \frac{d_n^2}{4} - d}.$$

In the case that $n = 2$, (19.13) is the equation of a circle, and has the form, in the unknowns X, Y,

$$X^2 + Y^2 + d_1X + d_2Y + d = 0$$

with $\frac{d_1^2}{4}+\frac{d_2^2}{4}-d>0$. The centre of the circle is $C(-\frac{d_1}{2},-\frac{d_2}{2})$, while the radius is

$$r=\sqrt{\frac{d_1^2}{4}+\frac{d_2^2}{4}-d}.$$

2. Let $\mathbf{E}$ be a Euclidean plane in which there is a given orientation, that is, an oriented Euclidean plane. Let ℓ and ℓ' be two half-lines starting at the same point O. Let $\mathbf{a}$ and $\mathbf{a}'$ be direction vectors of ℓ and ℓ' respectively. The angle $\widehat{\mathbf{aa}'}$ is called the *oriented angle between ℓ and ℓ'*, and is denoted $\widehat{\ell\ell'}$.

3. In an oriented Euclidean plane $\mathbf{E}$ fix a half-line with origin O. For each point $P\neq O$ consider the half-line ℓ_P with origin O and direction vector $\overrightarrow{OP}$. The positive scalar $\rho=\|\overrightarrow{OP}\|$ and the angle $\theta=\widehat{\ell\ell_P}$ are called *polar coordinates* of P, and respectively the *modulus* and *argument* of P with respect to the half-line ℓ. One says that ℓ defines a *system of polar coordinates* in $\mathbf{E}$.

   The polar coordinates (r,θ) of a point $P\neq O$ determine the point uniquely. Indeed, θ determines a half-line ℓ_θ with origin O, and P is the point of intersection of ℓ_θ with the circle with centre O and radius r — Fig. 19.3.

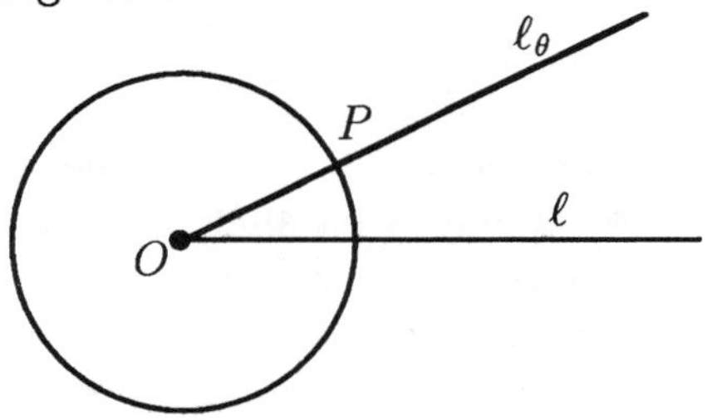

Fig. 19.3

Thus for every pair of real numbers (r,θ), with $r>0$, there is a unique point P whose polar coordinates are r and the angle determined by θ. It is convenient to extend polar coordinates to the point O as well, by giving it modulus $r=0$ and an indeterminate angle.

Consider the Cartesian frame $O\mathbf{ij}$ belonging to the given orientation in $\mathbf{E}$ with origin O, and such that $\mathbf{i}$ is a unit direction vector of ℓ ($O\mathbf{ij}$ is uniquely determined by these conditions). Let $P\neq O$ be a point with polar coordinates (r,θ) and Cartesian coordinates

(x, y). Then

$$x = r\cos\theta, \qquad y = r\sin\theta. \tag{19.16}$$

Conversely:

$$r = \sqrt{x^2+y^2}, \qquad \theta = \epsilon(y)\arccos\left(\frac{x}{\sqrt{x^2+y^2}}\right), \tag{19.17}$$

where $\epsilon(y) = \pm 1$ accordingly as $y \geq 0$ or $y < 0$ (the angle θ therefore lies between $-\pi$ and π).

Formulae (19.16) and (19.17) are the formulae for changing from polar coordinates to Cartesian coordinates and conversely; their verification is left to the reader.

4. Let $\mathbf{E}$ be an n-dimensional Euclidean space, $O\mathbf{e}_1 \ldots \mathbf{e}_n$ a Cartesian frame, and let $A_0(a_1^0, \ldots, a_n^0), A_1(a_1^1, \ldots, a_n^1), \ldots, A_n(a_1^n, \ldots, a_n^n) \in \mathbf{E}$ be independent points. The *volume of the parellelopiped determined by* $A_0, A_1, \ldots, A_n$ is defined to be $|\det(M)|$, where

$$M = \begin{pmatrix} a_1^1 - a_1^0 & a_2^1 - a_2^0 & \cdots & a_n^1 - a_n^0 \\ a_1^2 - a_1^0 & a_2^2 - a_2^0 & \cdots & a_n^2 - a_n^0 \\ \vdots & \vdots & & \vdots \\ a_1^n - a_1^0 & a_2^n - a_2^0 & \cdots & a_n^n - a_n^0 \end{pmatrix}.$$

In particular, for $n = 1, 2, 3$, one speaks of the *length, area* and *volume* of a segement, a parallelogram and a paralleliped, respectively. The length of a segment $A(a)B(b)$ in a Euclidean line is $|b - a|$, which coincides with $d(a, b)$.

The area of the parallogram determined by $A(a_1, a_2), B(b_1, b_2)$ and $C(c_1, c_2)$ in a Euclidean plane is

$$\begin{vmatrix} b_1 - a_1 & b_2 - a_2 \\ c_1 - a_1 & c_2 - a_2 \end{vmatrix}.$$

The volume of the parallopiped determined by $A(a_1, a_2, a_3)$, $B(b_1, b_2, b_2)$, $C(c_1, c_2, c_3)$ and $D(d_1, d_2, d_3)$ in a 3-dimensional Euclidean space is

$$\begin{vmatrix} b_1 - a_1 & b_2 - a_2 & b_3 - a_3 \\ c_1 - a_1 & c_2 - a_2 & c_3 - a_3 \\ d_1 - a_1 & d_2 - a_2 & d_3 - a_3 \end{vmatrix}.$$

Notice that the definition of volume of a parallelopiped is independent of the Cartesian frame $O\mathbf{e}_1 \ldots \mathbf{e}_n$. Indeed, the rows of the matrix M are the vectors $\overrightarrow{A_0A_1}, \overrightarrow{A_0A_2}, \ldots, \overrightarrow{A_0A_n}$ with respect to the basis $\{\mathbf{e}_1, \ldots, \mathbf{e}_n\}$; if $O'\mathbf{f}_1 \ldots \mathbf{f}_n$ is another Cartesian frame then the coordinates of $\overrightarrow{A_0A_1}, \overrightarrow{A_0A_2}, \ldots, \overrightarrow{A_0A_n}$ with respect to the basis $\{\mathbf{f}_1 \ldots \mathbf{f}_n\}$ are obtained from the previous ones by multiplying them by an orthogonal matrix. Then the analogue N of the matrix M in the frame $O'\mathbf{f}_1 \ldots \mathbf{f}_n$ is the product of M with this orthogonal matrix, which has determinant ± 1. Thus

$$|\det(N)| = |\det(M)|.$$

5. Let $\mathbf{E}$ be a Euclidean space. A subset $S \subset \mathbf{E}$ is said to be *bounded* if it is contained in some ball; that is if there exist $C \in \mathbf{E}$ and $r > 0$ such that $S \subset \mathbf{B}(C, r)$.

A *convex polyhedron* is a bounded subset of $\mathbf{E}$ which is not contained in any proper affine subspace of $\mathbf{E}$ and which is the intersection of finitely many half-spaces. A convex polyhedron is convex because any half-space is. The dimension of $\mathbf{E}$ is said to be the *dimension of the polyhedron.* If $\dim(\mathbf{E}) = 1$, a convex polyhedron is a segement. If $\dim(\mathbf{E}) = 2$ then a convex polyhedron is called a *convex polygon*, while if $\dim(\mathbf{E}) = 3$, convex polyhedra are called *convex solids.*

We leave the reader to define notions appropriate to convex polyhedra analogous to *vertex, side* and *angle*, to *regular n-gon*, and to *adjacent edges* for polygons in elementary Euclidean geometry.

Suppose $\dim(\mathbf{E}) = 3$ and that $\mathbf{\Pi} \subset \mathbf{E}$ is a convex solid. Suppose moreover that $\mathbf{\Pi}$ is contained in one of the half-spaces defined by the plane $\mathcal{P}$, then there are the following possibilities:

$\mathcal{P} \cap \mathbf{\Pi} = \emptyset$;
$\mathcal{P} \cap \mathbf{\Pi}$ is a point, called a *vertex* of $\mathbf{\Pi}$;
$\mathcal{P} \cap \mathbf{\Pi}$ is a segment, called an *edge* of $\mathbf{\Pi}$;
$\mathcal{P} \cap \mathbf{\Pi}$ is a polygon, called a *face* of $\mathbf{\Pi}$.

Since $\mathbf{\Pi}$ is the intersection of finitely many half-spaces, it follows that it has only a finite number V of vertices, E of edges and F of faces. It is easily seen that each edge is the edge of two faces, and that every vertex is the vertex of at least three faces and as many edges.

For any convex solid, the numbers of vertices, edges and faces satisfy the relation,

$$V - E + F = 2. \tag{19.18}$$

This famous identity was already known to Descartes in 1640, but was proved for the first time by Euler in 1752.

We give a proof of (19.18) similar to Euler's original one. To simplify the argument, we will suppose that it is possible to construct $\mathbf{\Pi}$ by starting with a face and adding one face at a time in such a way that each new face has only *adjacent* sides in common with the part of $\mathbf{\Pi}$ already constructed.

Note that necessarily $F \geq 2$. At each stage of the construction we put $\Phi = V - E + F - 1$. For a single face, we have $\Phi = 0$. We continue by induction on the number of faces inserted, proving that as long as the polyhedron is incomplete $\Phi = 0$. Suppose that indeed $\Phi = 0$ at some stage of the construction at which there are at least two more faces to be inserted. Now add a new face f having p sides, of which q consecutive ones will be attached to the preceding faces. Thus $q + 1$ vertices will be attached to the preceding faces. We have therefore added 1 new face, $p - q$ new edges and $p - q - 1$ new vertices. Denote by Φ' the new value of Φ. Then

$$\Phi' = \Phi + (p - q - 1) - (p - q) + 1 = \Phi = 0.$$

Thus, provided the last face has not been added, we still have $\Phi = 0$. Note that for the last face, neither V nor E is modified, while F is increased by 1. Thus for $\mathbf{\Pi}$ we have $\Phi = 1$; that is, (19.18).

The analogues in dimension 3 of the regular polygons are the 'regular solids'. A *regular solid* is a convex solid having identical regular polygons as faces. It is well-known that, unlike in 2 dimensions, in 3 dimensions there are only a finite number of similarity classes of regular solids, namely 5 (for a definition of similarity, see 20.10(2)): the *tetrahedron*, the *cube*, the *octahedron*, the *dodecahedron* and the *icosahedron* — see Fig. 19.4.

These solids have been known since antiquity. They are often called the *Platonic solids*, as the school of Plato was interested in

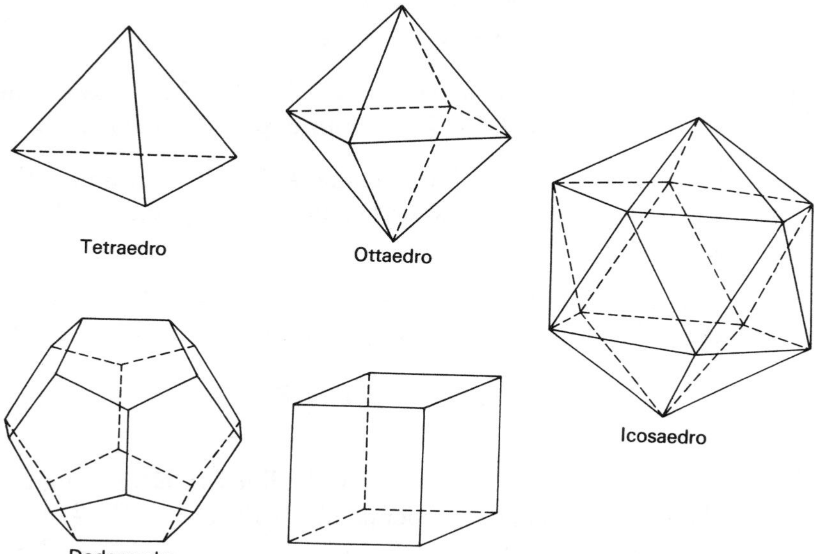

Fig. 19.4

them, and in particular Plato speaks of them in Thimeus. Thaetes studied them systematically around 380 B.C., and they are the subject of Book XIII of Euclid's *Elements*. For further details, the reader should consult [5].

EXERCISES

19.1 Let $\mathbf{E}$ be a Euclidean space with a given Cartesian frame $O\mathbf{e}_1 \ldots \mathbf{e}_n$, and let $\mathbf{H} \subset \mathbf{E}$ be a hyperplane with equation

$$a_1X_1 + \cdots + a_nX_n + c = 0.$$

A vector $\mathbf{m}$ is said to be *orthogonal* to $\mathbf{H}$ if it is orthogonal to every vector in its associated vector subspace. Prove that
a) two vectors orthogonal to $\mathbf{H}$ are proportional,
b) the vector $\mathbf{m}(a_1, \ldots, a_n)$ is orthogonal to $\mathbf{H}$.

19.2 For each of the following, calculate the distance from the point P to the line ℓ in $\mathbf{E}^2$:
a) $P = (1, -3)$, $\ell : 2X - Y + 1 = 0$,
b) $P = (1, 4)$, $\ell : 4X - 3Y + 7 = 0$.

19.3 Find the equation of the line ℓ' in $\mathbf{E}^3$ which meets the line $\ell : X + Y - 2 = 2Y - Z = 0$ and is perpendicular to it.

19.4 Find the equation of the plane $\mathcal{P}$ in $\mathbf{E}^3$ containing the line $\ell : \frac{X-1}{2} = \frac{Y-2}{3} = \frac{Z}{4}$, and orthogonal to the plane $\mathcal{Q} : 2X + 2Y + Z = 0$.

19.5 Find the Cartesian equations of the line ℓ in $\mathbf{E}^3$ passing through the point $P(1, 2, 1)$, meeting, and perpendicular to, the line

$$\ell' : X + 1 = Y - Z = X + Z.$$

19.6 Find the Cartesian equations of the line ℓ in $\mathbf{E}^3$ passing through the point $P(3, 2, 1)$, perpendicular to the line $\ell_1 : \frac{1}{3}(X + 1) = Y - 2 = -\frac{1}{2}Z$ and meeting the line $\ell_2 : X - 3Y - Z = X + 7Y + Z - 6 = 0$. Find the distance between ℓ and ℓ_1.

19.7 Find the cosine of the convex angle between the following two lines in $\mathbf{E}^3$:

$$\ell_1 : X - 3Y + Z - 2 = 0 = X - 5Y + 2Z + 2$$

$$\ell_2 : 2X - Y + Z + 1 = 0 = X - 2Y + 2Z - 3.$$

19.8 Find the Cartesian equations of the lines in $\mathbf{E}^3$ containing the point $P(1, 1, 1)$, parallel to the plane $\mathcal{P} : Y + \sqrt{2}Z - 1 = 0$, and whose convex angle with the X-axis is $\pi/3$.

19.9 In each of the following, find the distance between ℓ and ℓ_1 after first checking that they are skew:

a) $\ell : 2X - Y - Z - 1 = 0 = X + Y - 2Z$
$\ell_1 : 2X + Y - Z + 2 = 0 = Y + 3Z - 2$,
b) $\ell : Y + X - 3 = 0 = 2X - Z + 1$
$\ell_1 : x = 1 - t,\ y = 3 + 3t,\ z = 1 - 2t.$

19.10 Find the equations of the circles in $\mathbf{E}^2$ with the following centres and radii:
a) $C = (3, -4)$, $r = 1$
b) $C = (1, 2)$, $r = \sqrt{2}$
c) $C = (1, -1)$, $r = \frac{1}{2}$.

19.11 Find the centres and radii of the circles in $\mathbf{E}^2$ with equations:
a) $X^2 + Y^2 - 6X + 8Y = 0$
b) $X^2 + Y^2 + 8X - 10Y + 32 = 0$.

19.12 Let $\mathbf{E}$ be a Euclidean space. Prove that a ball in $\mathbf{E}$ is a convex set, but that a sphere is not convex. Prove also that the centre of a ball, or of a sphere, is its only centre of symmetry.

20

Unitary operators and isometries

Let $\mathbf{V}$ be a Euclidean vector space with scalar product $\langle\,,\rangle$. In this chapter we suppose that $\mathbf{V}$ is finite dimensional.

An operator $T : \mathbf{V} \to \mathbf{V}$ is said to be *unitary* if

$$\langle T(\mathbf{v}),\, T(\mathbf{w})\rangle = \langle \mathbf{v},\, \mathbf{w}\rangle \tag{20.1}$$

for every $\mathbf{v}, \mathbf{w} \in \mathbf{V}$. This condition is expressed by saying that T *preserves the scalar product.*

Theorem 20.1
Let $T : \mathbf{V} \to \mathbf{V}$ be a map. The following conditions are equivalent:
1) *T is a unitary operator.*
2) *T is an operator satisfying $\|T(\mathbf{v})\| = \|\mathbf{v}\|$ for all $\mathbf{v} \in \mathbf{V}$.*
3) *T satisfies $T(\mathbf{0}) = \mathbf{0}$ and $\|T(\mathbf{v} - \mathbf{w})\| = \|\mathbf{v} - \mathbf{w}\|$ for all $\mathbf{v}, \mathbf{w} \in \mathbf{V}$.*
4) *T is an operator, and for every orthonormal basis $\{\mathbf{e}_1, \ldots, \mathbf{e}_n\}$ of $\mathbf{E}$, $\{T(\mathbf{e}_1), \ldots, T(\mathbf{e}_n)\}$ is also an orthonormal basis.*
5) *T is an operator, and there exists an orthonormal basis $\{\mathbf{e}_1, \ldots, \mathbf{e}_n\}$ of $\mathbf{E}$ such that $\{T(\mathbf{e}_1), \ldots, T(\mathbf{e}_n)\}$ is also an orthonormal basis.*

Proof
The implication (1) $\Rightarrow$ (2) is obvious, since

$$\|T(\mathbf{v})\|^2 = \langle T(\mathbf{v}),\, T(\mathbf{v})\rangle = \langle \mathbf{v},\, \mathbf{v}\rangle = \|\mathbf{v}\|^2.$$

To prove (2) $\Rightarrow$ (1) observe that, for each $\mathbf{v}, \mathbf{w} \in \mathbf{V}$

$$4\langle \mathbf{v},\, \mathbf{w}\rangle = \langle \mathbf{v} + \mathbf{w},\, \mathbf{v} + \mathbf{w}\rangle - \langle \mathbf{v} - \mathbf{w},\, \mathbf{v} - \mathbf{w}\rangle, \tag{20.2}$$

and

$$4\langle T(\mathbf{v}),\,T(\mathbf{w})\rangle = \begin{array}{l}\langle T(\mathbf{v})+T(\mathbf{w}),\,T(\mathbf{v})+T(\mathbf{w})\rangle \\ -\,\langle T(\mathbf{v})-T(\mathbf{w}),\,T(\mathbf{v})-T(\mathbf{w})\rangle.\end{array} \tag{20.3}$$

From the linearity of T, it follows that the right-hand side of (20.3) is equal to

$$\langle T(\mathbf{v}+\mathbf{w}),\,T(\mathbf{v}+\mathbf{w})\rangle - \langle T(\mathbf{v}-\mathbf{w}),\,T(\mathbf{v}-\mathbf{w})\rangle,$$

which is equal to the right-hand side of (20.2) since T satisfies (2). Thus $4\langle T(\mathbf{v}),\,T(\mathbf{w})\rangle = 4\langle \mathbf{v},\,\mathbf{w}\rangle$; that is, T satisfies (1).
(2) $\Rightarrow$ (3) The proof is left to the reader.
(3) $\Rightarrow$ (1) For every $\mathbf{v} \in \mathbf{V}$ one has

$$\|T(\mathbf{v})\| = \|T(\mathbf{v}) - \mathbf{0}\| = \|T(\mathbf{v}) - T(\mathbf{0})\| = \|\mathbf{v} - \mathbf{0}\| = \|\mathbf{v}\|. \tag{20.4}$$

Expanding the equality

$$\|T(\mathbf{v}) - T(\mathbf{w})\|^2 = \|\mathbf{v} - \mathbf{w}\|^2$$

gives, for every $\mathbf{v}, \mathbf{w} \in \mathbf{V}$,

$$\|T(\mathbf{v})\|^2 - 2\langle T(\mathbf{v}),\,T(\mathbf{w})\rangle + \|T(\mathbf{w})\|^2 = \|\mathbf{v}\|^2 - 2\langle \mathbf{v},\,\mathbf{w}\rangle + \|\mathbf{w}\|^2.$$

Using (20.4) we have

$$\langle T(\mathbf{v}),\,T(\mathbf{w})\rangle = \langle \mathbf{v},\,\mathbf{w}\rangle \tag{20.5}$$

for every $\mathbf{v}, \mathbf{w} \in \mathbf{V}$.

It remains to show that T is linear. Let $\{\mathbf{e}_1, \ldots, \mathbf{e}_n\}$ be an orthonormal basis for $\mathbf{V}$. By (20.5) it follows that $\{T(\mathbf{e}_1), \ldots, T(\mathbf{e}_n)\}$ is an orthonormal basis, because it satisfies

$$\langle T(\mathbf{e}_i),\,T(\mathbf{e}_j)\rangle = \langle \mathbf{e}_i,\,\mathbf{e}_j\rangle = \delta_{ij} \quad \text{for every } 1 \le i, j \le n.$$

For every $\mathbf{v} = x_1\mathbf{e}_1 + \cdots + x_n\mathbf{e}_n$ one has

$$T(\mathbf{v}) = \sum_{i=1}^{n} \langle T(\mathbf{v}),\,T(\mathbf{e}_i)\rangle T(\mathbf{e}_i) = \sum_{i=1}^{n} \langle \mathbf{v},\,\mathbf{e}_i\rangle T(\mathbf{e}_i) = \sum_{i=1}^{n} x_i T(\mathbf{e}_i).$$

That is,

$$T\Big(\sum_{i=1}^{n} x_i\mathbf{e}_i\Big) = \sum_{i=1}^{n} x_i T(\mathbf{e}_i),$$

and so T is linear.

(1) $\Rightarrow$ (4) The proof is left to the reader.

(4) $\Rightarrow$ (5) Obvious.

(5) $\Rightarrow$ (1) Let $\{\mathbf{e}_1, \ldots, \mathbf{e}_n\}$ be an orthonormal basis with the property stated in (5). Let $\mathbf{v} = x_1\mathbf{e}_1 + \cdots + x_n\mathbf{e}_n$, $\mathbf{w} = y_1\mathbf{e}_1 + \cdots + y_n\mathbf{e}_n$. Then

$$\begin{aligned}\langle T(\mathbf{v}), T(\mathbf{w})\rangle &= \langle x_1T(\mathbf{e}_1) + \cdots + x_nT(\mathbf{e}_n), y_1T(\mathbf{e}_1) + \cdots + y_nT(\mathbf{e}_n)\rangle \\ &= \sum_{ij} \langle T(\mathbf{e}_i), T(\mathbf{e}_j)\rangle = \sum_i x_iy_i = \langle \mathbf{v}, \mathbf{w}\rangle.\end{aligned}$$

□

Condition (3) of Theorem 20.1 can be considered as a condition on the asociated affine space $\mathbf{V}_a$. It asserts that T leaves the vector $\mathbf{0}$ fixed, and preserves the distance between vectors in $\mathbf{V}$, without assuming that $\mathbf{V}$ is a linear map.

Condition (2) of the theorem asserts that T *preserves the norm* of vectors. From (2) it follows immediately that if T is unitary then $T(\mathbf{v}) = \mathbf{0}$ if and only if $\mathbf{v} = \mathbf{0}$. Thus, a unitary operator is invertible.

The inverse T^{-1} of a unitary operator T is again unitary, as is the composite of two unitary operators; the proofs are left to the reader. Thus *the unitary operators form a subgroup of* GL($\mathbf{V}$), called the *orthogonal group* of $\mathbf{V}$ and denoted O($\mathbf{V}$).

Suppose that $\boldsymbol{e} = \{\mathbf{e}_1, \ldots, \mathbf{e}_n\}$ is an orthonormal basis of $\mathbf{V}$, and let $A = M_{\boldsymbol{e}}(T)$ be the matrix of the operator T with respect to the basis $\boldsymbol{e}$. Since the columns $A_{(1)}, \ldots, A_{(n)}$ of A are the coordinates of $T(\mathbf{e}_1), \ldots, T(\mathbf{e}_n)$ with respect to $\boldsymbol{e}$, and since with respect to an orthonormal basis, the scalar product of two vectors is equal to the standard scalar product of their coordinates, it follows that by (5) of the theorem T is unitary if and only if

$$A_{(i)} \cdot A_{(j)} = \delta_{ij}$$

for every $1 \leq i, j \leq n$, that is if and only if $A^tA = \mathbf{I}_n$. Thus we have the following corollary.

Corollary 20.2

An operator $T : \mathbf{V} \to \mathbf{V}$ is unitary if and only if the matrix of T with respect to any orthonormal basis is orthogonal.

Therefore, any basis $\boldsymbol{e}$ of $\mathbf{V}$ defines a map

$$M_{\boldsymbol{e}} : \mathrm{GL}(\mathbf{V}) \to \mathrm{GL}_n(\mathbf{R}),$$

which associates to any operator T its matrix $M_{\mathbf{e}}(T)$ with respect to $\mathbf{e}$. *This map induces an isomorphism of the group* $\mathrm{O}(\mathbf{V})$ *onto the group* $\mathrm{O}(n)$.

Note that any unitary operator $T : \mathbf{V} \to \mathbf{V}$ satisfies

$$\det(T) = \pm 1 \tag{20.6}$$

since any orthogonal matrix has determinant equal to ± 1. The unitary operators with $\det(T) = 1$ form a subgroup of $\mathrm{O}(\mathbf{V})$ called the *special orthogonal group of* $\mathbf{V}$, and denoted $\mathrm{SO}(\mathbf{V})$. Elements of $\mathrm{SO}(\mathbf{V})$ are often called *rotations* of $\mathbf{V}$. From Corollary 20.2 it follows immediately that for a given orthonormal basis $\mathbf{e}$ of $\mathbf{V}$, the map $M_{\mathbf{e}}$ induces an isomorphism of $\mathrm{SO}(\mathbf{V})$ onto $\mathrm{SO}(n)$.

The eigenvalues of unitary operators have the following important property.

Proposition 20.3
If $T \in \mathrm{O}(\mathbf{V})$ and $\lambda \in \mathbf{R}$ is an eigenvalue of T then $\lambda = \pm 1$. Similarly, if $A \in \mathrm{O}(n)$ and $\lambda \in \mathbf{R}$ is an eigenvalue of A then $\lambda = \pm 1$.

Proof
By Corollary 20.2, the second statement follows from the first. Suppose that $\lambda \in \mathbf{R}$ is an eigenvalue of T, and let $\mathbf{v} \in \mathbf{V}$ be an associated eigenvector. Since T is unitary,

$$\|\mathbf{v}\| = \|T(\mathbf{v})\| = \|\lambda\mathbf{v}\| = |\lambda|\|\mathbf{v}\|$$

and so $|\lambda| = 1$. □

A useful characterization of unitary operators, equivalent to that of Corollary 20.2 but more intrinsic, can be obtained by introducing the concept of 'transpose' or 'adjoint' operator.

Proposition 20.4
For every operator $F \in \mathrm{End}(\mathbf{V})$ there is a unique operator $G \in \mathrm{End}(\mathbf{V})$ for which

$$\langle F(\mathbf{v}), \mathbf{w}\rangle = \langle \mathbf{v}, G(\mathbf{w})\rangle \tag{20.7}$$

for every $\mathbf{v}, \mathbf{w} \in \mathbf{V}$. The operator G is called the transpose *or* adjoint *of F.*

Proof
For every $\mathbf{w} \in \mathbf{V}$ the map $F_{\mathbf{w}} : \mathbf{V} \to \mathbf{R}$ defined by

$$F_{\mathbf{w}}(\mathbf{v}) = \langle F(\mathbf{v}), \mathbf{w}\rangle \quad \text{for all } \mathbf{v} \in \mathbf{V}$$

is a linear functional. Indeed

$$F_{\mathbf{w}} = b'_{\mathbf{w}} \circ F,$$

and since $b'_{\mathbf{w}}$ and F are linear, so is $F_{\mathbf{w}}$. Given that the scalar product $\langle , \rangle$ is a non-degenerate bilinear form, by Proposition 15.6 there is a unique $G(\mathbf{w}) \in \mathbf{V}$ for which there is an equality of linear functionals

$$F_{\mathbf{w}} = \langle -, G(\mathbf{w})\rangle,$$

i.e. such that

$$F_{\mathbf{w}}(\mathbf{v}) = \langle \mathbf{v}, G(\mathbf{w})\rangle, \quad \text{for all } \mathbf{v} \in \mathbf{V}.$$

The map $G : \mathbf{V} \to \mathbf{V}$ so defined satisfies condition (20.7) and is obviously unique; there remains only to show that G is linear. To this end, let $\mathbf{w}_1, \mathbf{w}_2 \in \mathbf{V}$ and $c_1, c_2 \in \mathbf{R}$. Then, for every $\mathbf{v} \in \mathbf{V}$,

$$\begin{aligned}
\langle \mathbf{v}, G(c_1\mathbf{w}_1 + c_2\mathbf{w}_2)\rangle &= F_{c_1\mathbf{w}_1+c_2\mathbf{w}_2}(\mathbf{v}) = \langle F(\mathbf{v}), c_1\mathbf{w}_1 + c_2\mathbf{w}_2\rangle \\
&= c_1\langle F(\mathbf{v}), \mathbf{w}_1\rangle + c_2\langle F(\mathbf{v}), \mathbf{w}_2\rangle \\
&= c_1 F_{\mathbf{w}_1}(\mathbf{v}) + c_2 F_{\mathbf{w}_2}(\mathbf{v}) \\
&= c_1\langle \mathbf{v}, G(\mathbf{w}_1)\rangle + c_2\langle \mathbf{v}, G(\mathbf{w}_2)\rangle \\
&= \langle \mathbf{v}, c_1 G(\mathbf{w}_1) + c_2 G(\mathbf{w}_2)\rangle,
\end{aligned}$$

and so $G(c_1\mathbf{w}_1 + c_2\mathbf{w}_2) = c_1 G(\mathbf{w}_1) + c_2 G(\mathbf{w}_2)$. Thus G is linear. □

The adjoint of an operator F is usually denoted F^t. It follows immediately from the symmetry of the scalar product that $(F^t)^t = F$, and so F and F^t are adjoints (or transposes) of each other.

The reason for the terminology just introduced is easily explained as follows.

Suppose that $\boldsymbol{e} = \{\mathbf{e}_1, \ldots, \mathbf{e}_n\}$ is an orthonormal bases of $\mathbf{V}$, and let $A = M_{\boldsymbol{e}}(F)$ and $B = M_{\boldsymbol{e}}(F^t)$. Then for $\mathbf{v} = x_1\mathbf{e}_1 + \cdots + x_n\mathbf{e}_n$ and $\mathbf{w} = y_1\mathbf{e}_1 + \cdots + y_n\mathbf{e}_n$ one has

$$\langle F(\mathbf{v}), \mathbf{w}\rangle = (A\mathbf{x}) \cdot \mathbf{y} = (A\mathbf{x})^t \cdot \mathbf{y} = \mathbf{x}^t A^t \mathbf{y} = \mathbf{x} \cdot (A^t \mathbf{y}).$$

and

$$\langle F(\mathbf{v}), \mathbf{w}\rangle = \langle \mathbf{v}, G(\mathbf{w})\rangle = \mathbf{x} \cdot (B\mathbf{y}).$$

Thus,

$$A^t\mathbf{y} = B\mathbf{y}. \tag{20.8}$$

Since (20.8) holds for any $\mathbf{y}$, it implies that $A^t = B$. In words, *the transpose of the matrix representing an operator with respect to an orthonormal basis is equal to the matrix representing the transpose of the operator.*

$F \in \mathrm{End}(\mathbf{V})$ is said to be *symmetric* or *self-adjoint* if $F^t = F$. It is said to be *skew-symmetric* if instead $F^t = -F$.

From what have just discussed, it follows that F is symmetric or skew-symmetric if and only if, with respect to an orthonormal basis, it is represented by a symmetric or skew-symmetric matrix, respectively.

We are now in a position to give a new characterization of unitary operators:

Proposition 20.5
An operator $T : \mathbf{V} \to \mathbf{V}$ is unitary if and only if $T^t \circ T = \mathbf{1}_\mathbf{V}$.

Proof
T is unitary if and only if, for every $\mathbf{v}, \mathbf{w} \in \mathbf{V}$

$$\langle \mathbf{v}, \mathbf{w}\rangle = \langle T(\mathbf{v}), T(\mathbf{w})\rangle = \langle \mathbf{v}, T^t(T(\mathbf{w}))\rangle = \langle \mathbf{v}, T^t \circ T(\mathbf{w})\rangle.$$

Since this holds for every $\mathbf{v}, \mathbf{w} \in \mathbf{V}$ it follows that

$$\mathbf{w} = (T^t \circ T)(\mathbf{w})$$

for every $\mathbf{w} \in \mathbf{V}$, that is $T^t \circ T = \mathbf{1}_\mathbf{V}$. □

The preceding discussion shows that the unitary operators are essentially the isomorphisms that preserve the Euclidean vector space structure. It is thus natural to use unitary operators to define a certain class of affine transformations that are compatible with the metric structure.

Definition 20.6
Let $\mathbf{E}$ be a Euclidean space over $\mathbf{V}$. An affine transformation $f : \mathbf{E} \to \mathbf{E}$ is said to be an *isometry* of $\mathbf{E}$ if the associated automorphism $\phi : \mathbf{V} \to \mathbf{V}$ is a unitary operator.

The identity map $\mathbf{1_E}$, and more generally any translation, is an isometry because the associated isomorphism is the identity on $\mathbf{V}$, which is a unitary operator. The composite of two isometries is an isometry because the associated isomorphism is the composite of two unitary operators, which is again unitary. Similarly, the inverse of an isometry is an isometry. Thus the isometries of $\mathbf{E}$ constitute a subgroup of Aff($\mathbf{E}$); that is a group of affine transformations of $\mathbf{E}$, denoted Isom($\mathbf{E}$), and called the *group of isometries of* $\mathbf{E}$.

The subgroups of Isom($\mathbf{E}$) are called *groups of isometries of* $\mathbf{E}$.

An *isometry* f, with associated isomorphism ϕ, is said to be *direct* if $\det(\phi) = 1$ and *inverse* if $\det(\phi) = -1$. The direct isometries form a subgroup $\mathrm{Isom}^+(\mathbf{E})$ of Isom($\mathbf{E}$).

The translations are a particular case of direct isometries, and so $T_{\mathbf{E}}$ is a group of direct isometries of $\mathbf{E}$.

Let $O \in \mathbf{E}$, and consider the stabilizer $\mathrm{Isom}(\mathbf{E})_O$ of O in Isom($\mathbf{E}$). The direct isometries belonging to $\mathrm{Isom}(\mathbf{E})_O$ are called *rotations with centre* O or, *about* O, and constitute the subgroup

$$\mathrm{Isom}^+(\mathbf{E}) \cap \mathrm{Isom}(\mathbf{E})_O \subset \mathrm{Isom}(\mathbf{E}).$$

The isomorphism

$$\Phi : \mathrm{Aff}(\mathbf{A})_O \to \mathrm{GL}(\mathbf{V})$$

(see Example 14.6(2)) induces isomorphisms

$$\begin{array}{rcl} \mathrm{Isom}(\mathbf{E})_O & \to & \mathrm{O}(\mathbf{V}) \\ \mathrm{Isom}(\mathbf{E})_O \cap \mathrm{Isom}^+(\mathbf{E}) & \to & \mathrm{SO}(\mathbf{V}). \end{array} \tag{20.9}$$

If we fix a basis $\boldsymbol{e} = \{\mathbf{e}_1, \ldots, \mathbf{e}_n\}$ of $\mathbf{V}$, then from Corollary 20.2 we deduce that the composite $M_{\boldsymbol{e}} \circ \Phi$ induces isomorphisms:

$$\begin{array}{rcl} \mathrm{Isom}(\mathbf{E})_O & \to & \mathrm{O}(n) \\ \mathrm{Isom}(\mathbf{E})_O \cap \mathrm{Isom}^+(\mathbf{E}) & \to & \mathrm{SO}(n). \end{array} \tag{20.10}$$

In particular, the second isomorphism identifies the group of rotations with centre O with the group of special orthogonal matrices of order n.

The following theorem follows immediately from Theorem 14.8 and Corollary 20.2.

Theorem 20.7
Let $\mathbf{E}$ *be a Euclidean space with a given Cartesian frame* $O\mathbf{e}_1 \ldots \mathbf{e}_n$*. Any* $f \in \mathrm{Isom}(\mathbf{E})$ *with associated automorphism* ϕ*, can be expressed in the form*

$$f(P(x_1, \ldots, x_n)) = Q(y_1, \ldots, y_n)$$

with

$$\mathbf{y} = A\mathbf{x} + \mathbf{c}, \tag{20.11}$$

where $\mathbf{c} = (c_1 \ldots c_n)^t \in \mathbf{R}^n$ *is the coordinate vector of* $f(O)$*, and* $A = M_{\mathbf{e}}(\phi) \in \mathrm{O}(n)$ *is the matrix of* ϕ *with respect to* $\mathbf{e}$*.*

Conversely, any transformation $f : \mathbf{E} \to \mathbf{E}$ *of the form (20.11) for some* $A \in \mathrm{O}(n)$ *and* $\mathbf{c} \in \mathbf{R}^n$ *is an isometry.*

In particular, the isometries, or direct isometries, of $\mathbf{E}^n$ *are precisely those affine transformations* $f_{A,\mathbf{c}}$ *for which* $A \in \mathrm{O}(n)$ *or, respectively,* $A \in \mathrm{SO}(n)$*.*

The following theorem provides a geometric characerization of isometries, analogous to that of condition (3) of Theorem 20.1.

Theorem 20.8
Let $\mathbf{E}$ *be a Euclidean space. A map* $f : \mathbf{E} \to \mathbf{E}$ *is an isometry if and only if*

$$d(f(P), f(Q)) = d(P, Q) \tag{20.12}$$

for every $P, Q \in \mathbf{E}$*.*

Proof
If f is an isometry, with associated isomorphism $\phi : \mathbf{V} \to \mathbf{V}$ then, since ϕ is a unitary operator and by Theorem 20.1(2),

$$d(f(P), f(Q)) = \|\overrightarrow{f(P)f(Q)}\| = \|\phi(\overrightarrow{PQ})\| = \|\overrightarrow{PQ}\| = d(P, Q).$$

Suppose conversely that f is such that (20.12) is satisfied. Choose any point $O \in \mathbf{E}$ and define the map $\phi : \mathbf{V} \to \mathbf{V}$ by

$$\phi(\overrightarrow{OP}) = \overrightarrow{f(O)f(P)}.$$

Since every vector $\mathbf{v} \in \mathbf{V}$ is of the form $\mathbf{v} = \overrightarrow{OP}$, the map ϕ is well defined and is such that $\phi(\mathbf{0}) = \phi(\overrightarrow{OO}) = \overrightarrow{f(O)f(O)} = \mathbf{0}$. Moreover, if $\mathbf{v} = \overrightarrow{OP}$ and $\mathbf{w} = \overrightarrow{OQ}$ then

$$\begin{aligned} \|\phi(\mathbf{v}) - \phi(\mathbf{w})\| &= \|\phi(\overrightarrow{OP}) - \phi(\overrightarrow{OQ})\| = \|\overrightarrow{f(O)f(P)} - \overrightarrow{f(O)f(Q)}\| \\ &= \|\overrightarrow{f(Q)f(P)}\| = \|\overrightarrow{QP}\| = \|\mathbf{v} - \mathbf{w}\|. \end{aligned}$$

By Theorem 20.1(3) it follows that ϕ is a unitary operator. Furthermore, for each $P, Q \in \mathbf{E}$,

$$\begin{aligned} \phi(\overrightarrow{PQ}) &= \phi(\overrightarrow{OQ} - \overrightarrow{OP}) = \phi(\overrightarrow{OQ}) - \phi(\overrightarrow{OP}) \\ &= \overrightarrow{f(O)f(Q)} - \overrightarrow{f(O)f(P)} = \overrightarrow{f(P)f(Q)} \end{aligned}$$

and so f is an affine transformation with associated automorphism ϕ. Consequently it is an isometry. □

Theorem 20.8 allows one to study isometries purely geometrically.

The most efficient method for finding the groups of isometries of a Euclidean space is to study the isometries of geometric figures.

Let $\mathbf{F} \subset \mathbf{E}$ be a geometric figure. An isometry $f \in \mathrm{Isom}(\mathbf{E})$ such that $f(\mathbf{F}) = \mathbf{F}$ is said to be an *isometry of* $\mathbf{F}$. It is clear that the set of isometries of a given geometric figure $\mathbf{F}$ forms a group of transformations of $\mathbf{F}$ which is a subgroup $\mathrm{Isom}(\mathbf{F})$ of $\mathrm{Isom}(\mathbf{E})$; it is called the *group of isometries of* $\mathbf{F}$, or its *isometry group*.

For example, if $O \in \mathbf{E}$ then

$$\mathrm{Isom}(O) = \mathrm{Isom}(\mathbf{E})_O.$$

If $\mathbf{S}(O, r)$ is the sphere centre O and radius $r > 0$, then

$$\mathrm{Isom}(\mathbf{S}(O, r)) = \mathrm{Isom}(\mathbf{E})_O. \tag{20.13}$$

Indeed, every $f \in \mathrm{Isom}(\mathbf{E})_O$ transforms $\mathbf{S}(O, r)$ into itself, since

$$r = d(O, P) = d(f(O), f(P)) = d(O, f(P)).$$

Thus $f(P) \in \mathbf{S}(O, r)$ for every $P \in \mathbf{S}(O, r)$; consequently

$$f \in \mathrm{Isom}(\mathbf{S}(O, r)).$$

To prove the converse, first note that for any pair of distinct points $P, Q \in \mathbf{S}(O, r)$ one has $d(P, Q) \leq d(O, P) + d(O, Q) = 2r$, with equality if and only if O, P, Q are collinear, that is, if and only if P and

Q are *diametrically opposite*. Suppose then that $f \in \text{Isom}(\mathbf{S}(O,r))$; given P, Q diametrically opposite, then $d(f(P), f(Q)) = 2r$ so $f(P)$ and $f(Q)$ must also be diametrically opposite. Therefore the midpoint of PQ, namely O, is transformed to the midpoint of $f(P)f(Q)$ which is also O. Thus $f(O) = O$, or $f \in \text{Isom}(\mathbf{E})_O$. This concludes the proof of (20.13).

The study of groups of isometries of Euclidean figures is one of the most classical and largest chapters in group theory. Intuitively, the notion of isometry of a geometric figure corresponds to the artistic and aesthetic notion of 'symmetry'. The larger the group $\text{Isom}(\mathbf{F})$, the more symmetry the figure has — or the more symmetric it is. Historically, the abstract notion of group was preceded by that of transformation group, and in particular by that of isometry group. Several examples of isometry groups are given in Chapter 21.

Definition 20.9

Two geometric figures $\mathbf{F}$ and $\mathbf{F}'$ are said to be *congruent* if there is an $f \in \text{Isom}(\mathbf{E})$ for which $f(\mathbf{F}) = \mathbf{F}'$. The properties of a figure $\mathbf{F}$ common to all figures congruent to it are said to be *Euclidean properties* of $\mathbf{F}$.

Every affine property of a figure $\mathbf{F}$ is also a Euclidean property, since $\text{Isom}(\mathbf{E}) \subset \text{Aff}(\mathbf{E})$, and so any figure that is congruent to $\mathbf{F}$ is also affinely equivalent to $\mathbf{F}$. In general, though, a Euclidean property is not an affine property. For example, the distance between two points $P, Q \in \mathbf{E}$ is a Euclidean property of the figure $\{P, Q\}$, but it is not an affine property since affine transformations do not necessarily map the pair P, Q into another pair the same distance apart.

Complements 20.10

1. Conditions (1), (2) and (3) of Theorem 20.1 also make sense if $\mathbf{V}$ is a Euclidean vector space that is not of finite dimension; the proof of their equivalence also makes no restrictions on the dimension of $\mathbf{V}$ provided one assumes that T is an operator. Thus (1), (2) and (3) are equivalent conditions for an operator defined on any Euclidean vector space.

2. In elementary geometry, one also studies properties of geometric figures which do not depend on their size, but only on their form

and proportions: the *properties of similarity.* These properties are preserved, not only by isometries, but also by homotheties (see 14.6(3)), and so by all affine transformations obtained by composing a finite number of isometries and homotheties in any way possible. Such affine transformations are naturally called *similarities*, and constitute a subgroup Simil(**E**) of Aff(**E**). Indeed, the identity is a homothety as well as an isometry, and so it is a similarity. Moreover, the composite of two similarities $\sigma_1 \circ \omega_1 \circ \cdots \circ \sigma_k \circ \omega_k$ and $\tau_1 \circ \zeta_1 \circ \cdots \tau_j \circ \zeta_j$ is $\sigma_1 \circ \omega_1 \circ \cdots \circ \sigma_k \circ \omega_k \circ \tau_1 \circ \zeta_1 \circ \cdots \tau_j \circ \zeta_j$, which is also a similarity. The inverse of the similarity $\sigma_1 \circ \omega_1 \circ \cdots \circ \sigma_k \circ \omega_k$ is $\omega_k^{-1} \circ \sigma_k^{-1} \circ \cdots \circ \omega_1^{-1} \circ \sigma_1^{-1}$, which is also a similarity. It is clear that Simil(**E**) contains both Isom(**E**) and the group of all homotheties.

We will denote the group of all direct similarities of **E** by Simil$^+$(**E**). It is of course a subgroup of Simil(**E**).

Identify $\mathbf{E}^2$ with **C** by associating to any point $(x, y) \in \mathbf{E}^2$ the complex number $z = x + iy \in \mathbf{C}$. With this identification, it is possible to give a simple desciption of Simil($\mathbf{E}^2$) and Simil$^+$($\mathbf{E}^2$) as follows.

Let $f(z)$ be an affine transformation of **C**, considered as a complex vector space of dimension 1:

$$f(z) = az + b, \quad a, b \in \mathbf{C}, \ a \neq 0. \tag{20.14}$$

Expression (20.14) can be interpreted as an affine transformation of the Euclidean plane $\mathbf{E}^2$. Writing $a = a' + ia''$ and $b = b' + ib''$, (20.14) becomes

$$f(x, y) = (a'x - a''y + b', \ a''x + a'y + b''),$$

and so is a direct similarity since its determinant is

$$a'^2 + a''^2 > 0.$$

With $b = 0$, (20.14) becomes

$$f(z) = az. \tag{20.15}$$

If $a \in \mathbf{R}$ then (20.15) represents a homothety of $\mathbf{R}^2$ since it has the form,

$$f(x + iy) = ax + iay.$$

If $|a| = 1$, so $a = \cos\theta + i\sin\theta$, (20.15) becomes

$$f(x+iy) = x\cos\theta - y\sin\theta + i(x\sin\theta + y\cos\theta),$$

which represents a rotation. Since $a = |a|u$ with $|u| = 1$, for every $a \in \mathbf{C}$, equation (20.15) is always the composite of a homothety and a rotation. From this it follows that the affine transformations (20.14) of $\mathbf{C}$ coincide with those direct affine transformations of the Euclidean plane $\mathbf{E}^2$ which are composites of translations, rotations and homotheties, i.e. which are direct similarities. Thus $\mathrm{Aff}_1(\mathbf{C})$ is identified with $\mathrm{Simil}^+(\mathbf{E}^2)$.

To obtain the inverse similarities of $\mathbf{E}^2$ as well, it suffices to compose all the direct affine transformations (20.14) with a single inverse isometry: for example *conjugation*, which associates to each $z = x + iy \in \mathbf{C}$ its conjugate $\bar{z} = x - iy$.

Thus, *the inverse similarities of* $\mathbf{E}^2$ *correspond to the transformations* σ *of* $\mathbf{C}$ *of the form*

$$z \to a\bar{z} + b, \quad a, b \in \mathbf{C},\ a \neq 0.$$

3. Let $\mathbf{V}$ be a Euclidean vector space, and let $\mathbf{v} \in \mathbf{V}$, $\mathbf{v} \neq \mathbf{0}$. The *reflexion* defined by $\mathbf{v}$ (Fig. 20.1) is the map $\rho_{\mathbf{v}} : \mathbf{V} \to \mathbf{V}$ defined by

$$\rho_{\mathbf{v}}(\mathbf{u}) = \mathbf{u} - 2\frac{\langle \mathbf{v}, \mathbf{u}\rangle}{\langle \mathbf{v}, \mathbf{v}\rangle}\mathbf{v}.$$

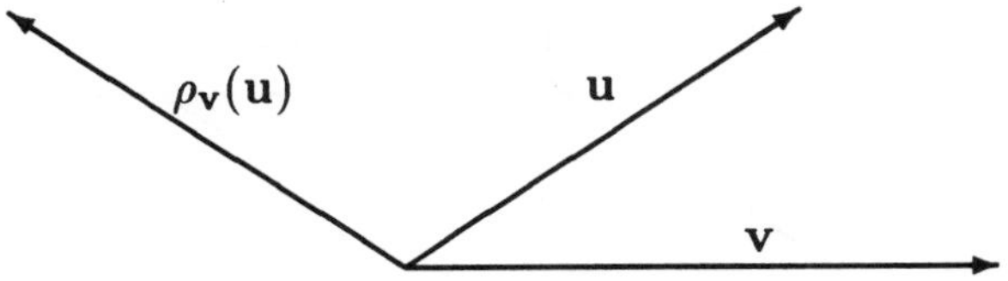

Fig. 20.1

Using the notion of Fourier coefficient $a_{\mathbf{v}}(\mathbf{u})$, this can be rewritten in the equivalent form,

$$\rho_{\mathbf{v}}(\mathbf{u}) = \mathbf{u} - 2a_{\mathbf{v}}(\mathbf{u})\mathbf{v}.$$

It is easily verified that $\rho_{\mathbf{v}}$ is linear; moreover, for any $\mathbf{u} \in \mathbf{V}$,

$$\begin{aligned}\|\rho_{\mathbf{v}}(\mathbf{u})\|^2 &= \|\mathbf{u} - 2a_{\mathbf{v}}(\mathbf{u})\mathbf{v}\|^2 \\ &= \|\mathbf{u}\|^2 - 4a_{\mathbf{v}}(\mathbf{u})\langle \mathbf{u}, \mathbf{v}\rangle + 4a_{\mathbf{v}}(\mathbf{u})^2\|\mathbf{v}\|^2 = \|\mathbf{u}\|^2,\end{aligned}$$

and so $\rho_{\mathbf{v}}$ is a unitary operator.

It follows immediately that $\rho_{\mathbf{v}}^2 = \mathbf{1}_{\mathbf{V}}$; that is, $\rho_{\mathbf{v}}^{-1} = \rho_{\mathbf{v}}$.

4. Suppose that, in a Euclidean space $\mathbf{E}$ over $\mathbf{V}$, we are given a Cartesian frame $O\mathbf{e}_1 \ldots \mathbf{e}_n$ and a hyperplane H with equation

$$a_1X_1 + \cdots + a_nX_n + c = 0.$$

Let $P(x_1, \ldots, x_n) \in \mathbf{E}$. The reflexion of P in the hyperplane H is the point $\rho_H(P)$ defined by the identity

$$\overrightarrow{N\rho_H(P)} = -\overrightarrow{NP},$$

or equivalently,

$$\overrightarrow{P\rho_H(P)} = -2\overrightarrow{NP},$$

where N is the foot of the perpendicular from P to H, i.e. the intersection of H with the line ℓ through P having direction vector $\mathbf{a}(a_1, \ldots, a_n)$ — see Fig. 20.2.

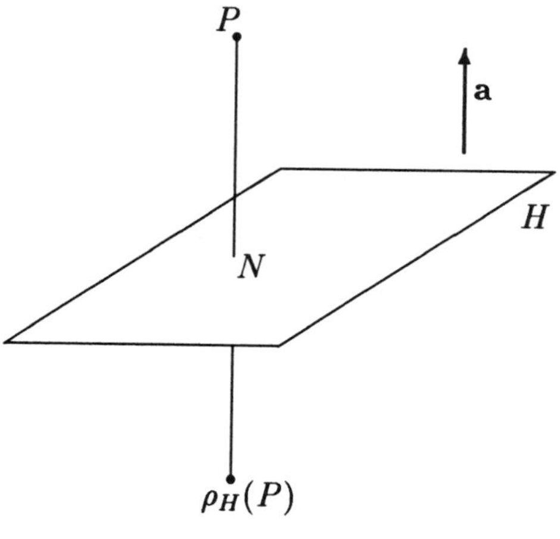

Fig. 20.2

Let $Q(q_1,\ldots,q_n) \in H$ be any point, then

$$\overrightarrow{P\rho_H(P)} = -2\langle\overrightarrow{QP}, \mathbf{a}\rangle/\langle\mathbf{a}, \mathbf{a}\rangle,$$

and so

$$\overrightarrow{Q\rho_H(P)} = \overrightarrow{QP} + \overrightarrow{P\rho_H(P)} = \rho_{\mathbf{a}}(\overrightarrow{QP}),$$

where $\rho_{\mathbf{a}}$ is the reflection defined by $\mathbf{a}$ — see (3).

Since $a_1q_1 + \cdots + a_nq_n = -c$, the *coordinates of* $\rho_H(P)$ are

$$\rho_H(P)_i = x_i - 2a_i\frac{a_1x_1 + \cdots + a_nx_n + c}{\sum_{j=1}^n a_j^2}, \quad i = 1,\ldots,n. \qquad (20.16)$$

It is easy to check that $\rho_H(\rho_H(P)) = P$ for every $P \in \mathbf{E}$, and that $\rho_H(P) = P$ if and only if $P \in H$.

Moreover, if $P, P' \in \mathbf{E}$,

$$\begin{aligned} d(\rho_H(P), \rho_H(P')) &= \|\overrightarrow{\rho_H(P)\rho_H(P')}\| \\ &= \|\overrightarrow{\rho_H(P)N} + \overrightarrow{NN'} + \overrightarrow{N'\rho_H(P')}\| \\ &= \|\overrightarrow{NP} + \overrightarrow{P'N'} + \overrightarrow{NN'}\| \\ &= \|\overrightarrow{NP} + \overrightarrow{P'N'} + \overrightarrow{N'N}\| \\ &= \|\overrightarrow{PP'}\| = d(P, P'), \end{aligned}$$

where we have used the fact that $\overrightarrow{NP} + \overrightarrow{P'N'}$ and $\overrightarrow{NN'}$ are perpendicular. The map $\rho_H : \mathbf{E} \to \mathbf{E}$ is thus an isometry. It is called the *reflexion defined by* H (or that *fixes* H).

In the case $n = 2$ one obtains the notion of *reflexion in an axis* which will be taken up again in Chapter 21.

From what we have seen, it follows that $\rho_H^2 = \mathbf{1_V}$, and that ρ_H fixes every point of H.

A subset $\mathbf{F}$ of $\mathbf{E}$ is said to be *symmetric about the hyperplane* H if $\rho_H(P) \in \mathbf{F}$ for every point $P \in \mathbf{F}$. In this case, H is said to be a *hyperplane of symmetry* of $\mathbf{F}$. If $n = 2$, it is usually called an *axis of symmetry* of $\mathbf{F}$.

5. Among the most interesting groups of isometries are the so-called 'discontinuous groups'.

Let $\mathbf{E}$ be a Euclidean space over the Euclidean vector space $\mathbf{V}$. A subgroup $\mathcal{G}$ of Isom($\mathbf{E}$) is said to be *discontinuous* if for each $P \in \mathbf{E}$ there exists $r > 0$ such that none of the points $g(P) \neq P$ with $g \in \mathcal{G}$ is contained in the ball $\mathbf{B}(P, r)$.

Every finite subgroup $\mathcal{G}$ of Isom($\mathbf{E}$) is a discontinuous group. Indeed, for $P \in \mathbf{E}$ any r with

$$0 < r < \min\{d(P, g(P)) \mid g \in \mathcal{G}, d(P, g(P)) \neq 0\},$$

satisfies the condition of the definition.

Given a non-zero vector $\mathbf{v} \in \mathbf{V}$, the set of all translations of the form $t_{h\mathbf{v}}$, with $h \in \mathbf{Z}$, is a discontinuous subgroup $T_{\mathbf{E}}(\mathbf{v})$ of Isom($\mathbf{E}$). Note that $T_{\mathbf{E}}(\mathbf{v})$ is an infinite group since $t_{h\mathbf{v}} = t_{k\mathbf{v}}$ if and only if $h = k$.

A finite group of isometries of $\mathbf{E}$ cannot contain a non-trivial translation, because if it contained the translation $t_{\mathbf{v}}$ for some non-zero $\mathbf{v} \in \mathbf{V}$ then it would also contain the infinite group $T_{\mathbf{E}}(\mathbf{v})$.

A discontinuous group of isometries which can be generated by a finite number of reflexions is called a *Coxeter group.*

6. Let $\mathbf{V}$ be a vector space over $\mathbf{K}$, and let $b : \mathbf{V} \times \mathbf{V} \to \mathbf{K}$ be a bilinear form. We say that the automorphism $f \in \mathrm{GL}(\mathbf{V})$ *preserves* b if

$$b(f(\mathbf{v}), f(\mathbf{w})) = b(\mathbf{v}, \mathbf{w}) \tag{20.17}$$

for every $\mathbf{v}, \mathbf{w} \in \mathbf{V}$.

The set of all automorphisms of $\mathbf{V}$ which preserve b is a linear group, called the *orthogonal group with respect to* b; it is denoted $\mathrm{O}_b(\mathbf{V})$. To verify that $\mathrm{O}_b(\mathbf{V})$ is indeed a group, note that for every $f \in \mathrm{O}_b(\mathbf{V})$ and $\mathbf{v}', \mathbf{w}' \in \mathbf{V}$ given by $\mathbf{v}' = f(\mathbf{v})$, $\mathbf{w}' = f(\mathbf{w})$, equation (20.17) can also be written

$$b(\mathbf{v}', \mathbf{w}') = b(f^{-1}(\mathbf{v}'), f^{-1}(\mathbf{w}')).$$

Thus $f^{-1} \in \mathrm{O}_b(\mathbf{V})$. It is clear that $\mathbf{1}_{\mathbf{V}} \in \mathrm{O}_b(\mathbf{V})$. Finally, if $f, g \in \mathrm{O}_b(\mathbf{V})$ then

$$\begin{aligned} b((g \circ f)(\mathbf{v}), (g \circ f)(\mathbf{w})) &= b(g(f(\mathbf{v})), g(f(\mathbf{w}))) \\ &= b(f(\mathbf{v}), f(\mathbf{w})) = b(\mathbf{v}, \mathbf{w}), \end{aligned}$$

for every $\mathbf{v}, \mathbf{w} \in \mathbf{V}$, so $f \circ g \in \mathrm{O}_b(\mathbf{V})$.

These ideas generalize those of unitary operator and orthogonal group, which correspond to taking $\mathbf{V}$ and b to be a Euclidean vector space and its scalar product.

Suppose that $\mathbf{V} = \mathbf{K}^n$ and b is the bilinear form associated to the matrix $A \in M_n(\mathbf{K})$; so b is defined by $b(\mathbf{x}, \mathbf{y}) = \mathbf{x}^t A\mathbf{y}$. Then $\mathrm{O}_b(\mathbf{K}^n)$ consists of the matrices $M \in \mathrm{GL}_n(\mathbf{K})$ satisfying

$$M^t AM = A. \tag{20.18}$$

Indeed, $M \in \mathrm{O}_b(\mathbf{K}^n)$ if and only if for each $\mathbf{x}, \mathbf{y} \in \mathbf{K}^n$

$$\mathbf{x}^t A\mathbf{y} = (M\mathbf{x})^t A(M\mathbf{y}) = \mathbf{x}^t(M^t AM)\mathbf{y}.$$

Since this identity holds for every $\mathbf{x}, \mathbf{y} \in \mathbf{K}^n$, M must satisfy (20.18).

Taking $\mathbf{K} = \mathbf{R}$ and $A = \mathbf{I}_n$, then (20.18) is the condition that $A \in \mathrm{O}(n)$. That is, $\mathrm{O}_b(\mathbf{R}^n) = \mathrm{O}(n)$ if b is the standard symmetric form.

If $\mathbf{V} = \mathbf{R}^n$ and b is the polar symmetric bilinear form of the quadratic form

$$q(\mathbf{x}) = x_1^2 + \cdots + x_p^2 - x_{p+1}^2 - \cdots - x_n^2$$

the orthogonal group of $\mathbf{V}$ with respect to b is denoted $\mathrm{O}(p, n-p)$. A particular case is $\mathrm{O}(3,1)$ which is the group of automorphisms of $\mathbf{R}^4$ preserving the Minkowski form; it is called the *Lorentz group.*

Another particularly important case is provided by the standard alternating form on $\mathbf{R}^{2k}$, $k \geq 1$:

$$b(\mathbf{x}, \mathbf{y}) = x_1 y_{k+1} + \cdots + x_k y_n - x_{k+1} y_1 - \cdots - x_n y_k = \mathbf{x}^t \mathbf{J}_k \mathbf{y},$$

where $n = 2k$, and

$$\mathbf{J}_k = \begin{pmatrix} \mathbf{0} & \mathbf{I}_k \\ -\mathbf{I}_k & \mathbf{0} \end{pmatrix}.$$

The corresponding orthogonal group is called the *symplectic group of order* $2k$ *over* $\mathbf{K}$. It is denoted $\mathbf{Sp}(2k, \mathbf{K})$. It follows from the preceding discussion that $M \in \mathrm{GL}_{2k}(\mathbf{K})$ belongs to $\mathbf{Sp}(2k, \mathbf{K})$ if and only if it satisfies the identity

$$M^t \mathbf{J}_k M = \mathbf{J}_k.$$

EXERCISES

20.1 In each of the following cases, find the isometry $f : \mathbf{E}^1 \to \mathbf{E}^1$ satisfying the following conditions
a) $f(1) = \frac{\pi}{2}$ and f is a direct isometry;
b) $f(\pi) = -2$ and f is an inverse isometry.

20.2 For each of the following, find the reflexion in $\mathbf{E}^2$ defined by the following lines

a) $X = 0$ b) $X + Y = 0$
c) $X - 2Y = 0$ d) $2X - 3Y = 0$
e) $X + Y - 1 = 0$.

20.3 In each of the following cases, prove that there is a unique isometry f of $\mathbf{E}^2$ satisfying the given conditions, and then find it:
a) $f(0,0) = (1,1)$, $f(1,0) = (2,1)$, and f is a direct isometry
b) $f(0,0) = (1,1)$, $f(1,0) = (2,1)$, and f is an inverse isometry
c) f fixes every point of the line $\ell : X - 2Y = 0$, and is not the identity
d) f fixes the points $(1,7)$ and $(-1,1)$, and is not the identity.

20.4 In each of the following, find the reflexion in $\mathbf{R}^3$ determined by the plane with the given equation:

a) $X - Y = 0$ b) $X + Y + Z = 0$
c) $X - Y + Z = 0$ d) $2X - Z + 1 = 0$
e) $2X - 2Y + Z - 4 = 0$.

20.5 In each of the following cases, prove that there is a unique isometry f of $\mathbf{E}^3$ satisfying the given conditions, and then find it:
a) f fixes every point of the X-axis and of the Y-axis and is a direct isometry
b) f fixes every point of the X-axis and of the Y-axis and is an inverse isometry.

20.6 Find the Cartesian equations of the line ℓ in $\mathbf{E}^3$ which is the reflexion in $\mathcal{P} : X + Y + Z = 0$ of the line given by

$$\tfrac{1}{2}X = Y = Z - 1.$$

21

Isometries of planes and three-dimensional spaces

This chapter treats isometries of 2- and 3-dimensional spaces in greater detail. These are particularly important cases because of their close relationship with elementary Euclidean geometry. For further information, the reader should consult [9], [1] and [11].

The elements of SO(2) are the matrices of the form

$$R_\theta = \begin{pmatrix} \cos\theta & -\sin\theta \\ \sin\theta & \cos\theta \end{pmatrix}, \qquad \theta \in \mathbf{R}.$$

To obtain the remaining elements of O(2), i.e. those of determinant -1, we can use the fact that, if $A, B \in \mathrm{O}(2)\setminus\mathrm{SO}(2)$, then $AB \in \mathrm{SO}(2)$, for

$$\det(AB) = \det(A)\det(B) = 1.$$

It follows that every element $A \in \mathrm{O}(2) \setminus \mathrm{SO}(2)$ can be written as

$$A = (AB)B^{-1},$$

that is, as the product of a matrix in SO(2) and a fixed matrix $B \in \mathrm{O}(2) \setminus \mathrm{SO}(2)$. Taking, for example, this fixed matrix to be

$$\begin{pmatrix} 1 & 0 \\ 0 & -1 \end{pmatrix},$$

the elements of $\mathrm{O}(2) \setminus \mathrm{SO}(2)$ can be written in the form

$$A_\theta = R_\theta \begin{pmatrix} 1 & 0 \\ 0 & -1 \end{pmatrix} = \begin{pmatrix} \cos\theta & -\sin\theta \\ \sin\theta & \cos\theta \end{pmatrix} \begin{pmatrix} 1 & 0 \\ 0 & -1 \end{pmatrix}$$

$$= \begin{pmatrix} \cos\theta & \sin\theta \\ \sin\theta & -\cos\theta \end{pmatrix}, \qquad \theta \in \mathbf{R}.$$

In particular,

$$A_0 = \begin{pmatrix} 1 & 0 \\ 0 & -1 \end{pmatrix}.$$

Lemma 21.1

1) *$A_\theta = R_\theta A_0 = A_0 R_{-\theta}$ for each $\theta \in \mathbf{R}$.*
2) *$A_\phi \circ A_\theta = R_{\phi-\theta}$ for every $\phi, \theta \in \mathbf{R}$.*
3) *For every $\theta \in \mathbf{R}$, the eigenvalues of A_θ are $\lambda = \pm 1$, and the eigenspaces, which are of dimension 1, are mutually orthogonal.*

Proof

1) This follows from a direct calculation.
2) By (1), we have

$$A_\phi A_\theta = (R_\phi A_0)(A_0 R_{-\theta}) = R_\phi(A_0 A_0)R_{-\theta} = R_\phi R_{-\theta} = R_{\phi-\theta}.$$

3) A direct calculation shows that the characteristic polynomial of A_θ is $T^2 - 1$, and hence the eigenvalues are $\lambda = \pm 1$. The eigenspaces are therefore 1-dimensional, and are given by the equations

$$\begin{aligned} (\cos\theta - 1)X + (\sin\theta)Y &= 0, \qquad \lambda = 1 \\ (\cos\theta + 1)X + (\sin\theta)Y &= 0, \qquad \lambda = -1. \end{aligned} \tag{21.1}$$

Since $(\cos\theta - 1)(\cos\theta + 1) + \sin^2\theta = 0$, these two eigenspaces are mutually orthogonal. □

Consider a Euclidean plane $\mathbf{E}$ with associated Euclidean vector space $\mathbf{V}$, and fix a Cartesian frame $O\mathbf{e}_1\,\mathbf{e}_2$. Let $C \in \mathbf{E}$, and let $\sigma : \mathbf{E} \to \mathbf{E}$ be a rotation with centre C. Via the isomorphism (20.10), we can associate to σ an element $R_\theta \in \mathrm{SO}(2)$. This θ is called the *angle* of the rotation σ. To distinguish it from the matrix R_θ, the rotation with centre C corresponding to R_θ will be denoted $R_{C,\theta}$. It can be represented as the composite

$$R_{C,\theta} = t_{\overrightarrow{OC}} \circ R_{O,\theta} \circ t_{-\overrightarrow{OC}}$$

and so the coordinates $\mathbf{y} = (y_1\, y_2)^t$ of the transform $R_{C,\theta}(P)$ of the point P with coordinates $\mathbf{x} = (x_1\, x_2)^t$ are

$$\mathbf{y} = R_\theta(\mathbf{x} - \mathbf{c}) + \mathbf{c},$$

where $\mathbf{c} = (c_1\, c_2)^t$ are the coordinates of the point C.

An isometry ρ_ℓ of $\mathbf{E}$ which fixes all points of a line ℓ but which is not the identity, is called a *reflexion* — see also 20.10(3). The line ℓ is the *axis* of the reflexion.

A reflexion is an inverse isometry whose square is equal to the identity. Figure 21.1 gives an example of a subset of the ordinary plane which is transformed into itself by the reflexion ρ_ℓ.

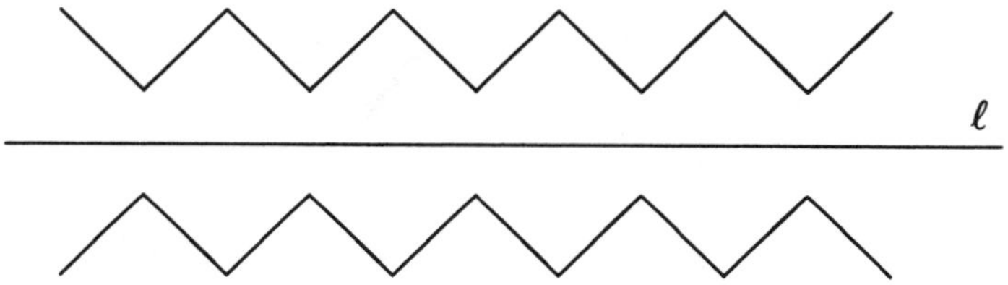

Fig. 21.1

A reflexion fixes every point of its axis. In particular, the reflexions whose axes pass through the origin are identified with the elements of $\mathrm{O}(2) \setminus \mathrm{SO}(2)$, because they fix the origin but are not rotations. Each one is therefore represented by a matrix A_θ for some $\theta \in \mathbf{R}$.

To distinguish it from the matrix A_θ, we denote the corresponding reflexion by $A_{O,\theta}$. The axis ℓ_θ of $A_{O,\theta}$ is the line through the origin with direction the eigenspace with eigenvalue $\lambda = 1$, that is, it is the line whose equation is the first of those in (21.1). A unit direction vector of ℓ_θ is given by $(\cos(\theta/2), \sin(\theta/2))$ — Fig. 21.2.

Lemma 21.2
1) Let ℓ be a line in $\mathbf{E}$, $C \in \ell$ a point and $R_{C,\theta}$ a rotation about C. There are lines ℓ' and ℓ'' containing C for which

$$R_{C,\theta} = \rho_\ell \circ \rho_{\ell'} = \rho_{\ell''} \circ \rho_\ell.$$

Conversely, for every pair of lines ℓ and ℓ' passing through a point C, the composite $\rho_\ell \circ \rho_{\ell'}$ is a rotation about C, and $\rho_\ell \circ \rho_{\ell'} = \mathbf{1_E}$ if and only if $\ell = \ell'$.
2) The composite $R_{C,\theta} \circ R_{D,\phi}$ of two rotations with centres C and D and angles θ and ϕ respectively, is a rotation with angle $\theta + \phi$, unless

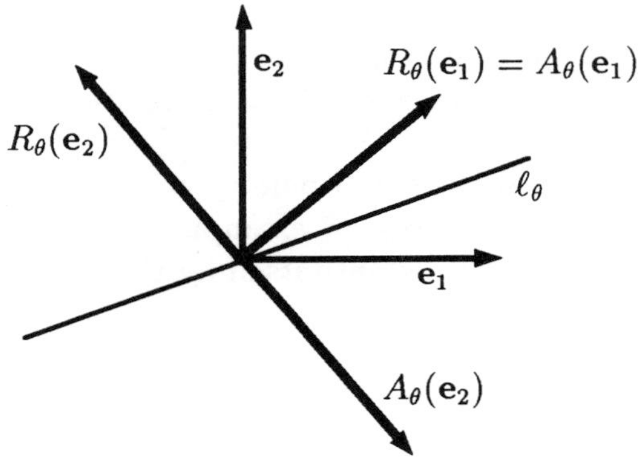

Fig. 21.2

$\theta + \phi = 2k\pi$ for some $k \in \mathbf{Z}$; in this case $R_{C,\theta} \circ R_{D,\phi}$ is a translation which is equal to the identity if and only if $C = D$.
3) If C and D are two distinct points and ℓ is the line containing them, and if $R_{C,\theta}$ and $R_{D,\phi}$ are non-trivial rotations with $\theta + \phi \neq 2k\pi$, then the centres of the rotations $R_{C,\theta} \circ R_{D,\phi}$ and $R_{C,-\theta} \circ R_{D,-\phi}$ are distinct and symmetric with respect to ℓ.

Proof
1) We can suppose that $C = O$ and so $\rho_\ell = A_{O,\alpha}$ for some $\alpha \in \mathbf{R}$. By Lemma 21.1(2) we have that $R_\theta = A_\alpha \circ A_{\alpha-\theta}$, and so $R_{O,\theta} = \rho_\ell \circ \rho_{\ell'}$ where ℓ' is the axis of the reflexion $A_{O,\alpha-\theta}$; similarly,

$$R_{O,\theta} = A_{O,\theta+\alpha} \circ A_{O,\alpha} = \rho_{\ell''} \circ \rho_\ell,$$

where ℓ'' is the axis of $A_{O,\theta+\alpha}$. If $C = O$ then the converse is a reformulation of Lemma 21.1(2).
2) If $C = D$ then obviously $R_{C,\phi} \circ R_{C,\theta} = R_{C,\theta+\phi}$. Suppose now that $C \neq D$, and that $\theta,\ \phi \neq 2k\pi$. Let ℓ be the line through C and D. By (1) there is a line ℓ'' containing C and a line ℓ' containing D for which

$$R_{C,\theta} = \rho_{\ell''} \circ \rho_\ell, \quad R_{D,\phi} = \rho_\ell \circ \rho_{\ell'}. \tag{21.2}$$

Thus,

$$R_{C,\theta} \circ R_{D,\phi} = (\rho_{\ell''} \circ \rho_\ell) \circ (\rho_\ell \circ \rho_{\ell'}) = \rho_{\ell''} \circ (\rho_\ell \circ \rho_\ell) \circ \rho_{\ell'} = \rho_{\ell''} \circ \rho_{\ell'} \tag{21.3}$$

If ℓ' and ℓ'' are parallel, then $\rho_{\ell''} \circ \rho_{\ell'}$ is a translation perpendicular to ℓ' and ℓ''; if they are not parallel, then by (1) $\rho_{\ell''} \circ \rho_{\ell'}$ is a rotation.

On the other hand, let $\mathbf{c} = (c_1\, c_2)^t$ and $\mathbf{d} = (d_1\, d_2)^t$ be the coordinates of C and D respectively. Using (21.3) we find that for every $P \in \mathbf{E}$ with coordinates $\mathbf{x} = (x_1\, x_2)^t$ the point $(R_{C,\theta} \circ R_{D,\phi})(P)$ has coordinates

$$\mathbf{y} = R_\theta[R_\phi(\mathbf{x} - \mathbf{d}) + \mathbf{d} - \mathbf{c}] + \mathbf{c} = R_{\theta+\phi}(\mathbf{x} - \mathbf{d}) + R_\theta(\mathbf{d} - \mathbf{c}) + \mathbf{c}. \tag{21.4}$$

This is a translation if and only if $\theta + \phi = 2k\pi$; otherwise, by what we have seen already, (21.4) is a rotation through an angle equal to $\theta + \phi$. If $\theta + \phi = 2k\pi$ then

$$\mathbf{y} = \mathbf{x} + [R_\theta(\mathbf{d} - \mathbf{c}) - (\mathbf{d} - \mathbf{c})],$$

which is not the identity as $\mathbf{d} - \mathbf{c} \neq \mathbf{0}$, and $R_\theta \neq \mathbf{I}_2$.
3) Let ℓ' and ℓ'' be the lines defined in the proof of (1) satisfying (21.2). Then

$$R_{C,-\theta} = (R_{C,\theta})^{-1} = \rho_\ell \circ \rho_{\ell''}$$
$$R_{D,-\phi} = (R_{D,\phi})^{-1} = \rho_{\ell'} \circ \rho_\ell$$

and, by (21.3)

$$R_{C,-\theta} \circ R_{D,-\phi} = \rho_\ell \circ \rho_{\ell''} \circ \rho_{\ell'} \circ \rho_\ell = \rho_\ell \circ R_{C,\theta} \circ R_{D,\phi} \circ \rho_\ell.$$

From this expression it follows that, if Q is the centre of rotation of $R_{C,\theta} \circ R_{D,\phi}$, then $\rho_\ell(Q)$ is transformed into itself by the rotation $R_{C,-\theta} \circ R_{D,-\phi}$, and so is its centre. Since $R_{C,\theta}$ and $R_{D,\phi}$ are non-trivial, we have $\ell \neq \ell'$ and $\ell \neq \ell''$ and so $Q = \ell' \cap \ell''$ does not belong to ℓ. Thus Q and $\rho_\ell(Q)$ are distinct. □

A *glidereflexion* is an isometry f of $\mathbf{E}$ obtained as the composite $f = t_\mathbf{v} \circ \rho_\ell$ of a reflexion ρ_ℓ about the axis ℓ and a translation $t_\mathbf{v} \neq \mathbf{1_E}$ such that the vecor $\mathbf{v} \neq \mathbf{0}$ is parallel to ℓ. The line ℓ is said to be the *axis* of f. It is easily checked that $f = \rho_\ell \circ t_\mathbf{v}$.

A glidereflexion is an inverse isometry of $\mathbf{E}$ which does not fix any point of $\mathbf{E}$.

Figure 21.3 gives an example of a subset of the ordinary plane which is transformed into itself by a glidereflexion with axis ℓ.

A classical theorem asserts that any isometry of $\mathbf{E}$ is one of the four types we have described:

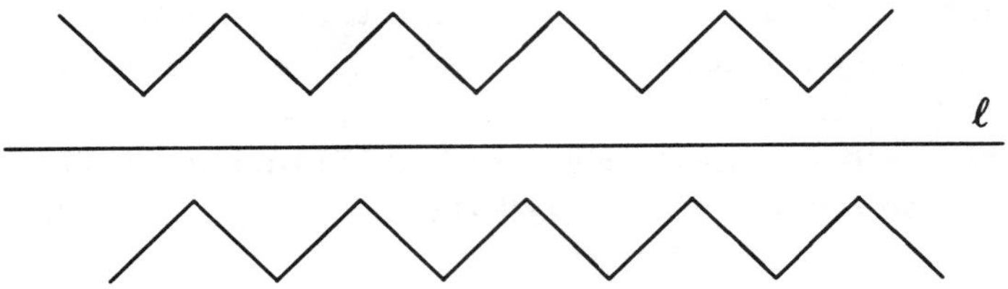

Fig. 21.3

Theorem 21.3 (Chasles, 1831)
An isometry of the Euclidean plane **E** *that fixes a point is either a rotation or a reflexion, accordingly as it is direct or inverse.*

An isometry of **E** *without any fixed points is either a translation or a glidereflexion, accordingly as it is direct or inverse.*

Proof
If $f \in \text{Isom}(\mathbf{E})$ fixes a point, then the result follows from the discussion preceding Lemma 21.2.

Suppose now that f is a direct isometry without any fixed points. Then f^2 also has no fixed points, because if $f^2(P) = P$ for some P, the segment $Pf(P)$ would be transformed by f into the segment

$$f(P)P = f(P)f^2(P),$$

that is, into the same segement with its end-points interchanged. The mid-point of the segment would therefore be fixed by f, which is not possible.

For every P, consider the three points $P, f(P), f^2(P)$. As we have just seen, these three points are distinct; we now show that they are collinear.

Were they not collinear (Fig. 21.4), the perpendicular bisectors of the two segments $Pf(P)$ and $f(P)f^2(P)$ would meet in a point Q. Since $d(P, f(P)) = d(f(P), f^2(P))$, one would also have

$$d(Q, P) = d(Q, f(P)) = d(Q, f^2(P)).$$

Since f preserves orientation, it follows that the triangle $QPf(P)$ is transformed into the triangle $Qf(P)f^2(P)$, and so $Q = f(Q)$ which is a contradiction.

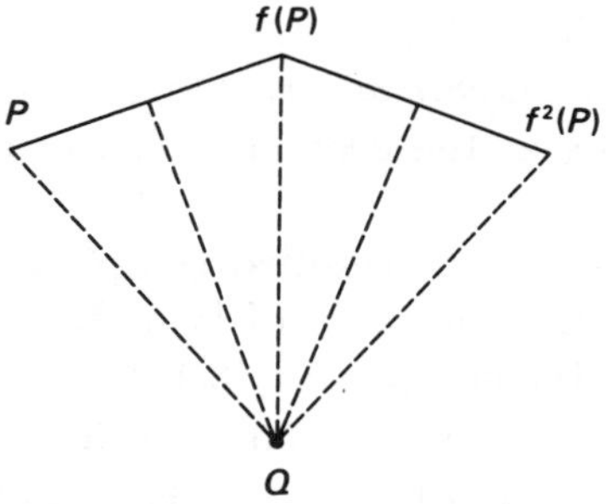

Fig. 21.4

It follows that the points P, $f(P)$, $f^2(P)$, $\dots$, $f^i(P)$, $\dots$ are all collinear, so that f acts on this line as a translation. However, since it is a direct isometry, f must act in the same way on the whole plane, and so is a translation.

Suppose finally that f is an inverse isometry. Then f^2 is a direct isometry, and reasoning as before, we conclude that $f^2 = t_{\mathbf{v}}$ for some $\mathbf{v}$.

Consider any point $P \in \mathbf{E}$. The lines $\ell_0 = \overline{Pf^2(P)}$ and $\ell_1 = \overline{f(P)f^3(P)}$ are parallel (possibly equal) and are interchanged by f. Thus f transforms the line ℓ into itself, where ℓ is the line parallel to ℓ_0 and ℓ_1 and equidistant between them — Fig. 21.5.

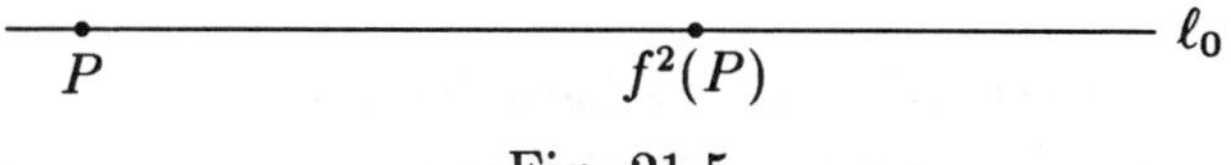

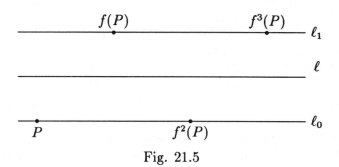

Fig. 21.5

Since f^2 acts on ℓ as a translation $t_{\mathbf{v}}$, it follows that f acts on ℓ as the translation $t_{\mathbf{v}/2}$. The composite $t_{-\mathbf{v}/2} \circ f$ thus fixes all points of ℓ, and so since it is an inverse isometry, it is a reflexion. It now follows that $f = t_{\mathbf{v}/2} \circ (t_{-\mathbf{v}/2} \circ f)$ is a glidereflexion. □

The discontinuous groups of isometries of the Euclidean plane $\mathbf{E}$ can be divided into three classes: the finite groups, the *frieze groups*, and the *plane crystallographic groups*, see for example [9]. Such groups are studied by looking at the geometric figures of which they are the isometry groups.

These groups have been classified completely; we limit ourselves here to the case of *finite subgroups* of $\mathrm{Isom}(\mathbf{E})$.

One such group is the group $\mathrm{Isom}(\Pi_n)$ of isometries of the regular n-gon Π_n, with $n \geq 3$ sides. We will suppose that this figure is inscribed in the circle $\mathbf{S}$ centre O and radius 1 in $\mathbf{E}$ — Fig. 21.6.

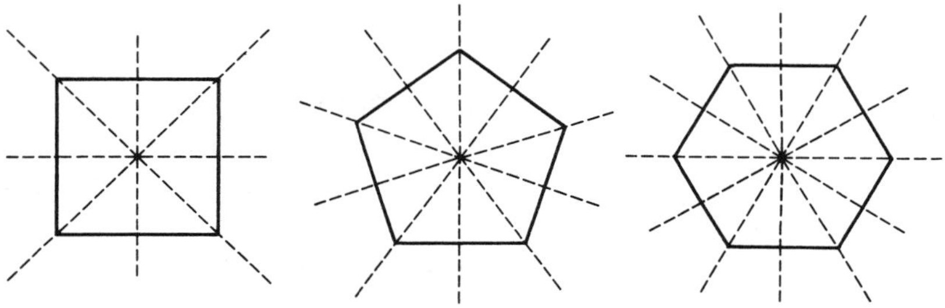

Fig. 21.6

$\mathrm{Isom}(\Pi_n)$ contains the rotation $\sigma = R_{2\pi/n}$, and so also the rotations

$$\sigma^0 = 1_{\mathbf{E}},\ \sigma,\ \sigma^2,\ \ldots,\ \sigma^{n-1}.$$

Since it is clear that every $f \in \mathrm{Isom}(\Pi_n)$ must fix O, and so must be either a rotation about O or a reflexion in an axis containing O, we can conclude that these are the only rotations in $\mathrm{Isom}(\Pi_n)$.

Furthermore, if for each side s of Π_n we consider the diameter ℓ_1 of $\mathbf{S}$ that bisects s, then the reflexion ρ_s with axis s belongs to $\mathrm{Isom}(\Pi_n)$. Similarly, for each vertex v of Π_n, the reflexion ρ_v with axis the diameter ℓ_v of $\mathbf{S}$ through v belongs to $\mathrm{Isom}(\Pi_n)$.

If n is odd, then for each vertex v we have $\rho_v = \rho_s$, where s is the side opposite v.

Let ρ be one of the reflexions we have described.

If $\alpha \in \mathrm{Isom}(\Pi_n)$ is an inverse isometry, then $\alpha \circ \rho$ is a direct isometry and so $\alpha \circ \rho = \sigma^i$ for some i. It follows that $\alpha = \sigma^i \circ \rho^{-1} = \sigma^i \circ \rho$, since $\rho = \rho^{-1}$.

Therefore, $\mathrm{Isom}(\Pi_n)$ consists of the following transformations:

$$\sigma^0 = 1,\ \sigma,\ \sigma^2,\ \ldots,\ \sigma^{n-1},$$
$$\rho, \sigma \circ \rho, \sigma^2 \circ \rho, \ldots, \sigma^{n-1} \circ \rho,$$

and it is easy to check that these are all distinct.

We can conlude that Isom(Π_n) is isomorphic to the *dihedral group* of order $2n$, denoted D_{2n}.

Note that Isom(Π_n) is generated by σ and ρ: this is clear from the way we have expressed the elements. However, it can also be generated by ρ and $\sigma \circ \rho$, since $\sigma = (\sigma \circ \rho) \circ \rho$. Since $\sigma \circ \rho$ is a reflexion it follows that Isom(Π_n) can be generated by 2 reflexions, and so is a Coxeter group.

Note that $\{1, \sigma, \sigma^2, \ldots, \sigma^{n-1}\}$ is a subgroup of Isom(Π_n) isomorphic to the cyclic group $\mathbf{Z}/n\mathbf{Z}$. It is easy to check that this can be identified with the group of isometries of a particular irregular polygon P_{2n} with $2n$ sides contained in Π_n. Figure 21.7 shows the cases $n = 3, 4$, where P_6 and P_8 are the shaded regions.

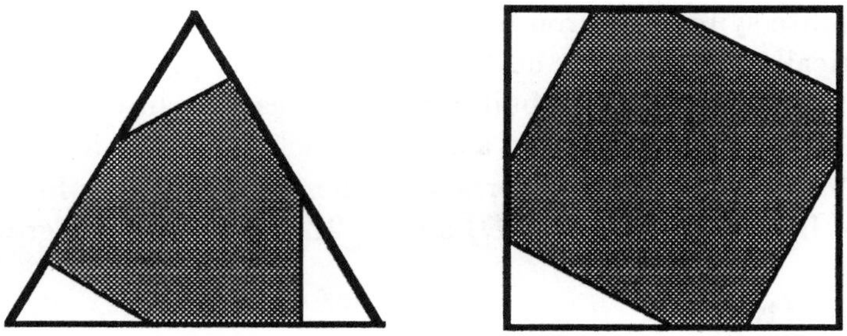

Fig. 21.7

Another finite subgroup of Isom(Π_n) is the group of isometries of an isosceles triangle $\mathbf{T}$ which is not equilateral: Isom($\mathbf{T}$) is isomorphic to $\mathbf{Z}/2\mathbf{Z}$, because it contains only the identity and a single reflexion — Fig. 21.8.

Consider a rectangle $\mathbf{F}$ which is not a square. The group Isom($\mathbf{F}$) is a finite subgroup of Isom($\mathbf{E}$) different from those we have already discussed: if $\mathbf{F}$ is centred on the origin, with sides parallel to the coordinate axes (Fig. 21.8), then Isom($\mathbf{F}$) is the subgroup of $\mathrm{O}(2)$ consisting of the identity, of the two reflexions ρ_1 and ρ_2 with axes the coordinate axes, and of the rotation $R_{O,\pi}$.

This group is called the *Klein 4-group*, or just the *4-group*; it is usually denoted V (for *Viergruppe*) and is isomorphic to D_4, the dihedral group of order 4.

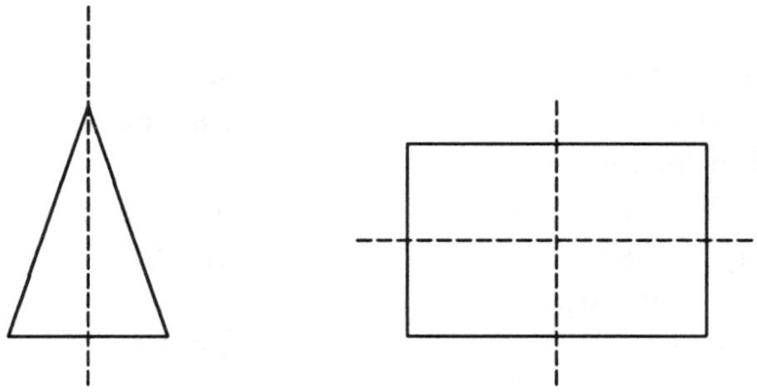

Fig. 21.8

It is clear that if we consider our polygons to be centred on some point $C \in \mathbf{E}$ other than the origin, then we would obtain subgroups of Isom($\mathbf{E}$) isomorphic to those described above, but contained in Isom$(\mathbf{E})_C$ rather than Isom$(\mathbf{E})_O$.

The finite symmetry groups that we have considered were studied systematically by Leonardo da Vinci in the course of his architectural work on the symmetries of buildings. He was also interested in the way this symmetry changed as apses and niches were added.

The following theorem asserts that there are no other finite subgroups of Isom($\mathbf{E}$). Note that D_2 and $\mathbf{Z}/2\mathbf{Z}$ are isomorphic groups.

Theorem 21.4

Every non-trivial finite subgroup of Isom($\mathbf{E}$) *is isomorphic either to* $\mathbf{Z}/n\mathbf{Z}$, *for some* $n \geq 3$, *or to a dihedral group* D_{2n} *for some* $n \geq 1$.

Proof

Let $\mathcal{G}$ be a non-trivial finite subgroup of Isom($\mathbf{E}$). As we have already seen in 20.10(5), $\mathcal{G}$ cannot contain any non-trivial translations, and consequently cannot contain any glidereflexions, since the square of a glidereflexion is a non-trivial translation. Thus, by Theorem 20.3, $\mathcal{G}$ can contain only rotations and reflexions.

Suppose first that $\mathcal{G}$ contains no reflexions.

Let $R_{C,\theta}, R_{D,\phi} \in \mathcal{G}$ be two rotations different from the identity, and such that $C \neq D$.

If $\theta + \phi = 2k\pi$ then $R_{C,\theta} \circ R_{D,\phi}$ is a non-trivial translation, which is impossible. Suppose, therefore, that $\theta + \phi \neq 2k\pi$. $\mathcal{G}$ contains the composite

$$f = (R_{C,\theta})^{-1} \circ (R_{D,\phi})^{-1} \circ R_{C,\theta} \circ R_{D,\phi}.$$

Since $(R_{C,\theta})^{-1} = R_{C,-\theta}$ and $(R_{D,\phi})^{-1} = R_{D,-\phi}$, it follows from Lemma 21.2 that $g_1 = (R_{C,\theta})^{-1} \circ (R_{D,\phi})^{-1}$ and $g_2 = R_{C,\theta} \circ R_{D,\phi}$ are rotations with angles $-\theta - \phi$ and $\theta + \phi$ respectively, and with different centres. Thus, again by Lemma 21.2, $f = g_1 \circ g_2$ is a non-trivial translation. Since this is impossible, it follows that $C = D$.

Thus the group $\mathcal{G}$ consists only of rotations with one and the same centre C.

Let $R = R_{C,\gamma} \in \mathcal{G}$ be such that γ has the smallest possible positive value. If $R_{C,\theta} \in \mathcal{G}$ with $\theta > 0$ it follows that $R_{C,\theta} = R^k$ for some integer $k > 0$, as otherwise for some k we would have $R_{C,\theta} \circ R^{-k} = R_{C,\theta-k\gamma} \in \mathcal{G}$, contradicting the minimality of γ. Thus the elements of $\mathcal{G}$ are all powers of R, and $\mathcal{G}$ is a cyclic group.

Suppose now that $\mathcal{G}$ contains at least one reflexion. The subset of $\mathcal{G}$ consisting of all direct isometries forms a subgroup of $\mathcal{G}$, which, by the first part of this proof, is a cyclic group consisting of all powers of a rotation R through a certain angle $2\pi/n$:

$$R,\ R^2, \ldots,\ R^n = 1$$

(if $n = 1$ then $\mathcal{G}$ contains no rotations other than the identity). Suppose that $\mathcal{G}$ contains m reflexions, and let ρ be one such.

The n products $\rho \circ R$, $\rho \circ R^2$, ..., $\rho \circ R^{n-1}$, ρ are n distinct inverse isometries, and so are all reflexions. Thus $m \geq n$.

On the other hand, if we multiply ρ on the right by each of the m distinct reflexions in $\mathcal{G}$, we obtain as many distinct rotations. Thus $n \geq m$.

In conclusion then, $m = n$ and the elements of $\mathcal{G}$ are

$$R,\ R^2, \ldots,\ R^{n-1},\ R^n = 1,$$

$$\rho \circ R,\ \rho \circ R^2, \ldots,\ \rho \circ R^{n-1},\ \rho,$$

and $\mathcal{G}$ is thus isomorphic to D_{2n}. □

We now proceed to the three-dimensional case.

Lemma 21.5

Every $R \in \mathrm{SO}(3), R \neq \mathbf{I}_3$, has $\lambda = 1$ as an eigenvalue with a 1-dimensional eigenspace.

Proof

Since R is a real square matrix of odd order, it has at least 1 eigenvalue λ (see 13.15(1)). By Proposition 20.3 we know that $\lambda = \pm 1$.

If $\lambda = -1$, let $\mathbf{x} \in \mathbf{R}^3$ be an eigenvector with eigenvalue -1. For every $\mathbf{y} \in \mathbf{x}^\perp$ one has

$$-R\mathbf{y} \cdot \mathbf{x} = R\mathbf{y} \cdot (-\mathbf{x}) = R\mathbf{y} \cdot R\mathbf{x} = \mathbf{y} \cdot \mathbf{x} = 0 \qquad (21.5)$$

and so $R\mathbf{y} \in \mathbf{x}^\perp$. It follows that R transforms $\mathbf{x}^\perp$ into itself, and defines a unitary operator R' on $\mathbf{x}^\perp$. Choosing an orthonormal basis of $\mathbf{R}^3$ whose first element is $\mathbf{x}/\|\mathbf{x}\|$, then $1 = \det(R) = (-1)\det(R')$. It follows that R' is a reflexion of the plane $\mathbf{x}^\perp$. Since R' has eigenvalue 1, so does R.

Let $\mathbf{W} \subset \mathbf{R}^3$ be the corresponding eigenspace. By hypothesis $\dim(\mathbf{W}) \leq 2$, as $R \neq \mathbf{I}_3$. As in (21.5) one checks that $R(\mathbf{W}^\perp) = \mathbf{W}^\perp$. If $\dim(\mathbf{W}) = 2$ then, since $\det(R) = 1$, R acts as the identity on $\mathbf{W}^\perp$, so that $\mathbf{W}^\perp \subset \mathbf{W}$, contradicting the supposition that $\dim(\mathbf{W}) = 2$. Thus $\dim(\mathbf{W}) = 1$. □

From Lemma 21.5 it follows that a rotation R leaves fixed each point of a line through $\mathbf{0}$, called the *axis of rotation*. Choose an orthonormal basis $\{\mathbf{n}, \mathbf{e}_1, \mathbf{e}_2\}$, with the same orientation as the standard basis $\{\mathbf{E}_1, \mathbf{E}_2, \mathbf{E}_3\}$, and such that $\mathbf{n}$ belongs to the axis. R induces on the plane $\langle \mathbf{e}_1, \mathbf{e}_2 \rangle$ a rotation through an angle θ with $0 \leq \theta < 2\pi$. Replacing the basis $\{\mathbf{n}, \mathbf{e}_1, \mathbf{e}_2\}$ with a basis $\{-\mathbf{n}, \mathbf{e}_1, \mathbf{e}_2\}$ changes the angle θ into $2\pi - \theta$, so by an appropriate choice of $\mathbf{n}$ we can suppose that $0 \leq \theta \leq \pi$. This θ is called the *convex angle of the rotation R*.

Note that every rotation $R \in \mathrm{SO}(3)$ determines uniquely a pair $(\mathbf{n}, \theta)$, unless $\theta = \pi$, in which case $\theta = 2\pi - \theta$ and both $\mathbf{n}$ and $-\mathbf{n}$ define the same angle.

Conversely, every pair $(\mathbf{n}, \theta) \in \mathbf{S}^2 \times [0, \pi]$, where $\mathbf{S}^2$ denotes the unit sphere in $\mathbf{E}^3$, determines an element $R \in \mathrm{SO}(3)$ defined by

$$\begin{aligned} R(\mathbf{n}) &= \mathbf{n} \\ R(\mathbf{e}_1) &= (\cos\theta)\mathbf{e}_1 + (\sin\theta)\mathbf{e}_2, \\ R(\mathbf{e}_2) &= -(\sin\theta)\mathbf{e}_1 + (\cos\theta)\mathbf{e}_2, \end{aligned} \qquad (21.6)$$

where $\mathbf{e}_1, \mathbf{e}_2$ are such that $\{\mathbf{n}, \mathbf{e}_1, \mathbf{e}_2\}$ is an orthonormal basis with the same orientation as the standard basis $\{\mathbf{E}_1, \mathbf{E}_2, \mathbf{E}_3\}$.

The above discussion can be summed up as follows:

Proposition 21.6
Equations (21.6) define a map $\rho : \mathbf{S}^2 \times [0, \pi] \to \mathrm{SO}(3)$ *whose restriction to* $\mathbf{S}^2 \times (0, \pi)$ *is bijective, and such that*

$$\begin{aligned} \rho(\mathbf{n}, 0) &= \mathbf{I}_3 \\ \rho(\mathbf{n}, \pi) &= \rho(-\mathbf{n}, \pi) \quad \textit{for every } \mathbf{n} \in \mathbf{S}^2. \end{aligned}$$

To compose two rotations about the same axis, one adds the angles. However, the composite of rotations about different axes is far more subtle; this fact makes the structure of the group SO(3) more complicated than that of SO(2).

There is a very explicit description of SO(3) due to Euler, obtained by using particular rotations.

For each $\theta \in \mathbf{R}$, let

$$\mathbf{X}_\theta = \begin{pmatrix} 1 & 0 & 0 \\ 0 & \cos\theta & -\sin\theta \\ 0 & \sin\theta & \cos\theta \end{pmatrix}$$

and

$$\mathbf{Z}_\theta = \begin{pmatrix} \cos\theta & -\sin\theta & 0 \\ \sin\theta & \cos\theta & 0 \\ 0 & 0 & 1 \end{pmatrix}.$$

These matrices represent rotations through an angle θ about the X-axis and the Z-axis respectively. Since we have fixed orientations of these two axes, it is not possible to reduce the angle θ to a convex angle, i.e. to one in $[0, \pi]$, as we did in the proof of Proposition 21.6. Thus, we need to consider all $\theta \in \mathbf{R}$.

Theorem 21.7 (Euler, 1776)
Every $R \in \mathrm{SO}(3)$ *is of the form*

$$R = \mathbf{Z}_\phi \circ \mathbf{X}_\theta \circ \mathbf{Z}_\psi$$

where $0 \leq \phi, \psi \leq 2\pi$, $0 \leq \theta \leq \pi$. *The angles* ϕ, θ, ψ *are uniquely determined by* R, *and are called its* Euler angles.

Proof
Since a rotation preserves orientation, it is completely determined by the images of the two vectors $\mathbf{e}_1 = (1, 0, 0)$ and $\mathbf{e}_3 = (0, 0, 1)$.

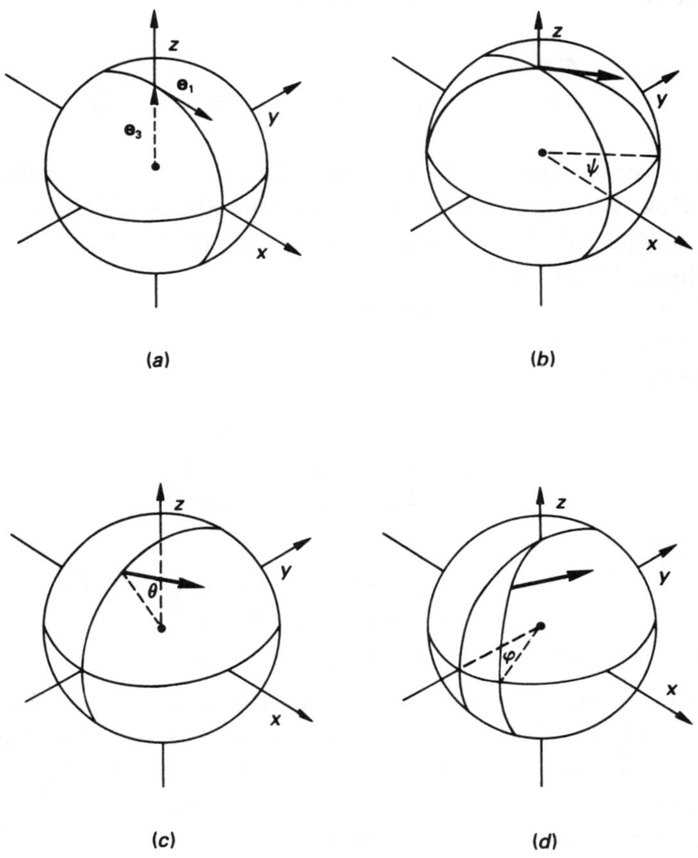

Fig. 21.9

To follow the argument, it helps to visualize the vector $\mathbf{e}_1$ as being based at the point $\mathbf{e}_3 = (0,0,1)$ — Fig. 21.9a — and $R(\mathbf{e}_1)$ as being based at $R(\mathbf{e}_3)$.

Applying first $\mathbf{X}_\theta$ and then $\mathbf{Z}_\phi$ for well-chosen θ and ϕ satisfying $0 \leq \theta \leq \pi$ and $0 \leq \phi < 2\pi$, we can form a transformation which transforms $\mathbf{e}_3$ into $R(\mathbf{e}_3)$: θ and ϕ are uniquely determined and correspond to the 'latitude' and 'longitude' of $R(\mathbf{e}_3)$.

The vector $(\mathbf{Z}_\phi \circ \mathbf{X}_\theta)(\mathbf{e}_1)$, based at $R(\mathbf{e}_3)$, forms an oriented angle ψ, say, with $R(\mathbf{e}_1)$. Applying the transformation $\mathbf{Z}_\psi$ before the composite $(\mathbf{Z}_\phi \circ \mathbf{X}_\theta)$, gives

$$(\mathbf{Z}_\phi \circ \mathbf{X}_\theta \circ \mathbf{Z}_\psi)(\mathbf{e}_1) = R(\mathbf{e}_1),$$

while,

$$(\mathbf{Z}_\phi \circ \mathbf{X}_\theta \circ \mathbf{Z}_\psi)(\mathbf{e}_3) = R(\mathbf{e}_3).$$

Thus $(\mathbf{Z}_\phi \circ \mathbf{X}_\theta \circ \mathbf{Z}_\psi) = R$.

Figures 21.9*b*, *c*, *d* illustrate the sequence of transformations. □

Let **E** be a 3-dimensional Euclidean vector space.

The classification of isometries of **E** is similar to that given in Theorem 21.3 for isometries of the plane. Apart from rotations, reflexions and translations, there are the following three types of isometry.

A *glidereflexion* is defined to be the composite of a reflexion in and a translation in a direction parallel to the plane of symmetry of the reflexion.

A *gliderotation* is the composite of a rotation with a translation in a direction parallel to the axis of the rotation.

A *rotation-reflexion* is the composite of a rotation with the reflexion in the plane perpendicular to the axis of rotation.

In 1776, Euler proved that *every symmetry of* **E** *is one of the six types we have described, namely rotations, reflexions, translations, glidereflexions, gliderotations and rotation-reflexions.*

We do not give a proof of this fact. It is curious that the analogous, but simpler, Theorem 21.3 was not proven until 1831, over 150 years later.

22

Diagonalizing symmetric operators

We have introduced two notions of equivalence for square matrices: similarity and congruency. Recall that two matrices $A, B \in M_n(\mathbf{K})$ are said to be *similar* if there exists $M \in \mathrm{GL}_n(\mathbf{K})$ for which $A = M^{-1}AM$; they are *congruent* if there exists $M \in \mathrm{GL}_n(\mathbf{K})$ for which $A = M^tAM$.

Similarity was introduced for matrices representing an operator on a vector space with respect to different bases; congruency, on the other hand, was introduced for matrices representing a bilinear form with respect to different bases.

Corresponding to the two types of equivalence, two different problems of diagonalizing a matrix arise. They can be stated thus: given a matrix $A \in M_n(\mathbf{K})$, find a diagonal matrix B similar or, respectively, congruent to A.

Theorem 16.1 tells us that it is always possible to find a diagonal matrix congruent to a given matrix provided the given matrix is symmetric.

As we know, simple examples show that the analogous problem for similarity is not in general soluble, that is, not every similarity class contains a diagonal matrix — see Example 13.5(3).

In this chapter we consider a related problem which is more restricted but is of considerable importance in Euclidean geometry: the diagonalization of symmetric operators by orthogonal matrices.

If $A \in M_n(\mathbf{R})$ and $M \in \mathrm{O}(n)$, then $M^{-1} = M^t$, so

$$M^{-1}AM = M^tAM \tag{22.1}$$

and the matrix (22.1) is simultaneously congruent and similar to A. It is therefore unnecessary to specify whether one is diagaonalizing symmetric operators by congruence or by similarity when using only orthogonal matrices. Restricting to the use of orthogonal matrices is equivalent to considering, in a finite dimensional Euclidean space, only orthogonal bases. Thus the diagonalizability of a symmetric matrix A by orthogonal matrices means that both the quadratic form defined by A and the operator with matrix A with respect to an orthonormal basis, are diagonalizable with respect to an *orthonormal basis* of $\mathbf{V}$.

A simple but fundamental property of real symmetric matrices is given by the following lemma.

Lemma 22.1
The characteristic polynomial of a symmetric matrix $A \in M_n(\mathbf{R})$ has only real roots.

Proof
We can consider A as a matrix over $\mathbf{C}$ (that is, with complex entries), and so as an operator $T_A : \mathbf{C}^n \to \mathbf{C}^n$. Let $\lambda \in \mathbf{C}$ be a root of the characteristic polynomial of A, and let $\mathbf{x} \in \mathbf{C}^n$ be a corresponding eigenvector. Then

$$A\mathbf{x} = \lambda\mathbf{x}. \tag{22.2}$$

Taking the complex conjugates of both sides gives

$$A\bar{\mathbf{x}} = \bar{\lambda}\bar{\mathbf{x}}.$$

Consider the scalar $\bar{\mathbf{x}}^t A\mathbf{x}$, and writing it in two different ways using (22.2) and (22.3) gives

$$\bar{\mathbf{x}}^t A\mathbf{x} = \bar{\mathbf{x}}^t(A\mathbf{x}) = \bar{\mathbf{x}}^t\lambda\mathbf{x} = \lambda\mathbf{x}^t\mathbf{x}, \tag{22.4}$$

$$\bar{\mathbf{x}}^t A\mathbf{x} = (\bar{\mathbf{x}}^t A)\mathbf{x} = (A\bar{\mathbf{x}})^t\mathbf{x} = (\bar{\lambda}\bar{\mathbf{x}})^t\mathbf{x} = \bar{\lambda}\bar{\mathbf{x}}^t\mathbf{x}. \tag{22.5}$$

Note that

$$\bar{\mathbf{x}}^t\mathbf{x} = \bar{x}_1 x_1 + \cdots + \bar{x}_n x_n$$

is a strictly positive real number, since $\mathbf{x} \neq \mathbf{0}$. We can therefore deduce from (22.4) and (22.5) that $\bar{\lambda} = \lambda$, that is that λ is real. □

Theorem 22.2 (Spectral Theorem)
Let $\mathbf{V}$ be a finite dimensional Euclidean vector space and $T : \mathbf{V} \to \mathbf{V}$ a symmetric operator. There is an orthonormal basis of $\mathbf{V}$ with respect to which the matrix of T is diagonal.

Proof

The proof is by induction on $n = \dim(\mathbf{V})$. If $n = 1$ there is nothing to prove. Suppose therefore that $n \geq 2$, and that the theorem holds for spaces of dimension $n - 1$. Since the operator T is symmetric, it has only real roots, by Lemma 22.1. Thus T has an eigenvalue λ; let $\mathbf{e}_1$ be a corresponding eigenvector, which we can take to be of length 1. Let $\mathbf{U} = \mathbf{e}_1^{\perp}$, the orthogonal complement of $\mathbf{e}_1$. For each $\mathbf{u} \in \mathbf{U}$,

$$\langle T(\mathbf{u}), \mathbf{e}_1 \rangle = \langle \mathbf{u}, T(\mathbf{e}_1) \rangle = \langle \mathbf{u}, \lambda \mathbf{e}_1 \rangle = \lambda \langle \mathbf{u}, \mathbf{e}_1 \rangle = \lambda 0 = 0,$$

and so $T(\mathbf{u}) \in \mathbf{U}$, that is T induces an operator $T_{\mathbf{U}} : \mathbf{U} \to \mathbf{U}$. Since $T_{\mathbf{U}}(\mathbf{u}) = T(\mathbf{u})$ for every $\mathbf{u} \in \mathbf{U}$, the operator $T_{\mathbf{U}}$ is symmetric. By the inductive hypothesis, $\mathbf{U}$ has an orthonormal basis $\{\mathbf{e}_2, \ldots, \mathbf{e}_n\}$ which diagonalizes $T_{\mathbf{U}}$. Thus $\{\mathbf{e}_1, \ldots, \mathbf{e}_n\}$ is an orthonormal basis of $\mathbf{V}$ which diagonalizes T. □

Theorem 22.3

For every real symmetric matrix $A \in M_n(\mathbf{R})$ there is an orthogonal matrix $M \in \mathrm{O}(n)$ such that $M^{-1}AM$ is diagonal.

Proof

A is the matrix of a symmetric operator T_A on $\mathbf{R}^n$ with respect to the canonical basis. By the Spectral Theorem, there is an orthonormal basis with respect to which the matrix of T_A is diagonal. □

There is an equivalent statement of the Spectral Theorem as follows.

Theorem 22.4

For every quadratic form $q : \mathbf{V} \to \mathbf{R}$ on a finite dimensional Euclidean space, there is a diagonalizing orthonormal basis.

The proof of Theorem 22.4 is similar to that of the preceding theorem and is left to the reader.

The principal geometric application of the spectral theorem is a classifiaction of Euclidean conics, and more generally of quadrics in Euclidean spaces of higher dimensions.

The following result, which holds for Euclidean spaces which are not finite dimensional, is implicit in Theorem 22.2 in the finite dimensional case.

Proposition 22.5
Let $T : \mathbf{V} \to \mathbf{V}$ be a symmetric operator on a Euclidean vector space, not necessarily finite of dimension. If λ, μ are two distict eigenvalues of T then every eigenvector with eigenvalue λ is orthogonal to every eigenvector with eigenvalue μ.

Proof
Let $\mathbf{v}, \mathbf{w} \in \mathbf{V}$ be eigenvectors of T with eigenvalues λ and μ respectively. Then,

$$\begin{aligned} \langle T(\mathbf{v}), \mathbf{w} \rangle &= \langle \lambda \mathbf{v}, \mathbf{w} \rangle = \lambda \langle \mathbf{v}, \mathbf{w} \rangle, \\ \langle \mathbf{v}, T(\mathbf{w}) \rangle &= \langle \mathbf{v}, \mu \mathbf{w} \rangle = \mu \langle \mathbf{v}, \mathbf{w} \rangle. \end{aligned}$$

Since T is symmetric, the right-hand sides are equal, and so $\lambda\langle \mathbf{v}, \mathbf{w} \rangle = \mu\langle \mathbf{v}, \mathbf{w} \rangle$. Thus, since $\lambda \neq \mu$, we must have $\langle \mathbf{v}, \mathbf{w} \rangle = 0$. □

Complement 22.7
As we said at the beginning of this chapter, not all matrices $A \in M_n(\mathbf{K})$ are similar to a diagonal matrix. This gives rise to the problem of finding a class of matrices which are as simple as possible, containing the diagonal matrices, and is such that every similarity class contains at least one element of this class. The matrices of such a class are called *canonical forms*. If such a class of matrices exists, its elements can be taken as representatives of the similarity classes, and so it provides an explicit classification of square matrices up to similarity. Of all the known solutions to this problem, the so-called *Jordan canonical form* is the most important. We give a brief description of this here, leaving the reader to consult more advanced texts for the proofs and a fuller treatment, for example [4].

A *Jordan block of order n* is an $n \times n$ matrix over $\mathbf{K}$ of the form

$$\begin{pmatrix} \lambda & 0 & \cdots & 0 & 0 \\ 1 & \lambda & \cdots & 0 & 0 \\ 0 & 1 & \cdots & 0 & 0 \\ \vdots & \vdots & & \vdots & \vdots \\ 0 & 0 & \cdots & 1 & \lambda \end{pmatrix}$$

for some $\lambda \in \mathbf{K}$. A Jordan block of this form is denoted $J_{n,\lambda}$.

A matrix $A \in M_n(\mathbf{K})$ is said to be in *Jordan canonical form* if it is in the following form:

$$A = \begin{pmatrix} J_{n_1,\lambda_1} & 0 & \cdots & 0 \\ 0 & J_{n_2,\lambda_2} & \cdots & 0 \\ \vdots & \vdots & & \vdots \\ 0 & 0 & \cdots & J_{n_k,\lambda_k} \end{pmatrix}$$

for some positive integers $n_1, \ldots, n_k$ with $n_1 + \cdots + n_k = n$, and $\lambda_1, \ldots, \lambda_k \in \mathbf{K}$. It is easy to prove that the scalars $\lambda_1, \ldots, \lambda_k$ are the eigenvalues of A.

In particular, if $k = n$, so $n_1 = \cdots = n_k = 1$, then $A = \mathrm{diag}(\lambda_1, \ldots, \lambda_k)$ is a diagonal matrix.

Jordan's Theorem

Suppose $\mathbf{K}$ *is algebraically closed. Let* $\mathbf{V}$ *be a finite dimensional* $\mathbf{K}$*-vector space, and* $T : \mathbf{V} \to \mathbf{V}$ *an operator. There is a basis* $\mathbf{e}$ *of* $\mathbf{V}$ *such that the matrix* $M_{\mathbf{e}}(T)$ *is in Jordan canonical form.*

An immediate consequence of Jordan's Theorem is that every matrix $M \in M_n(\mathbf{K})$ is similar to a matrix in Jordan canonical form.

EXERCISES

22.1 For each of the following, find the matrix $M \in SO(2)$ which diagonalizes the given symmetric matrix.

a) $\begin{pmatrix} 6 & 2 \\ 2 & 9 \end{pmatrix}$ b) $\begin{pmatrix} 5 & -13 \\ -13 & 5 \end{pmatrix}$

c) $\begin{pmatrix} 7 & -2 \\ -2 & 5/3 \end{pmatrix}$ d) $\begin{pmatrix} 1 & 1/2 \\ 1/2 & 1 \end{pmatrix}$.

22.2 For each of the following, find the orthogonal transformation of $\mathbf{R}^3$ which diagonalizes the given quadratic form.

a) $q(x_1, x_2, x_3) = 2x_1^2 + 2x_1x_2 + 2x_1x_3 + 2x_2^2 + 2x_2x_3 + 2x_3^2$

b) $q(x_1, x_2, x_3) = -2x_1x_2 + 2x_1x_3 - 2x_2^2 - 2x_2x_3 - 2x_3^2$

c) $q(x_1, x_2, x_3) = x_1^2 + 4x_1x_3 - x_2^2 + x_3^2$,

and give the corresponding diagonal form.

22.3 Prove that if $A \in M_n(\mathbf{R})$ is a skew-symmetric matrix, then every non-zero root of its characteristic polynomial is a purely imaginary complex number.

23

The complex case

We have seen how all the metric notions of geometry can be defined in a Euclidean vector space using the scalar product. In a field $\mathbf{K}$ different from $\mathbf{R}$, the property of being positive definite has no meaning, and so it is not possible to define a useful scalar product for general fields. However, in the case $\mathbf{K} = \mathbf{C}$ it is possible to modify the definition of symmetric bilinear form to that of 'Hermitian form', and we will see that many of the properties of Euclidean spaces can be extended, using positive definite Hermitian forms, to 'Hermitian spaces'.

Definition 23.1

Let $\mathbf{V}$ be a complex vector space. A map $h : \mathbf{V} \times \mathbf{V} \to \mathbf{C}$ is a *Hermitian form* on $\mathbf{V}$ if it satisfies the following conditions:

[HF1] $h(\mathbf{v} + \mathbf{v}', \mathbf{w}) = h(\mathbf{v}, \mathbf{w}) + h(\mathbf{v}', \mathbf{w})$

[HF2] $h(\mathbf{v}, \mathbf{w} + \mathbf{w}') = h(\mathbf{v}, \mathbf{w}) + h(\mathbf{v}, \mathbf{w}')$

[HF3] $h(c\mathbf{v}, \mathbf{w}) = ch(\mathbf{v}, \mathbf{w})$

[HF4] $h(\mathbf{w}, \mathbf{v}) = \overline{h(\mathbf{v}, \mathbf{w})}$.

HF1 and HF3 together imply that $h(\mathbf{v}, \mathbf{w})$ is $\mathbf{C}$-linear in $\mathbf{v}$, while HF2 implies that $h(\mathbf{v}, \mathbf{w})$ is additive in $\mathbf{w}$. From HF4 it follows that

$$h(\mathbf{v}, c\mathbf{w}) = \overline{h(c\mathbf{w}, \mathbf{v})} = \overline{ch(\mathbf{w}, \mathbf{v})} = \bar{c}\overline{h(\mathbf{w}, \mathbf{v})} = \bar{c}h(\mathbf{v}, \mathbf{w}). \qquad (23.1)$$

HF2 and HF4 therefore imply together that $h(\mathbf{v}, \mathbf{w})$ is antilinear in $\mathbf{w}$ — see Complement 11.14(3).

It also follows from HF4 that $h(\mathbf{v}, \mathbf{v}) \in \mathbf{R}$ for every $\mathbf{v} \in \mathbf{V}$.

A Hermitian form h is said to be

$$\begin{array}{rlll} \textit{positive definite} & \text{if} & h(\mathbf{v},\mathbf{v}) > 0 & \text{for all } \mathbf{v} \neq \mathbf{0} \\ \textit{positive semi-definite} & \text{if} & h(\mathbf{v},\mathbf{v}) \geq 0 & \text{for all } \mathbf{v} \in \mathbf{V} \\ \textit{negative definite} & \text{if} & h(\mathbf{v},\mathbf{v}) < 0 & \text{for all } \mathbf{v} \neq \mathbf{0} \\ \textit{negative semi-definite} & \text{if} & h(\mathbf{v},\mathbf{v}) \leq 0 & \text{for all } \mathbf{v} \in \mathbf{V}. \end{array}$$

Suppose that $\mathbf{V}$ is finite dimensional, and let $\boldsymbol{e} = \{\mathbf{e}_1, \ldots, \mathbf{e}_n\}$ be a basis. For every $1 \leq i, j \leq n$, put $h_{ij} = h(\mathbf{e}_i, \mathbf{e}_j)$. The matrix

$$H = (h_{ij}) \in M_n(\mathbf{C})$$

is called the *matrix representing the Hermitian form with respect to the basis* $\boldsymbol{e}$. By HF4, for each $1 \leq i, j \leq n$

$$h_{ij} = \overline{h_{ji}}.$$

In other words, $H = \overline{H}^t$.

A matrix $H \in M_n(\mathbf{C})$ satisfying $H = \overline{H}^t$ is said to be a *Hermitian matrix*. Thus, with respect to any basis, the matrix representing a Hermitian form is a Hermitian matrix. Note that in particular, if H is a Hermitian matrix, then for each $i = 1, \ldots, n$, $h_{ii} \in \mathbf{R}$. If H is symmetric with real entries, then it is Hermitian.

As in the case of bilinear forms, the matrix with respect to a given basis $\boldsymbol{e}$ determines the form. Indeed, for every

$$\mathbf{v} = x_1\mathbf{e}_1 + \cdots + x_n\mathbf{e}_n, \quad \mathbf{w} = y_1\mathbf{e}_1 + \cdots + y_n\mathbf{e}_n \tag{23.2}$$

one has

$$\begin{aligned} h(\mathbf{v},\mathbf{w}) &= h(x_1\mathbf{e}_1 + \cdots + x_n\mathbf{e}_n, y_1\mathbf{e}_1 + \cdots + y_n\mathbf{e}_n) \\ &= \sum_{ij} x_i\overline{y_j}h(\mathbf{e}_i,\mathbf{e}_j) = \mathbf{x}^t H \overline{\mathbf{y}}. \end{aligned}$$

Conversely, let $H \in M_n(\mathbf{C})$ be a Hermitian matrix, and $\boldsymbol{e}$ a basis of $\mathbf{V}$. Putting

$$h(\mathbf{v},\mathbf{w}) = \mathbf{x}^t H \overline{\mathbf{y}}$$

for every $\mathbf{v}, \mathbf{w} \in \mathbf{V}$ as in (23.2), defines a Hermitian form on $\mathbf{V}$. This is left to the reader to check.

In the particular case that H is the zero matrix, the corresponding Hermitian form is the *zero Hermitian form*: $h(\mathbf{v}, \mathbf{w}) = 0$ for all $\mathbf{v}, \mathbf{w} \in \mathbf{V}$.

Many of the ideas and results we established for symmetric bilinear forms extend to Hermitian forms.

Two vectors $\mathbf{v}, \mathbf{w} \in \mathbf{V}$ are said to be *orthogonal* or *perpendicular* if $h(\mathbf{v}, \mathbf{w}) = 0$. If $S \subset V$ define

$$S^{\perp} = \{\mathbf{w} \in \mathbf{V} \mid \mathbf{w} \text{ is orthogonal to every } \mathbf{v} \in S\}$$

and $S^{\perp}$ is called the subspace orthogonal to S.

A basis $\mathbf{e} = \{\mathbf{e}_1, \ldots, \mathbf{e}_n\}$ of $\mathbf{V}$ is said to be *diagonalizing* or *orthogonal* for h if the vectors are pairwise orthogonal.

Theorem 23.2
Let $\mathbf{V}$ be a finite dimensional complex vector space, and h a Hermitian form on $\mathbf{V}$. Then there is a diagonalizing basis for h.

The proof of this result is left to the reader; it is merely an adaptation of the corresponding result for symmetric bilinear forms, Theorem 16.1.

The most important case is that in which h is positive definite. A positive definite Hermitian form is also called a *Hermitian product* on $\mathbf{V}$. A complex vector space on which there is a given Hermitian product is called a *Hermitian vector space.*

Putting

$$h(\mathbf{x}, \mathbf{y}) = \mathbf{x} \cdot \overline{\mathbf{y}} = x_1\overline{y}_1 + x_2\overline{y}_2 + \cdots + x_n\overline{y}_n \tag{23.3}$$

defines a Hermitian product on $\mathbf{C}^n$, called the *standard Hermitian product.* It is easy to check that this does indeed define a Hermitian product on $\mathbf{C}^n$.

Hermitian vector spaces are the analogue in the complex case of Euclidean spaces in the real case, and the theory developed in the Euclidean case extends with little change to the Hermitian case. For example, the notions of *norm*, or *length*, of a vector, of *Fourier coefficient* and of *projection of a vector in the direction of a non-zreo vector* are defined in exactly the same way as for a Euclidean space.

Consequently, *orthonormal bases* can be defined in the same way as for Euclidean spaces. From Theorem 23.2 it follows immediately that there exist orthonormal bases in any finite dimensional Hermitian vector space, which can be obtained from an orthogonal basis by dividing each basis vector by its norm. The *Gram-Schmidt* process extends to Hermitian spaces, with no change.

Let $\mathbf{V}$ be a finite dimensional Hermitian vector space, and denote the Hermitian product of two vectors by $\langle \mathbf{v}, \mathbf{w} \rangle$.

If $\boldsymbol{e} = \{\mathbf{e}_1, \ldots, \mathbf{e}_n\}$ is an orthonormal basis of $\mathbf{V}$, then the matrix which represents the Hermitian product with respect to $\boldsymbol{e}$ is $\mathbf{I}_n$. The Hermitian product of two vectors $\mathbf{v} = x_1\mathbf{e}_1 + \cdots + x_n\mathbf{e}_n$ and $\mathbf{w} = y_1\mathbf{e}_1 + \cdots + y_n\mathbf{e}_n$ is thus

$$\langle \mathbf{v}, \mathbf{w} \rangle = \mathbf{x}^t\overline{\mathbf{y}},$$

i.e. it is equal to the standard Hermitian product of its coordinates.

The Schwartz inequality also extends to Hermitian vector spaces, but the proof is a little different to the Euclidean case.

Theorem 23.3 (Schwartz Inequality)
For every $\mathbf{v}, \mathbf{w} \in \mathbf{V}$,

$$|\langle \mathbf{v}, \mathbf{w} \rangle| \leq \|\mathbf{v}\|\|\mathbf{w}\|, \tag{23.4}$$

with equality if and only if $\mathbf{v}$ *and* $\mathbf{w}$ *are parallel.*

Proof
If $\mathbf{w} = \mathbf{0}$ then (23.4) is obvious, so we can suppose that $\mathbf{w} \neq \mathbf{0}$. For every $a, b \in \mathbf{C}$

$$\begin{aligned} 0 \leq \langle a\mathbf{v} + b\mathbf{w}, a\mathbf{v} + b\mathbf{w} \rangle &= \langle a\mathbf{v}, a\mathbf{v} \rangle + \langle a\mathbf{v}, b\mathbf{w} \rangle \\ &\quad + \langle b\mathbf{w}, a\mathbf{v} \rangle + \langle b\mathbf{w}, b\mathbf{w} \rangle \\ &= a\bar{a}\langle \mathbf{v}, \mathbf{v} \rangle + a\bar{b}\langle \mathbf{v}, \mathbf{w} \rangle \\ &\quad + \bar{a}b\langle \mathbf{w}, \mathbf{v} \rangle + b\bar{b}\langle \mathbf{w}, \mathbf{w} \rangle. \end{aligned}$$

Substituting $a = \langle \mathbf{w}, \mathbf{w} \rangle$ and $b = -\langle \mathbf{v}, \mathbf{w} \rangle$ gives

$$0 \leq \|\mathbf{w}\|^4\|\mathbf{v}\|^2 - 2\|\mathbf{w}\|^2\langle \mathbf{v}, \mathbf{w} \rangle\overline{\langle \mathbf{v}, \mathbf{w} \rangle} + \|\mathbf{w}\|^2\langle \mathbf{v}, \mathbf{w} \rangle\overline{\langle \mathbf{v}, \mathbf{w} \rangle}.$$

Since $\langle \mathbf{v}, \mathbf{w} \rangle\overline{\langle \mathbf{v}, \mathbf{w} \rangle} = |\langle \mathbf{v}, \mathbf{w} \rangle|^2$, we get

$$\|\mathbf{w}\|^2|\langle \mathbf{v}, \mathbf{w} \rangle|^2 \leq \|\mathbf{w}\|^4\|\mathbf{v}\|^2.$$

Dividing by $\|\mathbf{w}\|^2$ gives (23.4). Equality holds if and only if $a\mathbf{v}+b\mathbf{w} = \mathbf{0}$, which occurs if and only if $\mathbf{v}$ and $\mathbf{w}$ are parallel. □

The *norm* of vectors in a Hermitian space also satisfies properties N1, N2 and N3 — see Chapter 17. N1 is obvious, and N2 follows from the identity

$$\langle r\mathbf{v}, r\mathbf{v}\rangle = |r|^2\langle \mathbf{v}, \mathbf{v}\rangle.$$

N3 is the *triangle inequality*, and is proved as follows. Expanding,

$$\|\mathbf{v}+\mathbf{w}\|^2 = \langle \mathbf{v}+\mathbf{w}, \mathbf{v}+\mathbf{w}\rangle = \langle \mathbf{v}, \mathbf{v}\rangle + \langle \mathbf{v}, \mathbf{w}\rangle + \langle \mathbf{w}, \mathbf{v}\rangle + \langle \mathbf{w}, \mathbf{w}\rangle.$$

Observe that

$$\langle \mathbf{v}, \mathbf{w}\rangle + \langle \mathbf{w}, \mathbf{v}\rangle = \langle \mathbf{v}, \mathbf{w}\rangle + \overline{\langle \mathbf{v}, \mathbf{w}\rangle} \leq 2|\langle \mathbf{v}, \mathbf{w}\rangle|.$$

Using (23.4) we therefore deduce

$$\begin{aligned}\|\mathbf{v}+\mathbf{w}\|^2 &\leq \|\mathbf{v}\|^2 + 2|\langle \mathbf{v}, \mathbf{w}\rangle| + \|\mathbf{w}\|^2 \\ &\leq \|\mathbf{v}\|^2 + 2\|\mathbf{v}\|\|\mathbf{w}\| + \|\mathbf{w}\|^2 = (\|\mathbf{v}\| + \|\mathbf{w}\|)^2,\end{aligned}$$

that is, N3.

Definition 23.4
An operator $T : \mathbf{V} \to \mathbf{V}$ is said to be *unitary* if it satisfies

$$\langle T(\mathbf{v}), T(\mathbf{w})\rangle = \langle \mathbf{v}, \mathbf{w}\rangle, \quad \text{for every } \mathbf{v}, \mathbf{w} \in \mathbf{V}.$$

The following result is the analogue of Theorem 20.1, which characterizes unitary operators.

Theorem 23.5
Let $T : \mathbf{V} \to \mathbf{V}$ be a map. The following conditions are equivalent:
1) *T is a unitary operator.*
2) *T is an operator such that $\|T(\mathbf{v})\| = \|\mathbf{v}\|$ for all $\mathbf{v} \in \mathbf{V}$.*
3) *$T(\mathbf{0}) = \mathbf{0}$ and $\|T(\mathbf{v}) - T(\mathbf{w})\| = \|\mathbf{v} - \mathbf{w}\|$ for all $\mathbf{v}, \mathbf{w} \in \mathbf{V}$.*
4) *T is an operator, and for every orthonormal basis $\{\mathbf{e}_1, \ldots, \mathbf{e}_n\}$ of $\mathbf{V}$, $\{T(\mathbf{e}_1), \ldots, T(\mathbf{e}_n)\}$ is also an orthonormal basis.*
5) *T is an operator, and there exists an orthonormal basis $\{\mathbf{e}_1, \ldots, \mathbf{e}_n\}$ of $\mathbf{V}$ such that $\{T(\mathbf{e}_1), \ldots, T(\mathbf{e}_n)\}$ is also an orthonormal basis.*

The proof of this theorem follows exactly that of Theorem 20.1.

Note that from the preceding theorem, it follows that a unitary operator is necessarily invertible: $T \in \mathrm{GL}(\mathbf{V})$.

Corollary 23.6
Let $T : \mathbf{V} \to \mathbf{V}$ be a unitary operator.

1) *Every eigenvalue λ of T satisfies $|\lambda| = 1$.*
2) *If $\mathbf{v}$ and $\mathbf{w}$ are eigenvectors with distinct eigenvalues λ and μ respectively, then $\mathbf{v}$ and $\mathbf{w}$ are perpendicular.*

Proof
1) Let $\mathbf{v} \in \mathbf{V}$ be an eigenvector with eigenvalue λ. Then

$$\langle \mathbf{v}, \mathbf{v} \rangle = \langle T(\mathbf{v}), T(\mathbf{v}) \rangle = \langle \lambda \mathbf{v}, \lambda \mathbf{v} \rangle = \lambda \overline{\lambda} \langle \mathbf{v}, \mathbf{v} \rangle.$$

Since $\langle \mathbf{v}, \mathbf{v} \rangle \neq 0$, it follows that $\lambda\overline{\lambda} = 1$, that is, $|\lambda| = 1$.
2) We have,

$$\langle \mathbf{v}, \mathbf{w} \rangle = \langle T(\mathbf{v}), T(\mathbf{w}) \rangle = \langle \lambda \mathbf{v}, \mu \mathbf{w} \rangle = \lambda \overline{\mu} \langle \mathbf{v}, \mathbf{w} \rangle. \qquad (23.5)$$

If $\langle \mathbf{v}, \mathbf{w} \rangle \neq 0$, then it follows that $\lambda\overline{\mu} = 1$. But since $\lambda\overline{\lambda} = 1$ it follows that $\overline{\lambda} = \overline{\mu}$, whence $\lambda = \mu$, contradicting the hypothesis. □

Unitary operators correspond precisely to unitary matrices:

Corollary 23.7
An operator $T : \mathbf{V} \to \mathbf{V}$ is unitary if and only if the matrix representing T with respect to any orthonormal basis is unitary.

Proof
Let $\boldsymbol{e} = \{\mathbf{e}_1, \ldots, \mathbf{e}_n\}$ be an orthonormal basis of $\mathbf{V}$, and let $A = (a_{ij}) \in M_n(\mathbf{C})$ be the matrix representing T with repect to $\boldsymbol{e}$. Then T is unitary if and only if

$$\delta_{ij} = \langle \mathbf{e}_i, \mathbf{e}_j \rangle = \langle T(\mathbf{e}_i), T(\mathbf{e}_j) \rangle = A^t_{(i)} \overline{A}_{(j)}, \quad 1 \leq i, j \leq n,$$

where $A_{(1)}, \ldots, A_{(n)}$ are the n columns of A. Thus we have $A^t \overline{A} = \mathbf{I}_n$, or equivalently, $\overline{A}^t A = \mathbf{I}_n$. □

The principal difference between real and complex unitary operators is their diagonalizability. Indeed, the following result holds only for complex unitary operators, and not for their real counterparts.

Theorem 23.8
Let $T : \mathbf{V} \to \mathbf{V}$ be a unitary operator, and suppose that $\dim(\mathbf{V}) = n \geq 1$. *Then there is an orthonormal basis which diagonalizes T.*

Proof
We proceed by induction on n. If $n = 1$ there is nothing to prove, so we assume $n > 1$, and that the theorem holds for spaces of dimension less than n. Let $\mathbf{e}_1$ be an eigenvector of T with eigenvalue λ. We can suppose that $\|\mathbf{e}_1\| = 1$. Let $\mathbf{U} = \mathbf{e}_1^{\perp}$ Since $\lambda\overline{\lambda} = 1$, for every $\mathbf{u} \in \mathbf{U}$

$$\langle T(\mathbf{u}), \mathbf{e}_1\rangle = \langle T(\mathbf{u}), \lambda\overline{\lambda}\mathbf{e}_1\rangle = \lambda\langle T(\mathbf{u}), T(\mathbf{e}_1)\rangle = \lambda\langle \mathbf{u}, \mathbf{e}_1\rangle = \lambda 0 = 0.$$

Thus T transforms $\mathbf{U}$ to itself and so induces a unitary operator

$$T_{\mathbf{U}} : \mathbf{U} \to \mathbf{U}.$$

By the inductive hypothesis, $\mathbf{U}$ has an orthonormal basis $\{\mathbf{e}_2, \ldots, \mathbf{e}_n\}$ with respect to which $T_{\mathbf{U}}$ is diagonal. Then $\{\mathbf{e}_1, \ldots, \mathbf{e}_n\}$ is a diagonalizing basis for T. □

Theorem 23.8, together with Corollary 23.7, implies the diagonalizability of any unitary matrix using only unitary matrices. More precisely,

Corollary 23.9
For every $A \in \mathbf{U}(n)$ there is a a matrix $M \in \mathbf{U}(n)$ for which $M^{-1}AM$ is diagonal, or equuivalently, for which M^*AM is diagonal.

From the corollary, it follows that every matrix $A \in \mathrm{O}(n)$ is diagonalizable. A word of warning, however: an orthogonal matrix is *not* in general diagonalizable using real matries, because they do not in general have real eigenvalues. For example, the matrices $R_\theta \in \mathrm{O}(2)$, $0 < \theta < \pi$ do not have real eigenvalues.

Definition 23.10
An operator $T : \mathbf{V} \to \mathbf{V}$ is said to be *Hermitian* if it satisfies

$$\langle T(\mathbf{v}), \mathbf{w}\rangle = \langle \mathbf{v}, T(\mathbf{w})\rangle \quad \text{for every } \mathbf{v}, \mathbf{w} \in \mathbf{V}.$$

Suppose that $\boldsymbol{e} = \{\mathbf{e}_1, \ldots, \mathbf{e}_n\}$ is an orthonormal basis of $\mathbf{V}$, and let A be the matrix representing a Hermitian operator T with respect to this basis. Then, for every, $\mathbf{v} = x_1\mathbf{e}_1 + \cdots + x_n\mathbf{e}_n$ and $\mathbf{w} = y_1\mathbf{e}_1 + \cdots + y_n\mathbf{e}_n$, we have

$$\begin{aligned} \langle T(\mathbf{v}), \mathbf{w}\rangle &= (A\mathbf{x})^t\overline{\mathbf{y}} = \mathbf{x}^t A^t \overline{\mathbf{y}} \\ \langle \mathbf{v}, T(\mathbf{w})\rangle &= \mathbf{x}^t(\overline{A\mathbf{y}}) = \mathbf{x}^t\overline{A}\overline{\mathbf{y}} \end{aligned} \tag{23.6}$$

and so $\mathbf{x}^t A^t \overline{\mathbf{y}} = \mathbf{x}^t \overline{A}\overline{\mathbf{y}}$. Since this is true for every $\mathbf{x}, \mathbf{y} \in \mathbf{C}^n$, it follows that $A^t = \overline{A}$. That is, A is a Hermitian matrix.

Converely, if $A \in M_n(\mathbf{C})$ is a Hermitian matrix and T is an operator representing T with repect to an orthonormal basis $\boldsymbol{e}$ then from (23.6) it follows that T is Hermitian. We thus have the following proposition.

Proposition 23.11
An operator $T : \mathbf{V} \to \mathbf{V}$ is Hermitian if and only if the matrix which represents T with respect to any orthonormal basis is a Hermitian matrix.

Hermitian operators are to Hermitian vector spaces as symmetric operators are to Euclidean vector spaces. There is the following extension of Lemma 22.1.

Proposition 23.12
All eigenvalues of a Hermitian operator $T : \mathbf{V} \to \mathbf{V}$ are real.

Proof
Let $\lambda \in \mathbf{C}$ be an eigenvalue of T, and let $\mathbf{v} \in \mathbf{V}$ be an eigenvector with eigenvalue λ. Then

$$\lambda\langle \mathbf{v}, \mathbf{v}\rangle = \langle \lambda\mathbf{v}, \mathbf{v}\rangle = \langle T(\mathbf{v}), \mathbf{v}\rangle = \langle \mathbf{v}, T(\mathbf{v})\rangle = \langle \mathbf{v}, \lambda\mathbf{v}\rangle = \overline{\lambda}\langle \mathbf{v}, \mathbf{v}\rangle.$$

Since $\langle \mathbf{v}, \mathbf{v}\rangle \neq 0$, it follows that $\lambda = \overline{\lambda}$. □

The spectral theorem that we proved for symmetric operators also extends to Hermitian operators.

Theorem 23.13 (Spectral Theorem)
Let $T : \mathbf{V} \to \mathbf{V}$ be a Hermitian operator. There is an orthonormal basis of $\mathbf{V}$ which diagonalizes T.

The proof is identical to that of Theorem 22.2.

Complement 23.14

Let $h : \mathbf{V} \times \mathbf{V} \to \mathbf{C}$ be a Hermitian form on a complex vector space $\mathbf{V}$. Separating the real and imaginary parts, we can write, for every $\mathbf{v}, \mathbf{w} \in \mathbf{V}$

$$h(\mathbf{v}, \mathbf{w}) = s(\mathbf{v}, \mathbf{w}) + ia(\mathbf{v}, \mathbf{w})$$

where $s(\mathbf{v}, \mathbf{w})$, $a(\mathbf{v}, \mathbf{w}) \in \mathbf{R}$. It follows immediately from HF1 and HF2 that $s(\mathbf{v}, \mathbf{w})$ and $a(\mathbf{v}, \mathbf{w})$ are additive in $\mathbf{v}$ and $\mathbf{w}$. Moreover, for every $c \in \mathbf{R}$ and $\mathbf{v}, \mathbf{w} \in \mathbf{V}$,

$$s(c\mathbf{v}, \mathbf{w}) + ia(c\mathbf{v}, \mathbf{w}) = h(c\mathbf{v}, \mathbf{w}) = ch(\mathbf{v}, \mathbf{w}) = cs(\mathbf{v}, \mathbf{w}) + ica(\mathbf{v}, \mathbf{w}),$$

by HF3, and

$$s(\mathbf{v}, c\mathbf{w}) + ia(\mathbf{v}, c\mathbf{w}) = h(\mathbf{v}, c\mathbf{w}) = ch(\mathbf{v}, \mathbf{w}) = cs(\mathbf{v}, \mathbf{w}) + ica(\mathbf{v}, \mathbf{w}).$$

Thus $s(\mathbf{v}, \mathbf{w})$ and $a(\mathbf{v}, \mathbf{w})$ are two bilinear forms on $\mathbf{V}$ considered as a real vector space.

By HF4

$$s(\mathbf{w}, \mathbf{v}) + ia(\mathbf{w}, \mathbf{v}) = s(\mathbf{v}, \mathbf{w}) - ia(\mathbf{v}, \mathbf{w})$$

and so

$$\begin{aligned} s(\mathbf{w}, \mathbf{v}) &= s(\mathbf{v}, \mathbf{w}) \\ a(\mathbf{w}, \mathbf{v}) &= -a(\mathbf{v}, \mathbf{w}) \end{aligned}$$

for every $\mathbf{v}, \mathbf{w} \in \mathbf{V}$. Thus

$$s : \mathbf{V} \times \mathbf{V} \to \mathbf{R}$$

is a real symmetric bilinear form on $\mathbf{V}$, while

$$a : \mathbf{V} \times \mathbf{V} \to \mathbf{R}$$

is a real skew-symmetric bilinear form on $\mathbf{V}$.

Furthermore, expanding the identities $h(i\mathbf{v}, \mathbf{w}) = ih(\mathbf{v}, \mathbf{w})$ and $h(\mathbf{v}, i\mathbf{w}) = -ih(\mathbf{v}, \mathbf{w})$ gives

$$a(\mathbf{v}, \mathbf{w}) = -s(i\mathbf{v}, \mathbf{w}) = s(\mathbf{v}, i\mathbf{w}) \tag{23.7}$$

$$s(\mathbf{v}, \mathbf{w}) = a(i\mathbf{v}, \mathbf{w}) = -a(\mathbf{v}, i\mathbf{w}) \tag{23.8}$$

for every $\mathbf{v}, \mathbf{w} \in \mathbf{V}$.

Finally, expanding the identity $h(i\mathbf{v}, i\mathbf{w}) = h(\mathbf{v}, \mathbf{w})$ gives

$$s(i\mathbf{v}, i\mathbf{w}) = s(\mathbf{v}, \mathbf{w}) \tag{23.9}$$

$$a(i\mathbf{v}, i\mathbf{w}) = a(\mathbf{v}, \mathbf{w}) \tag{23.10}$$

for every $\mathbf{v}, \mathbf{w} \in \mathbf{V}$.

From (23.7) it follows that s determines a, while (23.8) on the other hand, shows that a determines s.

Conversely, given any real symmetric bilinear form $s : \mathbf{V} \times \mathbf{V} \to \mathbf{R}$ which satisfies (23.9), one can define a complex bilinear form h on $\mathbf{V}$:

$$h(\mathbf{v}, \mathbf{w}) = s(\mathbf{v}, \mathbf{w}) + is(\mathbf{v}, i\mathbf{w}).$$

This form is then Hermitian. We check properties HF3 and HF4 — the other two being obvious. For every $\mathbf{v}, \mathbf{w} \in \mathbf{V}$ and $c = a + ib \in \mathbf{C}$,

$$\begin{aligned} h(c\mathbf{v}, \mathbf{w}) &= s(a\mathbf{v}, \mathbf{w}) + is(a\mathbf{v}, i\mathbf{w}) + s(ib\mathbf{v}, \mathbf{w}) + is(ib\mathbf{v}, i\mathbf{w}) \\ &= ah(\mathbf{v}, \mathbf{w}) + b[s(i\mathbf{v}, \mathbf{w}) + is(i\mathbf{v}, i\mathbf{w})] \\ &= ah(\mathbf{v}, \mathbf{w}) + b[s(-\mathbf{v}, i\mathbf{w}) + is(\mathbf{v}, \mathbf{w})] \\ &= ah(\mathbf{v}, \mathbf{w}) + ibh(\mathbf{v}, \mathbf{w}) = ch(\mathbf{v}, \mathbf{w}), \end{aligned}$$

so HF3 is satisfied. Further,

$$\begin{aligned} h(\mathbf{w}, \mathbf{v}) &= s(\mathbf{w}, \mathbf{v}) + is(\mathbf{w}, i\mathbf{v}) = s(\mathbf{v}, \mathbf{w}) + is(i\mathbf{v}, \mathbf{w}) \\ &= s(\mathbf{v}, \mathbf{w}) + is(-\mathbf{v}, i\mathbf{w}) = \overline{h(\mathbf{v}, \mathbf{w})}, \end{aligned}$$

so HF4 is satisfied.

In a similar manner, one can prove that, given a skew-symmetric real bilinear form a on $\mathbf{V}$ satisfying (23.10), putting

$$h(\mathbf{v}, \mathbf{w}) = a(i\mathbf{v}, \mathbf{w}) + ia(\mathbf{v}, \mathbf{w})$$

defines a Hermitian from on $\mathbf{V}$.

Summarizing, we can say that giving a Hermitian form on a complex vector space $\mathbf{V}$ is equivalent to giving a real symmetric bilinear form on $\mathbf{V}$ which satisfies condition (23.9), or a real skew-symmetric bilinear form on $\mathbf{V}$ satisfying condition (23.10).

EXERCISES

23.1 Which of the following are Hermitian forms on $\mathbf{C}^2$?

a) $\langle \mathbf{x}, \mathbf{y} \rangle = x_1\overline{y}_1 + ix_1\overline{y}_2 + ix_2\overline{y_1}$

b) $\langle \mathbf{x}, \mathbf{y} \rangle = i|x_1||y_1|$

c) $\langle \mathbf{x}, \mathbf{y} \rangle = x_1\overline{y}_1 + 2ix_1\overline{y}_2 - 2ix_2\overline{y_1}$

d) $\langle \mathbf{x}, \mathbf{y} \rangle = 1 + x_1\overline{y}_1 + x_1\overline{y}_2$

e) $\langle \mathbf{x}, \mathbf{y} \rangle = x_1\overline{y}_1 + 2x_2\overline{y}_2$.

23.2 Which of the following matrices are Hermitian?

a) $\begin{pmatrix} 1 & 1+i \\ 1-i & -1 \end{pmatrix}$ b) $\begin{pmatrix} 0 & i \\ i & 0 \end{pmatrix}$

c) $\begin{pmatrix} 1 & 1 \\ 1 & 0 \end{pmatrix}$ d) $\begin{pmatrix} i & -i \\ i & 1 \end{pmatrix}$

e) $\begin{pmatrix} 0 & 1 & i \\ 1 & 2 & 1+i \\ -i & 1-i & 0 \end{pmatrix}$ f) $\begin{pmatrix} i & 1 & i \\ -1 & 2 & 2i \\ -i & -2i & 0 \end{pmatrix}$.

23.3 Using the Gram-Schmidt process, orthonormalize the following basis of $\mathbf{C}^3$ with respect to the standard Hermitian product:

$$\boldsymbol{b} = \{(i, -i, 0), (0, i, 0), (0, i, i)\}.$$

23.4 For each of the following Hermitian matrices A find a unitary matrix M for which M^*AM is diagonal:

a) $\begin{pmatrix} 1 & i \\ -i & 1 \end{pmatrix}$ b) $\begin{pmatrix} \frac{\sqrt{3}}{2} & -\frac{i}{2} \\ \frac{i}{2} & -\frac{\sqrt{3}}{2} \end{pmatrix}$.

Appendices

A

Domains, fields and polynomials

There are several agebraic results used in the main part of the text, and they are collected together in this appendix. Not all results are proved here, but proofs can be found in many algebra texts.

Definitions

The algebraic structures used most frequently in geometry are those of "field" and "domain".

A *field* is a triple $(\mathbf{K}, +, \cdot)$ consisting of a non-empty set $\mathbf{K}$ and of two binary operations on $\mathbf{K}$

$$+ : \mathbf{K} \times \mathbf{K} \to \mathbf{K}, \qquad \cdot : \mathbf{K} \times \mathbf{K} \to \mathbf{K}$$

called *sum* and *product.*

These associate to any pair $(a, b) \in \mathbf{K} \times \mathbf{K}$ an element $a + b \in \mathbf{K}$, called the "sum of a and b", and an element ab (or $a \cdot b$), called the "product of a and b", in such a way that the following axioms are satisfied:

[F1] (*Commutativity of summation*) $a+b = b+a$ for every $a, b \in \mathbf{K}$.

[F2] (*Associativity of summation*) $a+(b+c) = (a+b)+c$, for every $a, b, c \in \mathbf{K}$.

[F3] (*Existence of zero*) There is an element $0 \in \mathbf{K}$ for which $a+0 = a$ for every $a \in \mathbf{K}$.

[F4] (*Existence of opposites*) For every $a \in \mathbf{K}$ there is an element $a' \in \mathbf{K}$ for which $a + a' = 0$.

[F5] (*Commutativity of the product*) $ab = ba$, for every $a, b \in \mathbf{K}$.

[F6] (*Associativity of the product*) $a(bc) = (ab)c$, for every $a, b, c \in \mathbf{K}$.

[F7] (*Existence of a unity element*) There is an element $1 \in \mathbf{K}$ for which $1a = a$ for every $a \in \mathbf{K}$.

[F8] (*Existence of inverses*) For every $a \in \mathbf{K}$, $a \neq 0$ there is an element $a^* \in \mathbf{K}$ for which $aa^* = 1$.

[F9] (*Distributivity*) $a(b+c) = ab + ac$, for every $a, b, c \in \mathbf{K}$

[F10] (*Non-existence of divisors of zero*) If $ab = 0$ then either $a = 0$ or $b = 0$.

A set with two operations $(\mathbf{K}, +, \cdot)$ satisfying F1–F7, F9 and F10, but not necessarily F8, is called a *domain* (or *integral domain*).

When the operations $+$ and $\cdot$ on $\mathbf{K}$ are clear, the field, or domain, is denoted simply $\mathbf{K}$. The subset $\mathbf{K} \setminus \{0\}$ is denoted $\mathbf{K}^*$.

With the usual operations of sum and product, the set $\mathbf{Z}$ of integers is a domain. The sets $\mathbf{Q}$, $\mathbf{R}$ and $\mathbf{C}$ of rational numbers, real numbers and complex numbers, respectively, with their usual operations are fields. For any prime $p \geq 2$, the set $\mathbf{Z}/p\mathbf{Z}$ of residue classes of integers modulo p is a field.

Let $\mathbf{K}$ be a domain. The following properties follow easily from the axioms, and their proofs are omitted.

i) There is only one zero in $\mathbf{K}$: if $0' \in \mathbf{K}$ satisfies $0' + a = a$ for all $a \in \mathbf{K}$ then $0' = 0$.
ii) There is only one unity element in $\mathbf{K}$: if $1' \in \mathbf{K}$ satisfies $1'a = a$ for all $a \in \mathbf{K}$ then $1' = 1$.
iii) For any $a \in \mathbf{K}$, the element $a' \in \mathbf{K}$ whose existence is asserted in F4 is unique; it is called the *opposite* of a (or the *additive inverse of* a), and is usually denoted $(-a)$; one writes $b - a$ rather than $b + (-a)$.
iv) An element $a \in \mathbf{K}$ is said to be *invertible* (or a *unit*) if there is an element $a^* \in \mathbf{K}$ such that $aa^* = 1$. Such an element, if one exists, is uniquely determined by a; it is called the *inverse* of a and denoted a^{-1}. One often writes b/a rather than ba^{-1}.

Let $n \in \mathbf{N}$ and $a \in \mathbf{K}$. We write na for the sum

$$\overbrace{a + \cdots + a}^{n \text{ summands}},$$

and $0a = 0$ (in this last equation, the first 0 is an element of $\mathbf{N}$ while the second is an element of $\mathbf{K}$). If $n \in \mathbf{Z}$, $n < 0$ and $a \in \mathbf{K}$, then we define $na = |n|(-a)$.

If for every integer $n > 0$ we have $n1 \neq 0$ then we say the domain (or field) **K** has *characteristic* 0. If on the other hand, there is a $p > 0$ for which $p1 = 0$, the **K** has *positive characteristic*; the smallest positive integer p with this property is the *characteristic* of **K**.

Z, **Q**, **R**, and **C** all have characteristic 0. An example of a field of positive characteristic is provided by $\mathbf{Z}/p\mathbf{Z}$, for $p \geq 2$ prime.

A *subdomain* of a domain **D** is a subset **E** of **D** on which the operations of **D** are defined, and which is itself a domain with these operations. A *subfield* of a field **K** is a subdomain **F** of **K** which is also a field. If **F** is a subfield of **K**, one also says that **K** is an *extension* of **F**. For example, **Q** and **R** are subfields of **C**, and **Q** is a subfield of **R**. Other examples of subfields of **C** are:

$$\begin{aligned} \mathbf{Q}[i] &= \{a + ib \mid a, b \in \mathbf{Q}\} \\ \mathbf{Q}[\sqrt{n}] &= \{a + \sqrt{n}\, b \mid a, b \in \mathbf{Q}\}, \end{aligned}$$

where n is a positive integer not divisibe by a perfect square.

Z is a subdomain of **Q**, **R** and **C**.

A subset **F** of a field **K** is a subfield if and only if the following axioms are satisfied.

[SF1] $0, 1 \in \mathbf{F}$.
[SF2] If $a, b \in \mathbf{F}$ then $a - b \in \mathbf{F}$.
[SF3] If $a, b \in \mathbf{F}$ and $b \neq 0$ then $ab^{-1} \in \mathbf{F}$.

It is obvious that any subfield must satisfy SF1–SF3. Conversely, if **F** satisfies these three conditions, then from SF1 it follows that $F \neq \emptyset$; from SF2 with $a = 0$ it follows that if $b \in \mathbf{F}$ then its opposite $-b \in \mathbf{F}$, and from SF3 with $a = 1$ it follows that $b^{-1} \in \mathbf{F}$ if $b \neq 0$. The two operations of **K** therefore induce operations on **F** for which F3, F4, F7 and F8 are satisfied. The other axioms are satisfied because they are satisfied in **K**. Thus **F** is a subfield of **K**.

A subdomain of a domain **D** has the same characteristic as **D**. In particular, every subdomain of **C** has characteristic 0. From SF1, SF2 and SF3 it follows immediately that every subfield of **C** contains **Q**. Every domain containing **Q** has characteristic 0.

In a domain, the usual notation and conventions for exponents are used: for $a \in \mathbf{K}$ and $n \in \mathbf{N}$ one writes a^n for the product

$$\overbrace{aa \cdots a}^{n \text{ factors}},$$

and if $a \neq 0$ is invertible, then a^{-n} denotes $(a^{-1})^n$.

An important example of domain is provided by the set of polynomials in one indeterminate with coefficients in a domain or field.

Let $\mathbf{D}$ be a domain and X an indeterminate. For every finite sequence $a_0, \ldots, a_n$ of elements of $\mathbf{D}$, the expression

$$f(X) = a_0 + a_1 X + a_2 X^2 + \cdots + a_n X^n$$

defines a *polynomial in X with coefficients in* $\mathbf{D}$, a_0, $a_1, \ldots, a_n$ are the *coefficients* and a_0, $a_1 X, \ldots, a_n X^n$ the *terms* or *monomials*. Another expression $g(X) = b_0 + b_1 X + \cdots + b_m X^m$, b_0, $b_1, \ldots, b_n \in \mathbf{D}$, defines the same polynomial if and only if all the terms in $f(X)$ and $g(X)$ with non-zero coefficients are the same. The polynomial, all of whose coefficients are zero, is called the *zero polynomial*, and is denoted by 0. If $f(X) \neq 0$, then the *degree* of $f(X)$ is the largest integer d for which $a_d \neq 0$; it is denoted $\deg(f(X))$. A polynomial $f(X)$ of degree d is said to be *monic* if $a_d = 1$; $f(X)$ is said to be *constant* if either $f(X) = 0$ or $\deg(f(X)) = 0$. The set of all polynomials in X with coeffients in $\mathbf{D}$ is denoted $\mathbf{D}[X]$. The sum of two polyomials

$$f(X) = a_0 + a_1 X + \cdots + a_n X^n, \quad g(X) = b_0 + b_1 X + \cdots + b_m X^m,$$

is defined to be the polynomial

$$(f + g)(X) = (a_0 + b_0) + (a_1 + b_1)X + \cdots,$$

and their product is

$$(fg)(X) = a_0 b_0 + (a_0 b_1 + a_1 b_0)X + (a_0 b_2 + a_1 b_1 + a_2 b_0)X^2 + \cdots.$$

It follows immediately from the definition that if $f, g, f + g \neq 0$ then

$$\begin{aligned} \deg(f + g) &\leq \max\{\deg(f), \deg(g)\} \\ \deg(fg) &= \deg(f) + \deg(g). \end{aligned}$$

With the operations of sum and product we have just defined, $\mathbf{D}[X]$ is a domain. The proof is left to the reader.

Let $\mathbf{D}$ be a domain, and $X_1, \ldots, X_N$ be indeterminates. A *polynomial in $X_1, \ldots, X_N$ with coefficients in* $\mathbf{D}$ is defined inductively as a polynomial in X_N whose coefficients are polynomials in $X_1, \ldots, X_{N-1}$

with coeffiecients in $\mathbf{D}$. The set of all such polynomials is denoted $\mathbf{D}[X_1, \ldots, X_{N-1}]$. So in symbols, the defiition is

$$\mathbf{D}[X_1, \ldots, X_N] = \mathbf{D}[X_1, \ldots, X_{N-1}][X_N].$$

Every $f(X_1, \ldots, X_N) \in \mathbf{D}[X_1, \ldots, X_N]$ can be expressed in a unique way as a finite sum of *monomials*, that is of terms of the form

$$a_{i_1 i_2 \ldots i_N} X_1^{i_1} X_2^{i_2} \ldots X_N^{i_N},$$

where $a_{i_1 i_2 \ldots i_N} \in \mathbf{D}^*$ is the *coefficient* of the monomial, i_j is the *degree with respect to* X_j of the monomial, $i_1 + \cdots + i_N$ its *degree*, or *total degree*. If $a_{i_1 i_2 \ldots i_N} = 1$ then the monomial is said to be *monic*. The *degree with respect to* X_j of a non-zero polynomial $f(X_1, \ldots, X_N)$ is the largest of the degrees with respect to X_j of its monomials; the *degree*, or *total degree*, of $f(X_1, \ldots, X_N)$ is the largest of the degrees of its monomials, and is denoted $\deg(f(X_1, \ldots, X_N))$. The zero polynomial has undefined degree.

Let $\mathbf{D}$ and $\mathbf{D}'$ be two domains, with unities $1_{\mathbf{D}} \in \mathbf{D}$ and $1_{\mathbf{D}'} \in \mathbf{D}'$ respectively. A map $f : \mathbf{D} \to \mathbf{D}'$ is said to be a *homomorphism* if it satisfies the following conditions,

$$\begin{aligned} f(a+b) &= f(a) + f(b) \quad \text{for every } a, b \in \mathbf{D} \\ f(ab) &= f(a)f(b) \quad \text{for every } a, b \in \mathbf{D} \\ f(1_{\mathbf{D}}) &= 1_{\mathbf{D}'}. \end{aligned}$$

(*Note*: some authors do not insist that the last condition be satisfied.)

The identity map $\mathbf{1_D} : \mathbf{D} \to \mathbf{D}$ and the composition $g \circ f : \mathbf{D} \to \mathbf{D}''$ of two homomorphisms $f : \mathbf{D} \to \mathbf{D}'$ and $g : \mathbf{D}' \to \mathbf{D}''$ are homomorphisms.

An *isomorphism* is a homomorphism $f : \mathbf{D} \to \mathbf{D}'$ for which there is an inverse homomorphism. Two domains $\mathbf{D}$ an $\mathbf{D}'$ are said to be *isomorphic* if there exists an isomorphism $f : \mathbf{D} \to \mathbf{D}'$.

The identity, the inverse of an isomorphism and the composite of two isomorphisms are all isomorphisms; isomorphism is therefore an equivalence relation between domains.

Note that a homomorphism $f : \mathbf{D} \to \mathbf{D}'$ in which $\mathbf{D}$ is a field, is injective. Indeed, if there existed $x \in \mathbf{D}^*$ for which $f(x) = 0$, then $1_{\mathbf{D}'} = f(1_{\mathbf{D}}) = f(xx^{-1}) = f(x)f(x^{-1}) = 0f(x^{-1})$ which is absurd.

B

Permutations

In this appendix we discuss some of the properties of permutations of finite sets, particularly those used for the definition and study of the determinant.

Let F be a finite set. A *permutation* of F is a bijective map $p : F \to F$. Suppose that F consists of n elements. By numbering these elements, we can identify F with the set $\{1, 2, \dots, n\}$ of the first n natural numbers. We can therefore just consider permutations of $\{1, 2, \dots, n\}$. The set of all permutations of $\{1, 2, \dots, n\}$ is a group under composition, consisting of $n! = 1.2. \dots n$ elements (the proof is left to the reader); we denote this group of permutations by S_n. An element of S_n can be written in tabular form as

$$\begin{pmatrix} 1 & 2 & \cdots & n \\ p(1) & p(2) & \cdots & p(n) \end{pmatrix}$$

where under the element i is written its image $p(i)$. With this notation, the identity permutation $\mathbf{1}$ becomes

$$\begin{pmatrix} 1 & 2 & \cdots & n \\ 1 & 2 & \cdots & n \end{pmatrix}.$$

This notation does not in fact require that the top row is written in ascending order, which enables us to write the inverse $p^{-1} \in S_n$ of p as

$$\begin{pmatrix} p(1) & p(2) & \cdots & p(n) \\ 1 & 2 & \cdots & n \end{pmatrix}.$$

Let $a_1, \ldots, a_r$ be distinct elements of $\{1, 2, \ldots, n\}$. The permutation k defined by

$$k(a_1) = a_2, \quad k(a_2) = a_3, \quad \ldots \quad k(a_r) = a_1,$$

$$k(b) = b \qquad \text{for every } b \notin \{a_1, \ldots, a_r\}$$

is called a *cycle of length* r, and is denoted $(\, a_1 \quad a_2 \quad \cdots \quad a_r \,)$. In particular,

$$(\, 1 \quad 2 \quad \cdots \quad n \,) = \begin{pmatrix} 1 & 2 & \cdots & n-1 & n \\ 2 & 3 & \cdots & n & 1 \end{pmatrix}.$$

For example, $(\, 1 \quad 3 \quad 2 \quad 6 \,) \in S_6$ is the permutation

$$\begin{pmatrix} 1 & 2 & 3 & 4 & 5 & 6 \\ 3 & 6 & 2 & 4 & 5 & 1 \end{pmatrix}.$$

Every cycle of length 1 is the identity permutation. A cycle of length 2 is called a *transposition.* A transposition interchanges two elements, leaving the others fixed. In particular, note that a transposition is its own inverse.

Two cycles $(\, a_1 \quad a_2 \quad \cdots \quad a_r \,)$ and $(\, b_1 \quad b_2 \quad \cdots \quad b_s \,)$ are said to be *disjoint* if

$$\{a_1, \ldots, a_r\} \cap \{b_1, \ldots, b_s\} = \emptyset.$$

Proposition B.1

1) Every permutation $p \in S_n$ is a product of pairwise disjoint cycles.
2) Every permutation $p \in S_n$ is a product of transpositions.

Proof

Let $a_1 \in \{1, \ldots, n\}$ be any element, and let $a_2 = p(a_1)$, $a_3 = p(a_2)$, $a_4 = p(a_3), \ldots$. In the sequence

$$a_1, \; a_2, \; a_3, \; a_4, \; \ldots \tag{B.1}$$

the first element to repeat iteself is a_1, for if it were $a_r = a_k$ with $2 \leq k < r$, we would have $a_{r-1} = a_{k-1}$, and the repetition would not have been the first. The sequence (B.1) is therefore of the form

$$a_1, \; a_2, \; \ldots, \; a_r, \; a_1, \; a_2, \; \ldots$$

and so p gives a cyclic permutation of $a_1, \; a_2, \; \ldots, \; a_r$.

Consider the cycle $(a_1 \ a_2 \ \cdots \ a_r)$. If $r = n$ then p is the cycle $(a_1 \ a_2 \ \cdots \ a_r)$ and the assertion is true. Otherwise, there exists $b_1 \in \{1, \ldots, n\} \setminus \{a_1, \ldots, a_r\}$. Reasoning as before, we obtain a cycle $(b_1 \ b_2 \ \cdots \ b_r)$ disjoint from the preceding one. Proceeding in this way one gets a finite number of disjoint cycles $K_1, K_2, \ldots, K_i$ such that

$$p = K_1 \circ K_2 \circ \cdots \circ K_i. \tag{B.2}$$

2) By part (1), it is sufficient to prove the assertion for a cycle $p = (a_1 \ a_2 \ \cdots \ a_r)$. To this end, it suffices to note that

$$(a_1 \ a_2 \ \cdots \ a_r) = (a_1 \ a_r) \circ (a_1 \ a_{r-1}) \circ \cdots \circ (a_1 \ a_3) \circ (a_1 \ a_2).$$

□

Since the cycles $K_1, \ldots, K_i$ are pairwise disjoint, they commute, so that the expression (B.2) is independent of the order in which the cycles are written. Any cycles of length 1 that appear in (B.1) can be omitted, since they correspond to the identity permutation. Therefore, any permutation can be expressed in an irreducible manner as the product of cycles of length at least 2.

Note that the expression of a permutation as a product of transpositions is not unique: for example,

$$(1 \ 2 \ 3) = (1 \ 3) \circ (1 \ 2) = (2 \ 3) \circ (1 \ 3).$$

Theorem B.2
Let $p \in S_n$. Suppose that

$$p = T_1 \circ T_2 \circ \cdots \circ T_h = R_1 \circ R_2 \circ \cdots \circ R_k,$$

where $T_1, \ldots, T_h$, $R_1, \ldots, R_k$ are transpositions. Then $h \cong k \bmod 2$.

Proof
Since $\mathbf{1} = p \circ p^{-1} = R_1 \circ R_2 \circ \cdots \circ R_k \circ T_h \circ \cdots \circ T_2 \circ T_1$, it is enough to show that the identity permutation cannot be expressed as the product of an odd number of transpositions.

Let

$$\mathbf{1} = P_1 \circ P_2 \circ \cdots \circ P_m, \tag{B.3}$$

with $P_1, \ldots, P_m$ transpositions. Suppose first that $P_j = (1 \ a_j)$ for each $j = 1, \ldots, m$. Then, since $\mathbf{1}(a_j) = a_j$, the transposition

$(1\ \ a_j) = (a_j\ \ 1)$ must occur an even number of times in (B.3), and so the number m of factors is even. If, for some j, we have $P_j = (a_j\ \ b_j)$ with $a_j \neq 1 \neq b_j$, we can substitute P_j with

$$P_j = (1\ \ b_j) \circ (1\ \ a_j) \circ (1\ \ b_j).$$

Note that this substitution does not change the parity of the number of factors in [B.3]. By effecting such substitutions, we can reduce to the first case, for which the assertion has already been proved. □

Definition B.3
Let $p \in S_n$. If $p = T_1 \circ T_2 \circ \cdots \circ T_h$, then the *sign* of the permutation is $\varepsilon(p) = (-1)^h$.

It follows from Theorem B.2 that the sign of a permutation p is well-defined, since the parity of h depends only on p. The sign $\varepsilon(p)$ enjoys the following properties, which follow immediately from the definition.

Proposition B.4
1) $\varepsilon(\mathbf{1}) = 1$.
2) $\varepsilon(p^{-1}) = \varepsilon(p)$.
3) $\varepsilon(p \circ q) = \varepsilon(p)\varepsilon(q)$.
4) $\varepsilon(T) = -1$ *for every transposition* $T \in S_n$.

Selected solutions

Chapter 2

2.1 a) $\begin{pmatrix} 24+9\sqrt{2} \\ -8+5\sqrt{2} \end{pmatrix}$ b) $\begin{pmatrix} 14 \\ -21 \end{pmatrix}$ c) $\begin{pmatrix} 0 & 0 & 0 & 0 \\ 5 & 0 & 1 & 0 \\ 0 & 0 & 0 & 0 \\ 10 & 0 & 2 & 0 \end{pmatrix}$.

2.5 $(A+A^t)^t = A^t + A = A + A^t$, $(A-A^t)^t = A^t - A = -(A-A^t)$. One has $A = \frac{1}{2}(A+A^t)+\frac{1}{2}(A-A^t)$: the first matrix is symmetric and the second is skew-symmetric.

2.6 a) $\begin{pmatrix} 1 & \frac{1}{2} \\ \frac{1}{2} & 0 \end{pmatrix} + \begin{pmatrix} 0 & \frac{3}{2} \\ -\frac{3}{2} & 0 \end{pmatrix}$ b) $\begin{pmatrix} 3 & 1 \\ 1 & 0 \end{pmatrix} + \begin{pmatrix} 0 & 0 \\ 0 & 0 \end{pmatrix}$

c) $\begin{pmatrix} 1 & \frac{1}{2} & 1 \\ \frac{1}{2} & -1 & 0 \\ 1 & 0 & 0 \end{pmatrix} + \begin{pmatrix} 0 & -\frac{1}{2} & -1 \\ \frac{1}{2} & 0 & -1 \\ 1 & 1 & 0 \end{pmatrix}$.

2.10 c), d) and h) are not orthogonal; the others are.

Chapter 3

3.1 a) The general solution is $(7t, 4t, t)$.
b) $(-2t_1 - 5t_2 - 4t_3, \frac{1}{2}(t_1 - t_2 + t_3), t_1, t_2, t_3)$
c) Incompatible d) Incompatible
e) $(5 - 2t_1 - 2t_2, t_1, 3 - 2t_2, t_2)$.

3.4 c) $\begin{pmatrix} -\frac{1}{12} & -\frac{1}{6} \\ -\frac{1}{4} & 0 \end{pmatrix}$ e) $\frac{1}{9}\begin{pmatrix} 2 & 2-i \\ 2+i & -2 \end{pmatrix}$.

3.5 a) $(0,1,0)$ b) $(1,-i)$ c) $(\sqrt{2},1,0)$.

3.6 a) $\begin{pmatrix} 0 & 2 \\ \frac{1}{2} & 0 \end{pmatrix} = R_{21}R_1(\frac{1}{2})R_2(2) = \begin{pmatrix} 0 & 1 \\ 1 & 0 \end{pmatrix}\begin{pmatrix} \frac{1}{2} & 0 \\ 0 & 1 \end{pmatrix}\begin{pmatrix} 1 & 0 \\ 0 & 2 \end{pmatrix}$

b) $\begin{pmatrix} 1 & -1 \\ 2 & 0 \end{pmatrix} = R_2(2)R_{12}(1)R_1(-1)R_{12}$

$= \begin{pmatrix} 1 & 0 \\ 0 & 2 \end{pmatrix}\begin{pmatrix} 1 & 1 \\ 0 & 1 \end{pmatrix}\begin{pmatrix} -1 & 0 \\ 0 & 1 \end{pmatrix}\begin{pmatrix} 0 & 1 \\ 1 & 0 \end{pmatrix}$

c) $\begin{pmatrix} 3 & 5 \\ 1 & 2 \end{pmatrix} = R_{21}(\frac{1}{3})R_2(\frac{1}{3})R_{12}(5)R_1(3)$

d) $\begin{pmatrix} 2 & 1 & 0 \\ 1 & 1 & 0 \\ 0 & 0 & 2 \end{pmatrix} = R_1(2)R_{21}(1)R_{12}(1)R_2(\frac{1}{2})R_3(2)$

e) $\begin{pmatrix} 1 & 3 & 0 \\ 2 & 1 & 1 \\ 2 & -1 & 0 \end{pmatrix} = R_{32}R_{21}(2)R_2(-7)R_{12}(3)R_{31}(2)R_{32}(1)$.

Chapter 4

4.1 a) basis b) linearly dependent and not generating
c) basis d) linearly dependent and generating
e) the homogeneous system

$$\begin{pmatrix} 1 \\ 1 \\ 3 \end{pmatrix} a_1 + \begin{pmatrix} 2 \\ 2 \\ 0 \end{pmatrix} a_2 + \begin{pmatrix} 3 \\ 3 \\ -3 \end{pmatrix} a_3 = \begin{pmatrix} 0 \\ 0 \\ 0 \end{pmatrix}$$

has non-trivial solutions if and only if the vectors are linearly dependent. Solving this system, one finds that it has a 1-parameter family of solutions, and hence the vectors are linearly dependent. The subspace they generate has dimension 2, and is not $\mathbf{R}^3$.
f) independent g) dependent and generating
h) basis i) dependent and not generating.

4.3 a), i).

4.4 $\mathbf{U}$ contains $\mathbf{i} = \frac{1}{2}[(\mathbf{i}+\mathbf{j}) + (\mathbf{i}-\mathbf{j})]$ and $\mathbf{j} = \frac{1}{2}[(\mathbf{i}+\mathbf{j}) - (\mathbf{i}-\mathbf{j})]$; moreover $\mathbf{U}+\mathbf{W}$ contains $\mathbf{k} = (\mathbf{j}+\mathbf{k})-\mathbf{j}$. Thus $\mathbf{U}+\mathbf{W} = \mathbf{V}$ since it contains $\mathbf{i}, \mathbf{j}, \mathbf{k}$. The sum is not direct, because $\dim(\mathbf{U}) = \dim(\mathbf{W}) = 2$, and so by Grassman's formula, $\dim(\mathbf{U} \cap \mathbf{W}) = 1$.

4.6 The vectors $(1,1,0)$ and $(0,0,1)$ belong to $\mathbf{U}$ and are linearly independent. Since $\dim(\mathbf{U}) \leq 2$ it follows that $\{(1,1,0), (0,0,1)\}$

is a basis for $\mathbf{U}$. Now, $\mathbf{W} \cap \mathbf{U} = \langle \mathbf{0} \rangle$ since any non-zero vector belonging to $\mathbf{W}$ is of the form $(t, 0, t)$ with $t \neq 0$, which does not belong to $\mathbf{U}$ as its first two coordinates are different. Thus $\mathbf{U} + \mathbf{W} = \mathbf{U} \oplus \mathbf{W}$. Since $\mathbf{U}$ is a proper subspace of $\mathbf{U} \oplus \mathbf{W}$, it follows that $\dim(\mathbf{U} \oplus \mathbf{W}) = 3$, that is $\mathbf{U} \oplus \mathbf{W} = \mathbf{V}$.

4.11 $\mathrm{GL}_n(\mathbf{K})$ does not contain the zero matrix.

4.12 If we identify $M_n(\mathbf{K})$ with $\mathbf{K}^{n^2}$, then $\mathcal{T}_0$ is identified with the set of solutions of the linear homogeneous equation $a_{11} + \cdots + a_{nn} = 0$. $\mathcal{T}_0$ is thus a vector subspace. Since the above equation has an $(n^2 - 1)$-parameter family of solutions, it follows that $\dim(\mathcal{T}_0) = n^2 - 1$.

4.14 Suppose that there are n sequences $\mathbf{a}_1 = \{a_{10}, a_{11}, a_{12}, \ldots\}$, $\mathbf{a}_2 = \{a_{20}, a_{21}, a_{22}, \ldots\}$, $\ldots$, $\mathbf{a}_n = \{a_{n0}, a_{n1}, a_{n2}, \ldots\}$ which generate $S_{\mathbf{K}}$. Then for every $(b_0, b_1, \ldots, b_n) \in \mathbf{K}^{n+1}$ there exist $x_1, \ldots, x_n \in \mathbf{K}$ such that

$$\mathbf{b} = x_1\mathbf{a}_1 + x_2\mathbf{a}_2 + \cdots + x_n\mathbf{a}_n,$$

where

$$\mathbf{b} = \{b_0, b_1, \ldots, b_n, 0, 0, \ldots\}.$$

It follows that

$$\begin{aligned}(b_0, \ldots, b_n) &= x_1(a_{10}, \ldots, a_{1n}) + x_2(a_{20}, \ldots, a_{2n}) + \cdots \\ &\quad + x_n(a_{n0}, \ldots, a_{nn}).\end{aligned}$$

In other words, the n vectors

$$(a_{10}, \ldots, a_{1n}),\ (a_{20}, \ldots, a_{2n}),\ (a_{n0}, \ldots, a_{nn})$$

generate $\mathbf{K}^{n+1}$, which is absurd as $\mathbf{K}^{n+1}$ has dimension $n + 1$.

4.17 Every polynomial $f(X) \in \mathbf{R}[X]$ can be considered as an element of $C_{(a,b)}$, and so $\mathbf{R}[X]$ defines a subset of $C_{(a,b)}$. Since the operations of addition and scalar multiplication in $\mathbf{R}[X]$ correspond with those in $C_{(a,b)}$, $\mathbf{R}[X]$ is a subspace of $C_{(a,b)}$. Since $\mathbf{R}[X]$ is not finite dimensional, neither is $C_{(a,b)}$.

4.18 The polynomials 1, X, X^2, $\ldots$, X^d form a basis of $\mathbf{K}[X]_{\leq d}$.

Chapter 5

5.1 a) 3 b) 3 c) 2.

Chapter 6

6.2 a) The determinant of the matrix of coefficients is $m+2$, which vanishes for $m=-2$. When $m \neq -2$, the system is compatible, by Theorem 5.7, and has the unique solution $(1, 1-m)$, which can be found using either Cramer's rule or by elimination. When $m=-2$, the system becomes

$$\begin{aligned} 2X - Y &= -1 \\ -2X + Y &= 1, \end{aligned}$$

which is still compatible. It has the 1-parameter family of solutions $(t,\ 2t+1)$, $t \in \mathbf{R}$.
b) Compatible only if $m = \frac{1}{3}$, in which case the solution is $(\frac{1}{2}, 0)$.

6.3 b) If $m \neq 0, 2$ the system has the unique solution

$$\left(\frac{1-m}{2-m}, \frac{2}{2-m}, \frac{1-m}{2-m}\right).$$

If $m=0$ it has the 1-parameter family of solutions $(1+t,\ 1,\ -t)$, $t \in \mathbf{R}$; if $m=2$ it is incompatible.
c) If $m=-1$ the system is incompatible; if $m=2$ it has a 2-parameter family of solutions, given by $(\frac{1}{2}-s-t,\ s,\ t)$, $s, t \in \mathbf{R}$. If $m \neq -1, 2$ the system has the unique solution

$$\left(\frac{1}{2(m+1)}, \frac{1}{2(m+1)}, \frac{1}{2(m+1)}\right).$$

e) Incompatible for $m \neq 2$. If $m=2$ it has the 1-parameter family of solutions $(1,\ -2t,\ t)$, $t \in \mathbf{R}$.
f) Incompatible if $m=-1$. If $m \neq -1, 1$, it has the unique solution $\left(\frac{3m+2}{m+1}, \frac{m}{m+1}, \frac{m}{m+1}\right)$. If $m=1$, it has the 1-parameter family of solutions $(2+t,\ t,\ 1-t)$, $t \in \mathbf{R}$.
g) If $m=0$ the system has the 1-parameter family of solutions $(t, 0, 0)$, $t \in \mathbf{R}$; if $m=1$ it has the 1-parameter family of solutions $(-3t,\ t,\ t)$, $t \in \mathbf{R}$; if $m \neq 0, 1$ it has the unique solution $(0, 0, 0)$.

6.4 a) $(1, \sqrt{2}, \sqrt{2})$ b) $(1, i, -2i)$ c) $(2, 1, 1, -1)$.

6.6 The minors of maximal order of A taken with alternating signs are:

$$\alpha_i = (-1)^i \det[A(1\,2\,\cdots\,n-1 \mid 1\,2\,\cdots\hat{\imath}\,\cdots\,n)], \quad i = 1, \ldots, n.$$

The α_i are not all zero as $\mathrm{rk}(A) = n-1$.
For each $j = 1, \ldots, n-1$,

$$a_{j1}\alpha_1 + a_{j2}\alpha_2 + \cdots + a_{jn}\alpha_n = \det(B),$$

where

$$B = \begin{pmatrix} a_{j1} & a_{j2} & \cdots & a_{jn} \\ a_{11} & a_{12} & \cdots & a_{1n} \\ \vdots & \vdots & & \vdots \\ a_{n-1\,1} & a_{n-1\,2} & \cdots & a_{n-1\,n} \end{pmatrix}.$$

Since B has two equal rows, $\det(B) = 0$ and so $(\alpha_1, \alpha_2, \ldots, \alpha_n)$ satisfies the j-th equation of the system.

Chapter 8

8.1 a) $(\frac{1-y}{2}, y, z)$ b) $(\frac{1-y}{2}, y, z)$
c) $\frac{1}{5}(x - 2y + 2, -2x + 4y + 1, 2ix + iy + 5z - i)$
d) $(x, 1 - 2x, 4ix + 2iy + z - 2i)$.

Chapter 9

9.1 a), c).

9.2 a) $2X + 3Y = 6$ b) $\frac{X}{\sqrt{7}} + \frac{Y}{172} = 1$
c) The required line belongs both to the pencil determined by $\mathcal{S}$ and $\mathcal{S}'$, and to that determined by $\mathcal{T}$ and $\mathcal{T}'$. The general line of the first pencil has equation

$$(1 + 3t)X + 5Y + (6t - 8) = 0.$$

The condition that this line belong to the second pencil is

$$\begin{vmatrix} 1+3t & 5 & 6t-8 \\ 10 & -1 & -2 \\ 1 & -1 & -5 \end{vmatrix} = 0,$$

that is, $t = 7$. Thus the line we seek is $22X + 5Y + 34 = 0$.

9.3 a) $x = \frac{21}{13} + 2t, \quad y = -\frac{14}{13} + 4t$
b) $x = \frac{1}{2} - 5\sqrt{2}t, \quad y = \frac{1}{2} + 7t.$

9.4 $(\frac{13}{2}, \frac{3-\sqrt{5}}{2})$.

9.5 $(\frac{5}{4}, \frac{9}{8})$, $(\frac{3}{2}, \frac{13}{4})$, $(\frac{7}{4}, \frac{43}{8})$.

Chapter 10

10.1 a).

10.2 a) none b) $m = \frac{3}{4}$ c) none d) none.

10.3 a) $x = 2t + u, \quad y = \sqrt{2}t + u, \quad z = 1 + (\sqrt{2} - 1)u;$
$X - \sqrt{2}Y + Z = 1$
b) $x = 5 - 4t - 8u, \quad y = -1 + 2t + 2u, \quad z = \sqrt{5}t + \frac{\pi}{2}u;$
$(\pi - 2\sqrt{5})(X - 5) + (2\pi - 8\sqrt{5})(Y + 1) + 8Z = 0$
c) $x = 1 - 3t + u, \quad y = 1 + u, \quad z = 1 - t + u;$
$X + 2Y - 3Z = 0$
d) $x = t, \quad y = u, \quad z = 0; \quad Z = 0.$

10.4 a) $X + 2Y + 3Z - 9 = 0$ b) $2X - Y - i = 0$
c) $iY - 2Z + 3 + 2i = 0$ d) $Y - 1 = 0.$

10.5 a), b), c) No; d) Yes.

10.6 a) $x = 1 + 2t, \quad y = 1 - t, \quad z = \sqrt{2}t;$
$X + 2Y - 3 = 0, \quad X - \sqrt{2}Z - 1 = 0$
b) $x = -2 + t, \quad y = 2 + t, \quad z = -2t;$
$Z = -2, \quad X - Y + 4 = 0$
c) $x = 1 + t, \quad y = 2 + 2t, \quad z = 3 + 3t;$
$2X - Y = 0, \quad 3X - Z = 0$
d) $x = t, \quad y = 0, \quad z = 0;$
$Y = 0, \quad Z = 0$
e) $x = 1 + t, \quad y = 1 + t, \quad z = -t;$
$X - Y = 0, \quad X + Z - 1 = 0.$

10.7 a) $x = it, \quad y = t, \quad z = -1 - 2t;$
b) $x = t, \quad y = 4 + 3t, \quad z = 1 - 3t;$
c) $x = 1, \quad y = t, \quad z = 1;$
d) $x = \frac{1}{2}(1 - 7i) + (1 - i)t, \quad y = 2 + t, \quad z = -it.$

10.8 a) $x = 1 + it, \quad y = 1 + t, \quad z = 0;$
b) $x = 1, \quad y = 0, \quad z = t;$
c) $x = 2 + it, \quad y = 1, \quad z = -5 + t;$
d) $x = 3 + (\sqrt{2} + 5)t, \quad y = (3\sqrt{2} + 1)t, \quad z = 14t.$

10.9 a) $X + Z + 1 = 0,$ b) $X - Y + 2Z - 5 = 0,$
c) $X - 2Y + 3 = 0.$

10.10 a) skew b) skew c) parallel: $3X - 3Y + 3Z - 1 = 0$
d) skew e) incident: $2X - 3Z + 8 = 0.$

10.11 a) $\mathcal{R} \subset \mathcal{P}$ b) incident: $\mathcal{R} \cap \mathcal{P} = \{(-6, 17, 23)\}$
c) incident: $\mathcal{R} \cap \mathcal{P} = \{(-2, 1, 4)\}$ d) parallel.

10.12 a) $2X + Y - Z - 8 = 0$ b) $7X - Y + 10Z - 13 = 0$
c) $Y + 3Z - 6 = 0$ d) $3X + 2Y + Z - 12 = 0.$

10.13 a) $2X - Y + Z - 1 = 0, \quad X - Z - 1 = 0$
b) $X + Y + Z + 3 = 0, \quad X + 10Y - 7Z + 4 = 0$
c) $2X - Y = 0, \quad 2X + 2Y - Z - 3 = 0.$

10.14 a) $-13X + 25Y - 8Z + 4 = 0, \quad 3X + 7Y - 7Z + 4 = 0$
b) $3X + 7Y + 4Z + 2 = 0, \quad 13X + 2Y - 3Z - 32 = 0$
c) $4X + 6Y + Z + 3 = 0, \quad 2Y - 3Z - 1 = 0.$

10.16 $2X - 3Z - 2 = 0.$

Chapter 11

11.3 $p(x_1, x_2, x_3) = \left(x_1,\ \frac{1}{2}(-x_1 + x_2 - x_3),\ \frac{1}{2}(-x_1 - x_2 + x_3)\right).$

11.4 $p(x_1, x_2, x_3, x_4) = \left(-\frac{1}{2}x_3,\ 3x_4,\ x_3,\ x_4\right).$

11.5 a) $\{X_1,\ X_2,\ X_3\},$ b) $\left\{\frac{1}{2}X_1,\ \sqrt{2}X_2,\ -6X_3\right\}$
c) $\{2X_1 + X_2 - X_3,\ 2X_1 + 2X_2 - X_3,\ -X_1 - X_2 + X_3\}$
d) $\{X_1 - X_2,\ X_2 - X_3,\ X_3\}.$

Chapter 12

12.1 The columns of the required matrix are the coordinates with respect to the basis $\boldsymbol{b}'$ of the vectors $F(1,1) = (2,-1,1)$ and $F(0,-1) = (-1,2,0)$. Thus,

$$M_{\boldsymbol{b}',\boldsymbol{b}}(F) = \begin{pmatrix} x_1 & y_1 \\ x_2 & y_2 \\ x_3 & y_3 \end{pmatrix},$$

where $x_1, \ldots, y_3$ are the coefficients of the following linear combinations:

$$x_1 \begin{pmatrix} 1 \\ 1 \\ 1 \end{pmatrix} + x_2 \begin{pmatrix} 1 \\ -2 \\ 0 \end{pmatrix} + x_3 \begin{pmatrix} 0 \\ 0 \\ 1 \end{pmatrix} = \begin{pmatrix} 2 \\ -1 \\ 1 \end{pmatrix}$$

$$y_1 \begin{pmatrix} 1 \\ 1 \\ 1 \end{pmatrix} + y_2 \begin{pmatrix} 1 \\ -2 \\ 0 \end{pmatrix} + y_3 \begin{pmatrix} 0 \\ 0 \\ 1 \end{pmatrix} = \begin{pmatrix} -1 \\ 2 \\ 0 \end{pmatrix}.$$

Solving the two resulting systems of equations gives the solution

$$M_{\boldsymbol{b}',\boldsymbol{b}}(F) = \begin{pmatrix} 1 & 0 \\ 1 & -1 \\ 0 & 0 \end{pmatrix}.$$

12.2

$$M_{\boldsymbol{b}',\boldsymbol{b}}(F) = \begin{pmatrix} 1 & 0 \\ 1 & 1 \\ 0 & -1 \end{pmatrix}.$$

12.3 $\frac{i}{2}\begin{pmatrix} -3-2i & 4-2i & -4-i \\ -1-4i & 2i & -2-3i \end{pmatrix}.$

12.4 The vectors $\mathbf{v}_1$, $\mathbf{v}_2$, $\mathbf{v}_3$ are linearly dependent. If such an F were to exist, then their images $\mathbf{E}_1$, $\mathbf{E}_2$, $\mathbf{E}_3$ would also be linearly dependent. However, $\{\mathbf{E}_1, \mathbf{E}_2, \mathbf{E}_3\}$ is a basis of $\mathbf{R}^3$.

12.5 a) $M_{\boldsymbol{b},\boldsymbol{b}'} = \begin{pmatrix} 1 & \sqrt{3} \\ \sqrt{3} & 1 \end{pmatrix}.$

b) Letting $\boldsymbol{e}$ be the canonical basis, we have

$$M_{\boldsymbol{b},\boldsymbol{b}'} = M_{\boldsymbol{b},\boldsymbol{e}} M_{\boldsymbol{e},\boldsymbol{b}'} = \begin{pmatrix} 1 & 1 \\ -1 & 1 \end{pmatrix}^{-1} \begin{pmatrix} 1 & 1 \\ 0 & 1 \end{pmatrix} = \begin{pmatrix} \frac{1}{2} & 0 \\ \frac{1}{2} & 1 \end{pmatrix}.$$

c) $M_{\boldsymbol{b},\boldsymbol{b}'} = M_{\boldsymbol{b},\mathbf{e}} M_{\mathbf{e},\boldsymbol{b}'} = \begin{pmatrix} 2 & 2 \\ 1 & 2 \end{pmatrix}^{-1} \begin{pmatrix} \sqrt{5} & \sqrt{5} \\ -\sqrt{5} & \sqrt{5} \end{pmatrix}$
$= \begin{pmatrix} 2\sqrt{5} & 0 \\ -\frac{3\sqrt{5}}{2} & \frac{\sqrt{5}}{2} \end{pmatrix}.$

12.6 a) $M_{\boldsymbol{b},\boldsymbol{b}'} = \frac{1}{2}\begin{pmatrix} 2-i & 1-2i \\ 1-2i & 2-i \end{pmatrix}$ b) $M_{\boldsymbol{b},\boldsymbol{b}'} = \frac{1}{2}\begin{pmatrix} 1 & 1 \\ -i & i \end{pmatrix}.$

12.7 a) $M_{\boldsymbol{b},\boldsymbol{b}'} = \frac{1}{2}\begin{pmatrix} 1 & 0 & 1 \\ 1 & 0 & -1 \\ 1 & 2 & 1 \end{pmatrix}$

b) $M_{\boldsymbol{b},\boldsymbol{b}'} = -\frac{1}{2}\begin{pmatrix} 11 & -6 & 7 \\ 19 & 12 & -10 \\ -18 & 2 & 17 \end{pmatrix}.$

12.8 $x' = \frac{1}{2}(-x+y-1), \qquad y' = \frac{1}{2}(3x-y-1).$

12.9 $x' = \frac{1}{5}(-x+y+1), \qquad y' = -\frac{3}{5}(x+y).$

12.10 $x' = 2x - y - z - \frac{4\pi}{3}, \qquad y' = x - z, \qquad z' = -x + y + z + \pi.$

12.11 Let $M_{\mathbf{e},\boldsymbol{b}}(\mathbf{1_V}) = (n_{ij})$ and $M_{\boldsymbol{\beta},\boldsymbol{\eta}}(\mathbf{1_{V^*}}) = (m_{ij})$. Then

$$\mathbf{b}_j = \sum_{i=1}^{n} n_{ij}\mathbf{e}_j, \quad \boldsymbol{\eta}_l = \sum_{k=1}^{n} m_{kl}\beta_k,$$

for $j,l = 1,\ldots,n$. Evaluating the second equation on $\mathbf{b}_j$ gives $\eta_l(\mathbf{b}_j) = m_{jl}$, while evaluating η_l on the first gives $\eta_l(\mathbf{b}_j) = \eta_l(\sum_{i=1}^n n_{ij}\mathbf{e}_j) = n_{lj}$. Thus, for each $j,l = 1,\ldots,n$, $m_{jl} = n_{lj}$.

Chapter 13

13.1 Let $\boldsymbol{e}$ be the canonical basis. The required matrix is

$$M_{\boldsymbol{b}}(F) = M_{\boldsymbol{b},\mathbf{e}}(\mathbf{1}) M_{\mathbf{e}}(F) M_{\mathbf{e},\boldsymbol{b}}(\mathbf{1}).$$

We have

$$M_{\mathbf{e}}(F) = \begin{pmatrix} 1 & 1 & -1 \\ 0 & 1 & 1 \\ 2 & 0 & 0 \end{pmatrix}, \qquad M_{\mathbf{e},\boldsymbol{b}}(\mathbf{1}) = \begin{pmatrix} 1 & -1 & 1 \\ 1 & 0 & 1 \\ 0 & 1 & 1 \end{pmatrix}$$

$$M_{\boldsymbol{b},\mathbf{e}}(\mathbf{1}) = M_{\mathbf{e},\boldsymbol{b}}(\mathbf{1})^{-1} = \begin{pmatrix} -1 & 2 & -1 \\ -1 & 1 & 0 \\ 1 & -1 & 1 \end{pmatrix}.$$

Thus

$$M_{\boldsymbol{b}}(F) = \begin{pmatrix} -2 & 6 & 1 \\ -1 & 3 & 1 \\ 3 & -5 & 1 \end{pmatrix}.$$

13.2 Proceeding as in the previous exercise, one finds that

$$M_{\boldsymbol{b}}(F) = \begin{pmatrix} 1 & 1 & 1 \\ \frac{1}{2} & \frac{3}{2} & -\frac{1}{2} \\ -\frac{1}{2} & \frac{7}{2} & -\frac{5}{2} \end{pmatrix}.$$

Consequently,

$$M_{\boldsymbol{b}}(F^2) = M_{\boldsymbol{b}}(F)^2 = \begin{pmatrix} 1 & 6 & -2 \\ \frac{3}{2} & 1 & 1 \\ \frac{5}{2} & -4 & 4 \end{pmatrix}.$$

13.3 $M_{\boldsymbol{b}}(F) = \begin{pmatrix} 7 & -8i & 6i \\ 12i & 10 & -9 \\ 25i & 24 & -20 \end{pmatrix}.$

13.5 $M_{\boldsymbol{b}}(F) = \begin{pmatrix} 0 & 1 & -1 \\ -3 & 1 & -1 \\ 1 & 1 & -2 \end{pmatrix}.$

13.9 $\boldsymbol{b} = \{(1,1,-1),\ (1,1,0),\ (-1,0,1)\}.$

$$M_{\boldsymbol{b}}(F) = \begin{pmatrix} 2 & 0 & 0 \\ 0 & 1 & 0 \\ 0 & 0 & 1 \end{pmatrix}.$$

13.10 a) The characteristic polynomial of the given matrix A is $T^3 - 6T^2 + 12T - 8 = (T-2)^3$, and so A has the unique eigenvalue $\lambda = 2$ with algebraic multiplicity $h(2) = 3$. The matrix $A - 2\mathbf{I}_3$ has rank 2, and so $\dim(\mathbf{R}^3_2) = 3 - 2 = 1$. A is not diagonalizable since the sum of the dimensions of the eigenspaces of A is less than 3.
b) $\lambda = 2, 3$; $h(2) = 1, h(3) = 2$; $\dim(R^3_2) = \dim(\mathbf{R}^3_3) = 1$ so A is not diagonalizable.
c) $\lambda = 1, 2, 3$; $h(1) = h(2) = h(3) = 1$; $\dim(\mathbf{R}^3_1) = \dim(\mathbf{R}^3_2) = \dim(\mathbf{R}^3_3) = 1$; A is diagonalizable.
d) $\lambda = 1, -1$; $h(-1) = 2, h(1) = 1$; $\dim(\mathbf{R}^3_{-1}) = \dim(\mathbf{R}^3_1) = 1$; A is not diagonalizable.

e) $\lambda = 7, 0$; $h(7) = 1, h(0) = 2$; $\dim(\mathbf{R}^3_7) = \dim(\mathbf{R}^3_0) = 1$; A is not diagonalizable.
f) $\lambda = -4, 4$; $h(-4) = 2, h(4) = 1$; $\dim(\mathbf{R}^3_{-4}) = \dim(\mathbf{R}^3_4) = 1$; A is not diagonalizable.

13.11 ± 1, $\pm i$.

13.13 $|C - (a\lambda + b)\mathbf{I}_n| = |(aA + b\mathbf{I}_n) - (a\lambda + b)\mathbf{I}_n| = |a(A - \lambda\mathbf{I}_n)| = a^n|A - \lambda\mathbf{I}_n| = 0$.

13.15 The matrix representing F in the canonical basis is

$$A = \begin{pmatrix} 0 & 1 & -1 \\ -1 & 2 & -1 \\ 1 & -1 & 2 \end{pmatrix}.$$

For $B = M_{\boldsymbol{b}}(F)$, we have $A = MBM^{-1}$, where

$$M = M_{\boldsymbol{e},\boldsymbol{b}}(\mathbf{1}) = \begin{pmatrix} 1 & 1 & -1 \\ 1 & 1 & 0 \\ -1 & 0 & 1 \end{pmatrix},$$

and so $A^5 = MB^5M^{-1}$. Since

$$M^{-1} = \begin{pmatrix} -1 & 1 & -1 \\ 1 & 0 & 1 \\ -1 & 1 & 0 \end{pmatrix}, \quad \text{and} \quad B^5 = \begin{pmatrix} 32 & 0 & 0 \\ 0 & 1 & 0 \\ 0 & 0 & 1 \end{pmatrix},$$

it follows that

$$A^5 = \begin{pmatrix} -30 & 31 & -31 \\ -31 & 32 & -31 \\ 31 & -31 & 32 \end{pmatrix}$$

and so $F^5(x, y, z) = (-30x + 31y - 31z, -31x + 32y - 31z, 31x - 31y + 32z)$.

Chapter 14

14.1 The elements of $\mathbf{U}(1)$ can be identified with the complex numbers of modulus 1. Define $f : \mathbf{U}(1) \to \mathsf{SO}(2)$ by putting

$$f(a + ib) = \begin{pmatrix} a & -b \\ b & a \end{pmatrix}.$$

It easy to check that f is a group isomorphism.

14.2 a) $A = \overline{A} = A^t = A^*$.

b) $A = A^t$, $\overline{A} = A^* = \begin{pmatrix} -i & 0 \\ 0 & -i \end{pmatrix}$.

c) $A = A^t$, $\overline{A} = A^* = \begin{pmatrix} 1 & -i \\ -i & 1 \end{pmatrix}$.

d) $A = A^t$, $\overline{A} = A^* = \begin{pmatrix} \sqrt{2}/2 & -\sqrt{2}i/2 \\ -\sqrt{2}i/2 & \sqrt{2}/2 \end{pmatrix}$.

e) $A = A^t$, $\overline{A} = A^* = \begin{pmatrix} 3/5 & -4i/5 \\ -4i/5 & 3/5 \end{pmatrix}$.

f) $A^t = \overline{A} = \begin{pmatrix} 1 & 1-i \\ 1+i & -1 \end{pmatrix}$, $A^* = A$.

14.3 a), b), d), e).

14.7 a) $f(x, y) = (2x + y + 1,\ 3y - 1)$
b) $f(x, y) = (-x/2 + 3y/4 + 5/4,\ 3x/2 - 7y/4 + 3/4)$
c) $f(x, y) = (-x/4 - y/12 + 1/3,\ x/2 - y/6 - 1/3)$.

14.10 $\omega_{\mathbf{b},c}$ is determined by the condition $\mathbf{y} - \mathbf{b} = c(\mathbf{x} - \mathbf{b})$, where $\mathbf{y} = \omega_{\mathbf{b},c}(\mathbf{x})$, and so $\mathbf{y} = \mathbf{b}(1 - c) + c\mathbf{x}$. Thus, $\omega_{\mathbf{b},c} = T_{c\mathbf{1},\mathbf{b}(1-c)}$.

Chapter 15

15.1 c), e).

15.2 a) $3x_1x_2 - 4x_1y_2 - 4x_2y_1 - 3y_1y_2$,
b) $4x_1x_2 - \frac{9}{2}x_1y_2 - \frac{9}{2}x_2y_1 + 5y_1y_2$,
c) $4x_1x_2 - 2x_1y_2 - 2x_2y_1 + 7y_1y_2$,
d) $x_1x_2 - x_1y_2 - x_2y_1 + y_1y_2$,
e) $3x_1x_2 + 5x_1y_2 + 5x_2y_1 + 3y_1y_2$,
f) $3x_1y_2 + 3x_2y_1$.

15.3 a) $\begin{pmatrix} 3 & -4 \\ -4 & -3 \end{pmatrix}$, $r = 2$ b) $\begin{pmatrix} 4 & -\frac{9}{2} \\ -\frac{9}{2} & 5 \end{pmatrix}$, $r = 2$

c) $\begin{pmatrix} 4 & -2 \\ -2 & 7 \end{pmatrix}$, $r = 2$ d) $\begin{pmatrix} 1 & -1 \\ -1 & 1 \end{pmatrix}$, $r = 1$

e) $\begin{pmatrix} 3 & 5 \\ 5 & 3 \end{pmatrix}$, $r = 2$ f) $\begin{pmatrix} 0 & 3 \\ 3 & 0 \end{pmatrix}$, $r = 2$.

15.4 a) $\frac{1}{2}x_1z_2 + \frac{1}{2}x_2z_1 + \frac{1}{2}x_1y_2 + \frac{1}{2}x_2y_1 + \frac{1}{2}y_1z_2 + \frac{1}{2}y_2z_1$
b) $x_1y_2 + x_2y_1 + y_1y_2 - x_1z_2 - x_2z_1$

c) $x_1x_2 - x_1z_2 - x_2z_1 - y_1y_2 - z_1z_2$
d) $5x_1x_2 + 3y_1y_2 + \frac{1}{2}x_1z_2 + \frac{1}{2}x_2z_1$
e) $-x_1x_2 - 2x_1y_2 - 2x_2y_1 + 3y_1y_2 + 2z_1z_2$.

15.5 a) $\begin{pmatrix} 0 & \frac{1}{2} & \frac{1}{2} \\ \frac{1}{2} & 0 & \frac{1}{2} \\ \frac{1}{2} & \frac{1}{2} & 0 \end{pmatrix}, r = 3$ b) $\begin{pmatrix} 0 & 1 & -1 \\ 1 & 1 & 0 \\ -1 & 0 & 0 \end{pmatrix}, r = 3$

c) $\begin{pmatrix} 1 & 0 & -1 \\ 0 & -1 & 0 \\ -1 & 0 & -1 \end{pmatrix}, r = 3$ d) $\begin{pmatrix} 5 & 0 & \frac{1}{2} \\ 0 & 3 & 0 \\ \frac{1}{2} & 0 & 0 \end{pmatrix}, r = 3$

e) $\begin{pmatrix} -1 & -2 & 0 \\ -2 & 3 & 0 \\ 0 & 0 & 2 \end{pmatrix}, r = 3.$

Chapter 16

16.1 a) $\{(i, 0), (0, 1)\}$, $x = ix'$, $y = y'$
b) $\left\{\left(\frac{\sqrt{2}}{2} - \frac{\sqrt{2}}{2}i, 0\right), \left(0, \frac{\sqrt{2}}{2}i\right)\right\}, x = \left(\frac{\sqrt{2}}{2} - \frac{\sqrt{2}}{2}i\right)x'$, $y = \frac{\sqrt{2}}{2}iy'$
c) $\{(\frac{1}{2}, 0), (0, \frac{1}{3})\}$, $x = \frac{1}{2}x'$, $y = \frac{1}{3}y'$
d) $\{(i, 0), (0, 5i)\}$, $x = ix'$, $y = 5iy'$.

16.2 a) $\{(1/2, 0, 0), (0, 0, 1/2\sqrt{3}), (0, 1/\sqrt{5}, 0)\}$, $(2, 1)$
b) $\{(0, 0, 1/3), (1, 0, 0), (0, 1, 0)\}$, $(1, 1)$
c) $\{(0, 0, 1), (1, 0, 0), (0, 1, 0)\}$, $(1, 2)$
d) $\{(0, 1, 0), (0, 0, 1/4), (1, 0, 0)\}$, $(2, 0)$.

16.3 a) Following the procedure used in the second proof of Theorem 16.1, make the following change of coordinates: $x = x' + \frac{4}{3}y'$, $y = y'$. This transforms q into the diagonal form

$$q(x', y') = 3x'^2 - \frac{25}{3}y'^2.$$

The signature is thus (1,1).
b) $4x'^2 - \frac{1}{16}y'^2$, $x = x' + \frac{9}{8}y'$, $y = y'$, $(1, 1)$
c) $4x'^2 + 6y'^2$, $x = x' + \frac{1}{2}y'$, $y = y'$, $(2, 0)$
d) x'^2, $x = x' + y'$, $y = y'$, $(1, 0)$
e) $3x'^2 - \frac{16}{3}y'^2$, $x = x' - \frac{5}{3}y'$, $y = y'$, $(1, 1)$
f) In this case it is obvious that the change of coordinates $x = x' - y'$, $y = x' + y'$ transforms the quadratic form into the diagonal form $6x'^2 - 6y'^2$. The signature is therefore (1,1).

16.4 a) $\begin{pmatrix} 3 & 0 \\ 0 & -\frac{25}{3} \end{pmatrix} = \begin{pmatrix} 1 & 0 \\ \frac{4}{3} & 1 \end{pmatrix} \begin{pmatrix} 3 & -4 \\ -4 & -3 \end{pmatrix} \begin{pmatrix} 1 & \frac{4}{3} \\ 0 & 1 \end{pmatrix}$

b) $\begin{pmatrix} 4 & 0 \\ 0 & -\frac{1}{16} \end{pmatrix} = \begin{pmatrix} 1 & 0 \\ \frac{9}{8} & 1 \end{pmatrix} \begin{pmatrix} 4 & -\frac{9}{2} \\ -\frac{9}{2} & 5 \end{pmatrix} \begin{pmatrix} 1 & \frac{9}{8} \\ 0 & 1 \end{pmatrix}$

c) $\begin{pmatrix} 4 & 0 \\ 0 & 6 \end{pmatrix} = \begin{pmatrix} 1 & 0 \\ \frac{1}{2} & 1 \end{pmatrix} \begin{pmatrix} 4 & -2 \\ -2 & 7 \end{pmatrix} \begin{pmatrix} 1 & \frac{1}{2} \\ 0 & 1 \end{pmatrix}$

d) $\begin{pmatrix} 1 & 0 \\ 0 & 0 \end{pmatrix} = \begin{pmatrix} 1 & 0 \\ 1 & 1 \end{pmatrix} \begin{pmatrix} 1 & -1 \\ -1 & 1 \end{pmatrix} \begin{pmatrix} 1 & 1 \\ 0 & 1 \end{pmatrix}$

e) $\begin{pmatrix} 3 & 0 \\ 0 & -\frac{16}{3} \end{pmatrix} = \begin{pmatrix} 1 & 0 \\ -\frac{5}{3} & 1 \end{pmatrix} \begin{pmatrix} 3 & 5 \\ 5 & 3 \end{pmatrix} \begin{pmatrix} 1 & -\frac{5}{3} \\ 0 & 1 \end{pmatrix}$

f) $\begin{pmatrix} 6 & 0 \\ 0 & -6 \end{pmatrix} = \begin{pmatrix} 1 & 1 \\ -1 & 1 \end{pmatrix} \begin{pmatrix} 0 & 3 \\ 3 & 0 \end{pmatrix} \begin{pmatrix} 1 & -1 \\ 1 & 1 \end{pmatrix}.$

16.5 a) A first substitution of $x = x'$, $y = y' - x'$, $z = z'$ reduces q to the form $q(x', y', z') = -x'^2 + x'y' + y'z'$. Now proceed as in the proof of Theorem 16.1. With the substitution $x' = x'' + \frac{1}{2}y''$, $y' = y''$, $z' = z''$ the form becomes $q(x'', y'', z'') = -x''^2 + \frac{1}{4}y''^2 + y''z''$. Finally, the substitution $x'' = \tilde{x}$, $y'' = \tilde{y} - 2\tilde{z}$, $z'' = \tilde{z}$ reduces the quadratic form to

$$q(\tilde{x}, \tilde{y}, \tilde{z}) = -\tilde{x}^2 + \frac{1}{4}\tilde{y}^2 - \tilde{z}^2.$$

Exchanging $\tilde{x}$ and $\tilde{y}$, gives $q(\hat{x}, \hat{y}, \hat{z}) = \frac{1}{4}\hat{x}^2 - \hat{y}^2 - \hat{z}^2$, where $\hat{x} = \tilde{y}$, $\hat{y} = \tilde{x}$, $\hat{z} = \tilde{z}$. Signature (1,2).

b) $x'^2 + y'^2 - z'^2$; $x = z' - y'$, $y = x' + y' - z'$, $z = y'$; (2,1).

c) $x'^2 - y'^2 - 2z'^2$; $x = x' + z'$, $y = y'$, $z = z'$; (1,2).

d) $5x'^2 + 3y'^2 - \frac{1}{20}z'^2$; $x = x' - \frac{1}{10}z'$, $y = y'$, $z = z'$, (2,1).

e) $2x'^2 + 7y'^2 - z'^2$; $x = z'$, $y = y'$, $z = x' - 2y'$; (2,1).

16.6 a) The matrix M is the matrix corresponding to the successive changes of coordinates. Thus

$$M = \begin{pmatrix} 1 & 0 & 0 \\ -1 & 1 & 0 \\ 0 & 0 & 1 \end{pmatrix} \begin{pmatrix} 1 & \frac{1}{2} & 0 \\ 0 & 1 & 0 \\ 0 & 0 & 1 \end{pmatrix} \begin{pmatrix} 1 & 0 & 0 \\ 0 & 1 & -2 \\ 0 & 0 & 1 \end{pmatrix} \begin{pmatrix} 0 & 1 & 0 \\ 1 & 0 & 0 \\ 0 & 0 & 1 \end{pmatrix}$$

$$= \begin{pmatrix} \frac{1}{2} & 1 & -1 \\ \frac{1}{2} & -1 & -1 \\ 0 & 0 & 1 \end{pmatrix}.$$

Consequently,

$$\begin{pmatrix} \frac{1}{4} & 0 & 0 \\ 0 & -1 & 0 \\ 0 & 0 & -1 \end{pmatrix} = \begin{pmatrix} \frac{1}{2} & \frac{1}{2} & 0 \\ 1 & -1 & 0 \\ -1 & -1 & 1 \end{pmatrix} \begin{pmatrix} 0 & \frac{1}{2} & \frac{1}{2} \\ \frac{1}{2} & 0 & \frac{1}{2} \\ \frac{1}{2} & \frac{1}{2} & 0 \end{pmatrix} \begin{pmatrix} \frac{1}{2} & 1 & -1 \\ \frac{1}{2} & -1 & -1 \\ 0 & 0 & 1 \end{pmatrix}$$

b)

$$\begin{pmatrix} 1 & 0 & 0 \\ 0 & 1 & 0 \\ 0 & 0 & -1 \end{pmatrix} = \begin{pmatrix} 0 & 1 & 0 \\ -1 & 1 & 1 \\ 1 & -1 & 0 \end{pmatrix} \begin{pmatrix} 0 & 1 & -1 \\ 1 & 1 & 0 \\ -1 & 0 & 0 \end{pmatrix} \begin{pmatrix} 0 & -1 & 1 \\ 1 & 1 & -1 \\ 0 & 1 & 0 \end{pmatrix}$$

c)

$$\begin{pmatrix} 1 & 0 & 0 \\ 0 & -1 & 0 \\ 0 & 0 & -2 \end{pmatrix} = \begin{pmatrix} 1 & 0 & 0 \\ 0 & 1 & 0 \\ 1 & 0 & 1 \end{pmatrix} \begin{pmatrix} 1 & 0 & -1 \\ 0 & -1 & 0 \\ -1 & 0 & -1 \end{pmatrix} \begin{pmatrix} 1 & 0 & 1 \\ 0 & 1 & 0 \\ 0 & 0 & 1 \end{pmatrix}$$

d)

$$\begin{pmatrix} 5 & 0 & 0 \\ 0 & 3 & 0 \\ 0 & 0 & -\frac{1}{20} \end{pmatrix} = \begin{pmatrix} 1 & 0 & 0 \\ 0 & 1 & 0 \\ -\frac{1}{10} & 0 & 1 \end{pmatrix} \begin{pmatrix} 5 & 0 & \frac{1}{2} \\ 0 & 3 & 0 \\ \frac{1}{2} & 0 & 0 \end{pmatrix} \begin{pmatrix} 1 & 0 & -\frac{1}{10} \\ 0 & 1 & 0 \\ 0 & 0 & 1 \end{pmatrix}$$

e)

$$\begin{pmatrix} 2 & 0 & 0 \\ 0 & 7 & 0 \\ 0 & 0 & -1 \end{pmatrix} = \begin{pmatrix} 0 & 0 & 1 \\ -2 & 1 & 0 \\ 1 & 0 & 0 \end{pmatrix} \begin{pmatrix} -1 & -2 & 0 \\ -2 & 3 & 0 \\ 0 & 0 & 2 \end{pmatrix} \begin{pmatrix} 0 & -2 & 1 \\ 0 & 1 & 0 \\ 1 & 0 & 0 \end{pmatrix}.$$

Chapter 17

17.6 $\dim(\mathbf{W}) = 2$; orthonormal basis of $\mathbf{W}$ is given by:

$$\left\{\mathbf{v}_1(\tfrac{1}{\sqrt{3}}, \tfrac{1}{\sqrt{3}}, 0, \tfrac{1}{\sqrt{3}}),\ \mathbf{v}_2(\tfrac{2}{\sqrt{6}}, -\tfrac{1}{\sqrt{6}}, 0, -\tfrac{1}{\sqrt{6}})\right\}.$$

Vectors which complete this to a basis of $\mathbf{V}$ are:

$$\mathbf{v}_3(0,0,1,0),\ \mathbf{v}_4(0, -\tfrac{1}{2}, 0, \tfrac{1}{2}).$$

17.7 a) $\{(\frac{2}{\sqrt{5}}, 0, 0, \frac{1}{\sqrt{5}}), (-\frac{1}{\sqrt{13}}, \frac{2}{\sqrt{13}}, \frac{2}{\sqrt{13}}, \frac{2}{\sqrt{13}}), (\frac{\sqrt{33}}{2}, \frac{\sqrt{33}}{2}, \frac{3\sqrt{33}}{4}, -\sqrt{33})\}$
b) $\{(-\frac{1}{\sqrt{3}}, 0, \frac{1}{\sqrt{3}}, \frac{1}{\sqrt{3}}), (\frac{3}{\sqrt{19}}, \frac{1}{\sqrt{19}}, 0, \frac{3}{\sqrt{19}})\}$
c) $\{(\frac{1}{2}, \frac{1}{2}, -\frac{1}{2}, \frac{1}{2}), (\frac{1}{\sqrt{6}}, 0, \frac{2}{\sqrt{6}}, \frac{1}{\sqrt{6}}), (\frac{1}{\sqrt{2}}, 0, 0, -\frac{1}{\sqrt{2}})\}$.

17.8 $\{(0, \frac{\sqrt{2}}{2}, 0, \frac{\sqrt{2}}{2}), (1,0,0,0), (0, -\frac{\sqrt{2}}{2}, 0, \frac{\sqrt{2}}{2}), (0,0,1,0)\}$.

17.9 $\{(1,0,0,0), (-\frac{1}{2}, 1, 0, 0), (0,0,1,0), (0,0,-\frac{1}{2},1)\}$.

Chapter 19

19.2 a) $6/\sqrt{5}$, b) $1/5$.

19.3 $X - Y - 2Z - 3 = X + 3Y - Z - 2 = 0$.

19.4 $5X - 6Y + 2Z + 7 = 0$.

19.5 $X + Y - 3 = 0$, $Z = 1$.

19.6 $\ell : \frac{X-3}{3} = Y - 2 = \frac{Z-1}{5}$; $d(\ell, \ell_1) = \frac{4}{\sqrt{10}}$.

19.9 a) Common perpendicular: $\frac{3X}{2} + \frac{1}{2} = 6Y + 3 = -\frac{6Z}{5} + 1$;
$d(\ell, \ell_1) = \frac{5}{\sqrt{42}}$.

19.10 a) $X^2 + Y^2 - 6X + 8Y + 24 = 0$
b) $X^2 + Y^2 - 2X - 4Y + 3 = 0$
c) $X^2 + Y^2 - 2X + 2Y + \frac{7}{4} = 0$.

19.11 a) $C = (3, -4)$, $r = 5$, b) $C = (-4, 5)$, $r = 3$.

Chapter 20

20.1 a) Since f is a direct isometry, it is of the form $f(x) = x + c$ for some $c \in \mathbf{R}$. From the condition that $f(1) = \frac{\pi}{2}$, it follows that $c = \frac{\pi}{2} - 1$. Thus $f(x) = x + \frac{\pi}{2} - 1$.
b) $f(x) = -x + \pi - 2$.

20.2 a) $f(x, y) = (-x, y)$
b) $f(x, y) = (-y, -x)$
c) $f(x, y) = (\frac{3}{5}x + \frac{4}{5}y, \frac{4}{5}x - \frac{3}{5}y)$
d) $f(x, y) = (\frac{5}{13}x + \frac{12}{13}y, \frac{12}{13}x - \frac{5}{13}y)$
e) $f(x, y) = (-y + 1, -x + 1)$.

20.3 a) $f(x, y) = (x + 1, y + 1)$
b) $f(x, y) = (x + 1, -y + 1)$
c) $f(x, y) = (\frac{3}{5}x + \frac{4}{5}y, \frac{4}{5}x - \frac{3}{5}y)$
d) $f(x, y) = (-\frac{4}{5}x + \frac{3}{5}y - \frac{12}{5}, \frac{3}{5}x + \frac{4}{5}y + \frac{4}{5})$.

20.4 a) $f(x, y, z) = (y, x, z)$
b) $f(x, y, z) = (\frac{1}{3}x - \frac{2}{3}y - \frac{2}{3}z, -\frac{2}{3}x + \frac{1}{3}y - \frac{2}{3}z, -\frac{2}{3}x - \frac{2}{3}y + \frac{1}{3}z)$
c) $f(x, y, z) = (\frac{1}{3}x + \frac{2}{3}y - \frac{2}{3}z, \frac{2}{3}x + \frac{1}{3}y + \frac{2}{3}z, -\frac{2}{3}x + \frac{2}{3}y + \frac{1}{3}z)$
d) $f(x, y, z) = (-\frac{3}{5}x + \frac{4}{5}z - \frac{4}{5}, y, \frac{4}{5}x + \frac{3}{5}z + \frac{2}{5})$
e) $f(x, y, z) = (\frac{1}{9}x + \frac{8}{9}y - \frac{4}{9}z + \frac{16}{9}, \frac{8}{9}x + \frac{1}{9}y + \frac{4}{9}z - \frac{16}{9},$
$-\frac{4}{9}x + \frac{4}{9}y + \frac{7}{9}z + \frac{8}{9})$.

20.5 a) f is the identity b) $f(x,y,z)=(-x,y,z)$.

20.6 The given line meets $\mathcal{P}$ at the point $P=(-\frac{1}{2},-\frac{1}{4},\frac{3}{4})$. The line ℓ contains $\mathcal{P}$ and the reflexion of any point of the given line, say of $(0,0,1)$ whose reflexion is $Q=(-\frac{2}{3},-\frac{2}{3},\frac{1}{3})$. The line ℓ is the line with Cartesian equations

$$5X-2Y+2=0,\ Y-Z+1=0.$$

The solution can also be found by calculating directly the transform of the given line with respect to the reflexion in $\mathcal{P}$ (cf. Exercise 4(b)). With this approach, the equations one finds are $5X-4Y+2Z=0$, $Y-Z+1=0$.

Chapter 22

22.1 a) $\begin{pmatrix} \frac{2}{\sqrt{5}} & \frac{1}{\sqrt{5}} \\ -\frac{1}{\sqrt{5}} & \frac{2}{\sqrt{5}} \end{pmatrix}$ b) $\begin{pmatrix} \frac{1}{\sqrt{2}} & -\frac{1}{\sqrt{2}} \\ \frac{1}{\sqrt{2}} & \frac{1}{\sqrt{2}} \end{pmatrix}$

c) $\begin{pmatrix} \frac{1}{\sqrt{10}} & -\frac{3}{\sqrt{10}} \\ \frac{3}{\sqrt{10}} & \frac{1}{\sqrt{10}} \end{pmatrix}$ d) $\begin{pmatrix} \frac{1}{\sqrt{2}} & -\frac{1}{\sqrt{2}} \\ \frac{1}{\sqrt{2}} & \frac{1}{\sqrt{2}} \end{pmatrix}$.

22.2 a) The desired orthogonal transformation is one which changes from the canonical basis to an orthonormal basis consisting of eigenvectors of the matrix of q. The eigenvalues are $\lambda=1,4$. The eigenspace $\mathbf{R}^3_1$ has dimension 2, and an orthonormal basis is $\{(0,1/\sqrt{2},-1/\sqrt{2}),(-2/\sqrt{6},1/\sqrt{6},1/\sqrt{6})\}$. The eigenspace $\mathbf{R}^3_4$ has dimension 1, and contains the unit vector

$$(1/\sqrt{3},1/\sqrt{3},1/\sqrt{3}).$$

Thus the orthogonal transformation is

$$\begin{pmatrix} x_1 \\ x_2 \\ x_3 \end{pmatrix} = \begin{pmatrix} 0 & -\frac{2}{\sqrt{6}} & \frac{1}{\sqrt{3}} \\ \frac{1}{\sqrt{2}} & \frac{1}{\sqrt{6}} & \frac{1}{\sqrt{3}} \\ -\frac{1}{\sqrt{2}} & \frac{1}{\sqrt{6}} & \frac{1}{\sqrt{3}} \end{pmatrix} \begin{pmatrix} y_1 \\ y_2 \\ y_3 \end{pmatrix}.$$

The corresponding diagonal form is $q(y_1,y_2,y_3)=y_1^2+y_2^2+4y_3^2$.
b) Diagonal form: $q(y_1,y_2,y_3)=-2y_1^2-3y_2^2+y_3^2$. Orthogonal transformation:

$$\begin{pmatrix} x_1 \\ x_2 \\ x_3 \end{pmatrix} = \begin{pmatrix} \frac{1}{\sqrt{3}} & 0 & -\frac{2}{\sqrt{6}} \\ \frac{1}{\sqrt{3}} & \frac{1}{\sqrt{2}} & \frac{1}{\sqrt{6}} \\ -\frac{1}{\sqrt{3}} & \frac{1}{\sqrt{2}} & -\frac{1}{\sqrt{6}} \end{pmatrix} \begin{pmatrix} y_1 \\ y_2 \\ y_3 \end{pmatrix}.$$

c) Diagonal form: $q(y_1, y_2, y_3) = -y_1^2 - y_2^2 + 3y_3^2$. Orthogonal transformation:

$$\begin{pmatrix} x_1 \\ x_2 \\ x_3 \end{pmatrix} = \begin{pmatrix} \frac{1}{\sqrt{2}} & 0 & \frac{1}{\sqrt{2}} \\ 0 & 1 & 0 \\ -\frac{1}{\sqrt{2}} & 0 & \frac{1}{\sqrt{2}} \end{pmatrix} \begin{pmatrix} y_1 \\ y_2 \\ y_3 \end{pmatrix}.$$

Chapter 23

23.1 c), e).

23.2 a), c), e).

23.3 $\{(i/\sqrt{2}, -i/\sqrt{2}, 0), (i/\sqrt{2}, i/\sqrt{2}, 0), (0, 0, i)\}$.

23.4 a) A has eigenvalues $\lambda = 0, 2$. The corresponding eigenvectors $(i/\sqrt{2}, -1/\sqrt{2})$ and $(1/\sqrt{2}, i/\sqrt{2})$ form an orthonormal basis of $\mathbf{C}^2$ with respect to the standard Hermitian product. The required matrix is thus,

$$M = \begin{pmatrix} \frac{i}{\sqrt{2}} & \frac{1}{\sqrt{2}} \\ -\frac{1}{\sqrt{2}} & \frac{i}{\sqrt{2}} \end{pmatrix}.$$

b) $M = \begin{pmatrix} \frac{\sqrt{3}}{2} & \frac{i}{2} \\ \frac{i}{2} & \frac{\sqrt{3}}{2} \end{pmatrix}$.

Bibliography

[1] C.T. Benson and L.C. Grove, *Finite Reflection Groups*, Springer, New York, 1985.

[2] M. Berger, *Geometry I, II*, Springer, Berlin-London, 1987.

[3] H.W. Guggenheimer, *Plane Geometry and its Groups*, Holden Day, San Francisco, 1967.

[4] I. Herstein, *Topics in Algebra*, Wiley, London, 1976.

[5] D. Hilbert and S. Cohn-Vossen, *Geometry and the Imagination*, Chelsea, New York, 1952.

[6] F. Klein, *Lectures on the Icosahedron*, Dover, New York, 1956.

[7] F. Klein, *Elementary Mathematics from an Advanced Standpoint: Geometry*, Dover, New York, 1948.

[8] S. Lang, *Linear Algebra*, Addison-Wesley, Reading, Mass., 1966.

[9] R.C. Lyndon, *Groups and Geometry*, Cambridge University Press, Cambridge, 1985.

[10] V.V. Nikulin and I.R. Shafarevic, *Geometries and Groups*, Springer, Berlin-London, 1987.

[11] H. Weyl, *Symmetry*, Princeton University Press, Princeton, 1952.

Index of notation

Index